Books are to be returned on or before
the last date below.

17 MAY 1996

2 9 OCT 2001

- 9 DEC 1996 A

17 DEC 2003

16 DEC 1997 A

2 3 FEB 2004

27 APR 1998 M

29 NOV 1999

LIBREX —

2 4 OCT 2000

- 7 NOV 2000

BRICKS
THEIR PROPERTIES
AND USE

BRICKS
THEIR PROPERTIES
AND USE

The Brick Development Association

THE CONSTRUCTION PRESS LTD

ISBN   0 904406 04 0

© Brick Development Association

Published in 1974 by

The Construction Press Ltd.
(formerly MTP Construction)
P.O. Box 99
Lancaster
England

CONTENTS

Publisher's Note

This book is edited from material made available by the Brick Development Association which was originally published in booklet format by either the Clay Products Technical Bureau or the Brick Development Association itself. The resultant volume provides an unrivalled source of facts, figures and information about bricks and brickwork which are commonly required by members of the various construction and engineering professions.

Acknowledgement

We would like to gratefully acknowledge the full co-operation of the Brick Development Association in granting us permission to publish the material contained in this volume and in assisting with its preparation.

INTRODUCTION
CLAY AND CALCIUM SILICATE BRICKS

Part 1
Clay bricks

Introduction

For hundreds of years, the word 'brick' was exclusively associated with building units made of burnt clay. In modern usage, however, the word tends to be descriptive of the form regardless of the material. Generally speaking, the form may be defined as a rectangular prism of a size that can be handled conveniently with one hand. Thus, 'bricks' made from a variety of materials are available.

This publication is only concerned with the two most important types: Clay Bricks and Calcium Silicate Bricks.

History

First there were bricks made from mud dried in the sun. Then at least 5,000 years ago, mad discovered the strength and durability of bricks fired in a kiln. From the Middle East to Mediterranean Europe, the remains of countless ancient buildings still stand as a testimony to the endurance of kiln-burnt clay.

Bricks were first introduced into Britain by the Romans. Following a period of disuse in the Dark Ages, the material began to re-appear in the Early Middle Ages when bricks salvaged from the remains of the first Christian churches and cathedrals. Thereafter, the development of brick construction was unbroken, and followed the lines which had their parallels in most countries in Europe. Brick was first used for ecclesiastical buildings, then for the castles and houses of the nobility, and later for the dwellings of the emerging middle classes. Although it is difficult to isolate individual strands in the web of social/economic/architectural history, there can be no doubt that the trend towards the use of brickwork received a series of powerful boosts over the centuries from the urgent need for a fire resistant material that would obviate the frequent conflagrations which ravaged towns and cities throughout Europe.

The era of the Industrial Revolution saw the final emergence of brickwork as the dominant structural material. Throughout the nineteenth century, bricks were used in every kind of building and for a wide range of engineering structures. From the turn of the century to the present day, and despite the challenge from other materials and the demands of new architectural forms, brickwork has maintained its position as the leading structural medium.

Although the first patent for a brickmaking machine was granted as long ago as 1619, the brick manufacturing industry was largely based on manual processes until the Industrial Revolution. Then in response both to the explosion of demand for bricks and the new sources of power that were becoming available, machines rapidly supplanted hands. The turning point in this period of evolution came in 1858 with the introduction of the Hoffman kiln, which enabled all the stages of firing to be carried out concurrently and continuously. Since that time, continual research and development have led to the creation in Britain of one of the most efficient brickmaking industries in the world. In modern British plants, the skill and knowledge derived from centuries of brickmaking are blended into fully automated processes. Production in 1971 was 7,000 million bricks.

Materials

The materials used for clay brickmaking range from soft and plastic surface deposits to hard mudstone, shales, marls, and even some of the softer varieties of slate. The essential requirements are that after being ground, and tempered with water, the material should be capable of taking a good shape – either by moulding, extrusion or pressure – and that the shape should be retained without undue shrinkage, warping or cracking when the bricks are dried and fired.

Great Britain is exceptionally well provided with excellent deposits of clay, shales and other materials which meet these requirements. In consequence, brickworks are widely distributed throughout the country.

Properties

The many kinds of clay bricks available vary considerably in appearance and functional properties, according to the purposes for which they are intended.

Nevertheless, clay bricks generically may be said to perform extremely well in the following respects:
Attractive appearance which mellows instead of deteriorates with age
Economy
Durability
Structural strength
Fire resistance
Sound insulation
Low porosity
Low thermal and moisture movement
Resistance to adverse climatic conditions
Flexibility in application
Freedom from maintenance requirements
Reliability in performance is the first requirement in all building materials and components. Clay bricks in conformity with the functional and dimensional requirements of BS 3921* are readily available throughout the country.

Dimensions

Brick sizes may be specified in either or both of two ways:
the work size which is the actual size of the brick with allowances for manufacturing tolerances

Below In addition to the British Standard 225 mm format (centre foreground) various alternative brick sizes are widely available.

Right Brick paving, Harvey Court, Gonville & Caius College, Cambridge (Architects: Sir Leslie Martin and Colin St John Wilson). Paving is an increasingly popular application of brickwork – and is a good example of a situation where care should be exercised in selecting the right brick for the job.

the format size which is the actual size plus the thickness of the mortar joints

Roman bricks varied considerably in size but, typically, were about 300 x 150 x 33 mm (12 x 6 x 1¼ in). In the 13th century a 9 x 4½ x 2 in brick was introduced. Thereafter, although sizes continued to vary in different parts of the country, the principal variable was the thickness.

Through the publication of BS 3921:1965, Britain became the first country in Europe to adopt a single standard size brick:
Work size 8⅝ x 4⅛ x 2⅝ in.
Format size 9 x 4½ x 3 in (ie assuming a ⅜ in joint).

In 1969, to meet the impending change to metric dimensions and the trend towards dimensional co-ordination, BS 3921 defined the single standard metric brick size:
Work size 215 x 102.5 x 65 mm
Format size 225 x 112.5 x 75 mm (ie assuming a 10 mm joint)

Many manufacturers are also able to supply alternative sizes and special shapes, including 200 and 300 mm metric modular formats.

Applications

Clay bricks are used for an extremely wide range of applications in an equally extensive range of buildings and engineering structures. Among the most common applications are:
External and internal loadbearing walls
Loadbearing piers and columns
Partition walls
Party walls
Claddings and facings
Foundations
Perimeter and garden walls
Pavings and floorings

However, the sheer variety of applications attributable to bricks generically sometimes militates to their disadvantage in that users occasionally fail to exercise sufficient discrimination in selecting the brick best suited to a specific function. The Brick Development Association and its member companies are always ready to advise on the correct type of brick for a particular application.

*BS 3921 Bricks and blocks of fired brickearth, clay or shale
Part 1 : 1965 Imperial units
Part 2 : 1969 Metric units

Nomenclature

In common with some other traditional building materials and processes, the nomenclature of clay bricks exhibits marked local and regional peculiarities. No systematic nomenclature has been evolved, and the designations still in use to-day are based on at least six modes of classification:

1 Place of origin – Leicester red, Staffordshire blue, London stock
2 Raw material – marl, gault, shale, blaes
3 Method of manufacture – hand-made, wire-cut, stiff plastic, pressed
4 Use – facings, engineering, common, foundation, paviour
5 Colour – often associated with origin (above) – yellow, red, multi, brindle
6 Surface texture – smooth, glazed, sandfaced, rustic

Two designations in general use should be noted:
Flettons This name is widely applied to the pressed bricks made from the Lower Oxford Clay which occurs in economically accessible deposits in the Peterborough, Bedford and Buckinghamshire regions. Bricks of this kind form a substantial proportion of the total United Kingdom output. The high carbonaceous content, which is a unique feature of the Lower Oxford Clay, has a most beneficial effect on the cost of firing Fletton bricks.

Stocks Nowadays, this term is often loosely used to refer to the stock or usual brick in a particular district. Nevertheless, it is fairly generally understood to refer to the yellow London stock brick – in itself a paradoxical designation in that they are not made from London clay, but from the deposits of clay and chalk which occur in Kent and Essex close to the Thames. Originally 'stock bricks' referred to a brick made by hand on a stock – a piece of wood fixed to the moulder's table and which served to locate the mould – but they are now generally machine-made.

Classification

In BS 3921, Bricks are classified according to *variety, quality* and *type*:

1 Varieties

1.1 Common Common bricks are defined as 'Suitable for general building work but having no special claim to give an attractive appearance'.

Thus, the term 'commons' is applied to the many varieties of clay bricks which fall outside the categories of Facings and Engineering bricks, and which are used externally and internally for general construction. All have good loadbearing properties and are available in various crushing strengths. They are widely used for foundations and are increasingly being specified for calculated loadbearing construction. Common bricks generally make an ideal backing for rendering, plaster and colour wash.

1.2 Facing According to the Standard, Facing bricks are 'Specially made or selected to give an attractive appearance when used without rendering or plaster or other surface treatment of the wall'.

In general, Facing bricks may be said to combine attractive appearance with structural strength and good resistance to exposure. They are available in a very wide range of strengths, colours and textures. Facings constitute a steadily increasing proportion of the output of the clay brick industry.

1.3 Engineering This variety is described as 'Having a dense and strong semi-vitreous body conforming to defined limits for absorption and strength'. Engineering bricks are the most precisely defined variety in that they are required to possess performance characteristics ascertainable by test:

Class	Average compressive strength MN/m²* not less than	Average absorption boiling or vacuum % weight not greater than
A	69.0 (10,000 lbf/in²)	4.5
B	48.5 (7,000 lbf/in²)	7.0

Although BS 3921 makes no specific reference to the distinction, Engineering bricks are generally further sub-divided into Facings and Commons according to their appearance properties.

It must be stressed that to qualify as Engineerings, bricks must possess both the required strength *and* absorption properties. Thus, for example, there are many excellent loadbearing common and facing bricks which can meet or exceed the compressive strength requirements for Class A, but are not classified as Engineerings.

Engineering bricks were first developed in the nineteenth century in response to the demand for structural material with very high loadbearing capacity. To this day, they remain one of the strongest, most reliable and durable materials available to architects and civil engineers.

2 Quality

BS 3921 recognises three qualities of bricks:

2.1 Internal quality Bricks suitable for internal use only

2.2 Ordinary quality Less durable than special quality, but normally durable in the external face of the building.

2.3 Special quality Bricks that are durable even when used in situations of extreme exposure where the structure may become saturated and be frozen e.g. retaining walls, sewerage plants or pavings. Such bricks have clearly defined limits for soluble salts content.

Engineering bricks normally attain this standard of durability. Many facings and commons do also, but this should not be assumed in any particular instance unless it is claimed by the manufacturer.

3 Types

The Standard also distinguishes between types of bricks, according to their physical form:

3.1 Solid In which holes passing through or nearly through a brick do not exceed 25% of its volume, or in which frogs (depressions in the bed face of a brick) do not exceed 20% of its volume. A small hole is defined as less than 20 mm wide or less than 500 mm² in area. Up to three larger holes, not exceeding 3250 mm² each may be incorporated as handling aids within the total of 25%.

3.2 Perforated In which small holes (as defined above) passing through the brick exceed 25% of its volume. Again, up to three larger holes not exceeding 3250 mm² each may be incorporated as handling aids.

Bricks of this type have become widely accepted in recent years for a variety of applications. Tests have shown that their strength and absorption characteristics are comparable to those of similar solid bricks.

3.3 Hollow In which holes larger than the small holes defined above (3.1) exceed 25% of its volume.

3.4 Cellular Bricks in which holes closed at one end exceed 20% of the volume.

3.5 Special shapes Shapes other than the normal rectangular prism.

*Numerically equal to N/mm²

3.6 Standard specials Special shapes that are in
general use – e.g. squints, radials, bullnose – and may
be available from stock. These are treated in greater
detail in BS 4729: 1971 *

Manufacture

British clay bricks are unsurpassed throughout the
world for the range of types, colours and textures
available, and for their quality and economy. Perhaps
because they are so inexpensive and have become the
most widely used building material, there is a tendency to undervalue them – to overlook the fact that
what appears to be a simple geometrical solid is, in
reality, a highly complex package of carefully controlled functional and aesthetic properties. Efficient
volume production of such units calls for great skill
and experience and the best resources of modern
technology.

Methods of manufacture vary according to the
nature of the material, but all follow the same basic
principles. The raw material is won from the ground;
it is prepared and formed into brick shape; finally, it
is burnt in a kiln to produce a tough and immensely
durable building unit.

Winning the raw material

Clays and shales are normally won from quarries,
holes or pits in deposits specifically selected for brick
making. However, they are sometimes obtained as the
residual by-products of other mining or quarrying
operations.

At the majority of works, the raw material is

extracted from the ground by means of mechanical
diggers, excavators or planers, driven by diesel or
electric power. The clay or shale must then be transported to the brickmaking plant, and this is normally
effected by conveyor belts, aerial buckets, trucks or
heavy earthmoving equipment, according to circumstances.

The tonnages involved in these operations are
extremely large. The manufacture of a thousand
bricks whose finished weight may be 2¼ to 3 tons
(2.286 to 3.048 tonnes), depending on type, requires
the excavation, transportation and processing of
between 4 and 5 tons (4.064 and 5.08 tonnes) of raw
material. Thus, a works with the quite moderate
capacity of, say, a million bricks per week, must move
about 5,000 tons (5,080 tonnes) of raw material to
maintain this output. Similarly, a very large works
producing 20 million bricks per week must win, move
and process something of the order of 100,000 tons
(101605 tonnes).

Preparation and making

Before the clay or shale can be made into brick shapes,
it must be transformed from raw earth to a carefully
refined, entirely homogeneous material of predictable
performance.

Each of the many stages of clay preparation is of
critical importance. Together, they largely predeter-
mine the chemical and physical changes that occur in
the kiln, and the performance characteristics of the
end-product. Although the nature of the raw material
affects every aspect of the brickmaking process, it is
here, perhaps, in these critical preparation and making
phases that the most significant variations in technique occur. For this reason, the individual methods
will be considered separately.

There are five methods of making clay bricks:

1 The wire-cut process, for plastic clays
2 The stiff plastic process. Mainly for colliery shales
and certain other shales and clays which do not easily
develop a high degree of plasticity
3 The semi-dry press process, for the manufacture of
bricks from clays and shales with low natural plasticity
4 The soft mud process
5 Handmaking

The wire-cut process

The clays used in this method are normally fairly soft
and of fine texture.

As in all clay brickmaking processes, the raw material
obtained from the workings is mechanically and
evenly fed to the preparation plant. If the clay contains large hard pieces, the first stage of processing
will entail the elimination of these in primary crushing
rolls, known as 'kibblers'. It is then passed to the wet
pan, which is a revolving roller mill in which most of
the water required to induce the correct degree of
plasticity is added. This is called 'tempering' the clay.
It then travels through one or more pairs of rolls, set
increasingly close together – the final pair of which
may be rotated at high and differential speed to
produce fine grinding by direct crushing and tearing
action.

After grinding, the tempered clay is delivered to a
pug mill – sometimes via auxiliary cylindrical double-
bladed mixers, which may incorporate steam or
vacuum treatments, to increase homogeneity. The pug

---

* *BS 4729: 1971 Specification for Shapes and Dimensions of Special Bricks.*

mill consolidates the clay and extrudes it in a column whose breadth and depth are calculated to ensure the correct length and width in the finished bricks after drying and firing contractions have taken place. If required, the column of clay can be sandfaced or textured before being cut by wires into brick units.

Facing or engineering wire-cut bricks are sometimes pressed, after being cut into units, to give them smoother faces and sharper arrises.

Bricks made by the wire-cut process contain from 15 to 25% of moisture and must be partly pre-dried before being placed in the kiln for firing. **Drying is** carried out in chamber or tunnel dryers, **both of** which utilise waste heat from the kilns. Precise control of temperature and humidity conditions is essential to prevent drying cracks, and to ensure that the 'green' bricks reach the kiln in the optimum state for firing.

The semi-dry press process
This is a very important technique, used in the manufacture of fletton bricks, which accounts for about 50% of the annual UK clay brick production.

The clays and shales used in the semi-dry press process are normally ground in dry pans in the condition in which they are won, although primary crushing in kibbler rolls is sometimes necessary. After grinding, the fine material is passed through screens which may consist of rotary perforated plates, rotary wire screens, piano wire screens, or stationary perforated plates set at an angle. Jig and vibrating screens are also used in some cases. The fine clay dust is then delivered in a relatively dry condition to the making machine where it is pressed in moulds into brick shape. Bricks may be re-pressed in the same mould, or in a synchronised de-press die. Most fletton bricks, for example, are pressed four times. After pressing, the faces of the bricks can be sandfaced or textured.

As with all other processes, great care and precision is required to control the moisture content, which may vary from about 8% for shales to 19% for Oxford clays. Owing to their lower moisture content, semi-dry press bricks do not require pre-drying, and can be despatched direct to the kiln from the making machine.

The stiff plastic process
Here again, the clays and shales used are normally ground dry in the condition in which they are won. The screening process also is very similar to that used in the semi-dry press technique. However, because of the inherently low plasticity of the raw material, the clay dust is then passed to rotating knife mixers in a trough where water is added. The tempered clay is delivered to an extrusion pug which forces rough brick-size clots into moulds. A press die gives the final shape and compaction to each brick – a further press being given to facings, in some cases.

The addition of water to the mix does not raise the moisture content to a level that necessitates pre-drying, so this type of brick also is transferred to the kiln immediately after making.

The soft mud process
The clays used in this process are almost invariably obtained from shallow surface deposits. The material may be used in its natural state, but, should a buff or yellow colour be required, lime or chalk will be added.

The clay, or clay and chalk, is normally mixed in a wet pan. Breeze or other material may be added to improve the appearance properties and/or to provide combustible material for firing. The clay is then prepared in a similar manner to the wire-cut process – but with a much higher moisture content – then discharged into hoppers and mechanically forced into pre-sanded moulds.

Due to their relatively high moisture content, bricks made by the soft mud process must be pre-dried

Left In the wire-cut process, the tempered clay is extruded as a column which may be sandfaced or textured before being cut into brick units by wires.

Below far left A battery of fletton brick presses. After pressing, the bricks may be sandfaced or textured.

Below left 'Green' bricks in a chamber dryer. Bricks with a relatively high moisture content – eg wire-cut, soft mud – have to be partly pre-dried before delivery to the kiln.

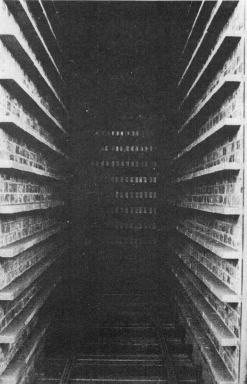

before firing and, in modern plants, this is carried out in tunnel or chamber driers.

This process has largely superseded the old hand moulding method used for the manufacture of yellow stock bricks – which were usually dried in the open air. The multi-stock bricks made in the south-east of England are also produced by similar means to those outlined above, but without the admixture of chalk or lime.

Handmaking

Good quality clays are normally a prerequisite for hand moulding. They are made up to a much softer consistency than is employed for wire-cut bricks, and the tempered clays may be left to stand or 'sour' for a time before use. The hand moulder throws a clot or lump of clay, previously rolled in sand, into a mould, taking care to fill all the corners. Surplus clay is struck from the top of the mould with a framed wire, flat trowel or other device.

Hand moulding is still used for making high-grade facings and some special shapes, and is characterised by the special texture produced on the faces of the bricks by the sand and the creasing of the clay in the moulds. Although more expensive than machine-made bricks, there is always a considerable demand for high quality handmades. No other building material can match their outstanding aesthetic merits at anything like a comparable cost.

Below A hand moulder throwing a pre-sanded clot of clay into a wooden mould.

Centre Pre-dried wire-cut bricks set on refractory-decked cars ready for firing in a tunnel kiln. The bricks are stacked in a special formation which ensures an even distribution of the hot kiln gases.

Firing
Setting

Firing constitutes the final stage in the transmutation of raw earth into aesthetically satisfying, dimensionally accurate, structural units. Before bricks can be fired however, they must be 'set' – (ie) stacked in such a way as to ensure an even distriubtion of the hot kiln gases. This used to be carried out by skilled craftsmen in the kiln itself. Nowadays, the green bricks are more usually built up into specially shaped packs at a setting station, and then transferred to the kiln by a fork-lift truck. In the growing number of plants where tunnel kilns are used (see below), the bricks are set on pallets or cars, either before or after the pre-drying process. Machine setting is possible in such cases, and is in limited use.

The nature of firing

Firing produces a large number of complicated chemical and physical changes in clay, and these must take place under very carefully controlled conditions to ensure consistent functional and appearance characteristics in the end-product. Although carried out as a continuous process, firing consists of four quite distinct stages: drying, pre-heating, firing and cooling. As noted earlier in the description of the wire-cut process, bricks with a relatively high moisture content are only *partly* pre-dried before being delivered to the kiln. The final stage of drying must be effected as an integral part of the firing process.

The firing of such diverse materials as are used in the manufacture of clay bricks does not readily lend itself to generalisations in regard to temperatures. However, the maximum temperature reached might be said to vary between 940 and 1180°C. Modifications in the properties of the finished bricks may be induced by varying the firing conditions. Thus, for example, colours can be modified very considerably by varying the temperature, or by adjustments to the amount of air admitted to the kiln during the later stages of firing.

Kilns

Four basic types of kiln are in use:
1 Clamps

This is a centuries old firing method which is still used for some stock and hand-made bricks. A clamp is formed by building up green bricks, in special close formations, on a layer of fuel or rough breeze laid on a base of burned bricks. The fuel or breeze is ignited by small fires started in flues arranged in the brick base, and this in turn eventually sets light to the fuel mixed in the green bricks (see the soft mud process, page 7). The clamp is usually then left to burn itself out and cool down.

2 Intermittent kilns

Prior to the introduction of the Hoffman continuous kiln in 1858, all bricks – except those fired in clamps – were burned in intermittent kilns of various kinds. Intermittent kilns are still used for the firing of special bricks and other requirements which do not justify large-scale production. They are heated by fires in grates in the outer walls – the bricks being set when

the kiln is cold, and the temperature gradually raised to cover each stage of the firing cycle in succession. The kiln is then allowed to cool slowly until the bricks can be handled and withdrawn.

3 The continuous kiln

As noted at the beginning of this publication, the invention of the Hoffman continuous kiln was a turning point in the evolution of the modern brick industry. The basic idea consists of a closed circuit of from 12 to 20 separate chambers, connected in such a way that the fire can be led from one to another. By this means, all the stages of firing – from setting and drying to cooling and drawing – can take place concurrently *and continuously* in different parts of the same kiln. The fire is made to progress around the kiln by feeding fuel – usually coal, oil, or gas – to zones via equi-distant openings in the crown.

The original Hoffman kiln was circular in design, but modern kilns usually consist of two parallel sections connected at the ends so that the chambers form a continuous tunnel. Where large output is required, or when even closer control over the firing is needed, the chambers are built as a series of arched structures, with the arches arranged transversely to the main direction of gas flow. Inter-chamber openings are provided to maintain the continuity of the system.

The modern continuous kiln is a highly efficient unit which combines excellent quality and large output with economy of operation.

4 Tunnel kilns

Tunnel kilns are a more recent innovation, and their use is steadily increasing in the clay brick industry. In tunnel kilns it is the bricks that move while the fire stays still. Green bricks are set on special cars, decked with refractories, which pass through a long straight tunnel, 300 ft or more in length, with a central firing zone. On their journey through the tunnel, the bricks pass successively through drying, pre-heating, firing and cooling zones. Temperatures and track speeds are controlled to provide the optimum conditions in each zone, and may be varied to produce specific functional and/or appearance characteristics.

Tunnel kilns are mostly gas or oil fired, although some utilise solid fuel which is normally fed to fire-mouths at the top or side.

Research and development

The centre of clay brick research in this country is the Heavy Clay Division of the British Ceramic Research Association, which is sponsored by the Brick Development Association and is one of the largest and best-equipped organisations of its kind in the world. The work carried out at its Mellor-Green Laboratories in Stoke-on-Trent, covers all aspects of the manufacture, application and performance of clay bricks.

A major and constant pre-occupation is the improvement of manufacturing techniques, and BCRA plays a particularly vital role in the siting and development of new brickworks. Potential sites for new clay workings are extensively test bored, and the samples subjected to rigorous laboratory scrutiny.

Top The control panel of a tunnel kiln. Modern brickmaking is highly automated and utilises the best resources of science and technology.

Above A top fired oil-burning tunnel kiln in a wire-cut plant. Operating on a 105 hour cycle at a temperature of approximately 104°C it has a rated output of 360,000 bricks per week. With a length of 157m (515ft) it is one of the longest tunnel kilns in the country.

The scientific equipment available to the Heavy Clay Division enables miniature bricks made from the potential clay deposits to be completely processed in a scaled-down replica of the production cycle in a brick factory. Through trials of this kind, BCRA scientists are able to predict with great accuracy the precise manufacturing conditions that will be required to produce the optimum performance and appearance characteristics in the future output of the projected factory.

In addition to this authoritative work on manufacturing techniques, the Brick Development Association has gained an international reputation for its promotion of advanced work in the field of user research, of which the investigations into gas explosions in loadbearing brick structure afford an excellent example. Although the Ronan Point disaster occurred in a precast concrete panel structure, the Brick Development Association was the first organisation in the world to initiate a series of trial gas explosions in a full-scale building. This imaginative and far-sighted programme of research yielded information whose relevance and importance stretches far beyond the field of loadbearing brick construction.

Communications and education represent another important aspect of the work undertaken by the Brick Development Association. Through the constant interchange of information with similar establishments throughout the world, through the publication of technical papers, and by organising conferences, symposia, lectures and courses, the Association plays an important part in the dissemination of new knowledge to the professional, constructional and manufacturing sectors of the building industry.

Well-equipped research laboratories are also maintained by many individual member companies. These are closely concerned with quality control, with user research in respect of specific products, and with new product development.

Mention must also be made of the comprehensive and authoritative work on many aspects of brick construction carried out by the Building Research Station, and of the special research projects sponsored by the Brick Development Association in various universities throughout the country.

Right Cladding failing during one of the trial gas explosions sponsored by the Brick Development Association. Among many other important results, this research showed that doors, windows and most kinds of infill panels fail at very low applied pressures and act as a safety valve for the rest of the structure.

Below Many member companies of the Brick Development Association have been concerned with the evolution of integrated systems for the mechanical handling of bricks from the point of delivery to the point of laying. The system illustrated here was evolved by a BDA member company in conjunction with the Building Research Station. The site packs of 339 or 330 bricks may be further sub-divided into 5 hand-barrow packs.

Part 2
Calcium silicate bricks

Calcium silicate bricks may be less familiar to the general population than clay bricks, but they are certainly no less economical and useful.

History

As long ago as 1866, an American named Van Derburgh obtained British Patent No. 2740 for 'artificial stone, etc.' to be obtained by the application of steam to a mixture of caustic lime and sand, with or without other ingredients. The process was never taken up because the patent did not envisage autoclaving (see later), and the use of live-steam would have made it unduly long. However, this early development was followed in 1881 by a German patent obtained by Michaelis for making 'artificial sandstone' by reacting an intimate mixture of lime, or similar substances with sand or siliceous compound with steam under pressure at $130^{\circ} - 300^{\circ}C$. This process was first developed in Germany in 1894. Since German clay deposits are inferior to those in the U.K., calcium silicate bricks made rapid progress in that country where, currently, they are used in about the same quantities as clay bricks.

Industrial production did not start in Britain until about ten years later, largely due to the limited availability of suitable sands and the high quality and low cost of clay bricks. No exact figures are available for the early years of the calcium silicate brick industry in this country, but it is believed that there were four factories in 1905 producing about 16 million bricks a year between them. Nevertheless, there has certainly been a steady increase in the number of factories and their average output ever since. In 1968, for example, the total output was 346 million calcium silicate bricks from factories producing on average about 14 million bricks each.

Some other countries with significant volumes of calcium silicate brick production are, firstly the USSR, then Holland, East Germany and Poland. Somewhat smaller outputs are achieved in Bulgaria, the USA, Taiwan and many other countries. It will be evident, therefore, that calcium silicate bricks have won a permanent place in the world constructional scene.

Materials

The main raw materials are silica sand, lime and water, crushed or uncrushed siliceous gravel or crushed siliceous rock are sometimes used instead of sand or in combination with it.

Not all sands are available and, since sand constitutes over 90% of the solids in the mix, it is common practice for calcium silicate brick factories to be located close to appropriate sand deposits. Lime is usually obtained from outside sources. However, the quality of this constituent is again critical and stringent acceptance tests are applied.

Properties

Calcium silicate bricks have broadly similar functional appearance and economic properties to those listed on page 4.

The method of manufacture produces bricks which are exceptionally uniform in appearance and which have somewhat sharper arrises and corners, and smoother faces than is normally the case with the majority of clay bricks. Because colour is a stable property added during manufacture — instead of being a consequence of the process — it is variable within very wide limits to suit users' requirements.

Calcium silicate brick manufacturers are no less

concerned with reliability than their counterparts in the clay brick industry, and their products are in strict conformity with the functional and dimensional requirements of BS 187.*

Dimensions

The *work size* and *format size* modes of specification (see page 4) are equally applicable to calcium silicate bricks.

In 1967, BS 187 defined the single Imperial size as:

Work size $8\frac{5}{8}$ x $4\frac{1}{8}$ x $2\frac{5}{8}$ in
Format size 9 x $4\frac{1}{2}$ x 3 in

Amendment 341 provided the following single standard metric size in November 1969:

Work size 215 x 103 x 65 mm
Format size 225 x 112.5 x 75 mm

Many manufacturers are also able to supply alternative sizes, including 200 & 300 mm modular formats.

Applications

Being substantially similar in application to clay bricks, calcium silicate bricks are used for an equally wide range of loadbearing and non-loadbearing purposes in all kinds of buildings and engineering structures (see page 4).

Nomenclature

The generic name is 'calcium silicate bricks'. Depending on the materials used, however, the more specific terms 'sandlime' and 'flintlime' are acceptable designations recognised by BS 187.

Classification

Six classes of calcium silicate bricks are recognised in BS 187, graded largely according to compressive strength:

Class 1	7.0 MN/m²	(1,000 lbf/in²)
Class 2A	14.0 MN/m²	(2,000 lbf/in²)
Class 2B	14.0 MN/m²	(2,000 lbf/in²)
Class 3A	20.5 MN/m²	(3,000 lbf/in²)
Class 3B	20.5 MN/m²	(3,000 lbf/in²)
Class 4	27.5 MN/m²	(4,000 lbf/in²)
Class 5	34.5 MN/m²	(5,000 lbf/in²)
Class 7	48.5 MN/m²	(7,000 lbf/in²)

The A and B variants in classes 2 and 3 differ only in drying shrinkage – a slightly higher percentage of the original wet length measurement being permitted in the B grades.

All the above classes are suitable for loadbearing purposes, and all except Class 1 are suitable for facing purposes.

Manufacture

Principles of the process

As noted earlier, the manufacture of calcium silicate bricks only became a practical proposition when the use of steam under pressure was introduced. This technique, which is based on the same principles as the familiar domestic pressure cooker, is known as autoclaving.

Under autoclaving conditions, silica – the main constituent of sand or flint reacts slowly with hydrated lime (calcium hydroxide) and water to form calcium hydrosilicates. There are many of these, and

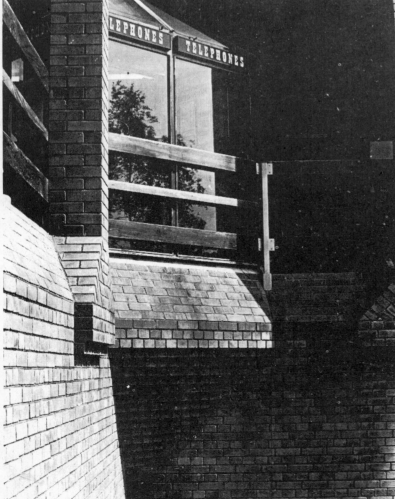

*BS 187: 1967 Calcium Silicate (Sandlime and Flintlime) bricks

they vary in complexity and properties. Ideally, all the autoclaving time is used up by this reaction so that the resultant calcium hydrosilicates envelop each grain of sand more or less completely and form bridges between them. The process is, in fact, very similar to the embedding of sand (and coarser aggregate) in a matrix of cement in concrete. As with concrete, an excess of binder would lower the strength. Under optimum conditions, all the spaces between the sand grains would just be completely filled, as the grains impart the strength and the hydrosilicates act as the glue that binds them together —the thinnest glue line being the best. In actual practice, however, there are always some empty spaces between the grains.

Other factors also affect strength development. If the mix is thoroughly compacted before moulding, fewer empty spaces have to be filled and the strength will be greater. Similarly, if all the sand grains are the same size, there will be more voids than if the sand is well graded so that smaller grains fill the spaces between larger ones. Furthermore, not all calcium hydrosilicates are equally strong, and some of the strong ones can change and become weaker through excessive autoclaving. As with many other complex industrial processes, the precise manufacturing conditions at all stages represent a compromise between desirable final properties, mechanical feasibility and acceptable cost.

Method

The whole manufacturing process is highly mechanised and, in many plants, very largely automated. Rigid quality control of raw materials and products is exercised at all stages.

The raw materials – sand and/or crushed flint, lime and water – are intimately mixed together, accurate proportioning being of critical importance. Colour is also added at this stage, if required.

The carefully blended mix is then stored to allow the quicklime to hydrate. This part of the process can, of course, be omitted if slaked (hydrated) lime is used instead of quicklime. When fully hydrated, the mix is compressed to ensure optimum compaction, and shaped into brick units by presses.

After pressing, the brick units are stacked on bogies which – like the cars in clay brick tunnel kilns – will enter the autoclave with their loads. In this 'pressure cooking' phase of the process, the brick units are subjected to compressed saturated steam for several hours, and this will cause them to harden into durable building components capable of bearing loads and withstanding the rigours of our climate. By varying the autoclaving time and the steam pressure, the performance characteristics of the bricks can be adjusted to suit requirements.

Hot bricks from the autoclave are normally stored in the open air. If necessary, however, they can be used as soon as they are cool enough to handle.

Research and development

Co-operative research for the great majority of calcium silicate brick manufacturers is carried out by the Welwyn Hall Research Association. Research is largely directed towards the investigation of the factors discussed in the paragraphs on the principles of the process, and to determining the best plant, machinery and techniques to be used in the light of existing and new knowledge. Questions of testing, application, standardisation and various aspects of

user research also form an important part of the programme, which is effectively rationalised through close co-operation with the corresponding German and Dutch research organisations.

The strength and value of the work carried out at Welwyn Hall is exemplified by the fact that whereas in 1955 BS 187 called for a compressive strength of 3,000 lbf/in² (20.5 MN/m²) in the best special purpose calcium silicate bricks, the 1967 version of the Standard specifies bricks with compressive strengths up to 7,000 lbf/in² (48.5 MN/m²). There is every reason for expecting that the progress of the Association, in enabling better and stronger bricks to be produced, will be maintained in the future☐

Opposite page Details of the shopping arcade at Chandler's Ford, Hants (Architects: Julian Keable & Partners, now Triad Architects). Calcium silicate bricks are substantially similar in application to clay bricks and are used for an equally wide range of loadbearing and non-loadbearing purposes.

This page
Left A modern hydraulic press capable of producing 4000 calcium silicate bricks per hour.

Below Autoclaves in which the pressed bricks are subjected to compressed saturated steam for several hours.

In recent years, the sharp revival in brick construction has reflected the disenchantment of designers with other materials and forms of construction which have proved neither quicker nor cheaper than brickwork and which deteriorate at a wholly unacceptable speed.

Here are four pictures which demonstrate the flexibility of brickwork to meet differing architectural requirements. **(Below)** Julio Lafuente's masterly external staircase at the Villa Mariotti near Rome. **(Below right)** Alvar Aalto's great wedge-shaped lecture hall at Otaniemi University, Finland. **(Right)** Nelson & Parker's very original private house at Scarisbrick, Lancs, which is entirely made up of U-shaped brick piers. **(Far right)** Alexandra Square, part of the superb environmental essay by Shepheard & Epstein which is Lancaster University. Shepheard himself has said: 'Our attachment to brick as a facing material starts from a wish to see our buildings looking better with age, and a conviction that the contrast provided by brick and white paint is a very useful theme, as Georgian London stands to witness, in our not-too-sunny climate. We find bricks in the higher price ranges an economic proposition; and even if much more were spent on buildings than the present cost limits permit, we would still probably use a great deal of brick'.

Section 1
BASIC INFORMATION

BRICK DATA

WEIGHT

The weight of bricks will vary according to the type of clay and method of manufacture, including the presence of one or two frogs or of perforations. In addition, the moisture content will affect their weight as used. The following figures are therefore approximate:

1 Standard Brick	5½—8½ lb.
1,000 Standard Bricks	2½—4 tons
1 cu. ft. brickwork (Commons)	125 lb.
1 cu. ft. brickwork (Dense Engineering)	150 lb.
1 sq. yd. 4½ in. thick... ...	approx. 410 lb.
1 sq. yd. 9 in. thick	approx. 820 lb.
1 sq. yd. 13½ in. thick ...	approx. 1,230 lb.

STRENGTH

In practice what matters is not merely the strength of the bricks but the combined strength of bricks and mortar. Bricks vary widely in strength according to their type and also within some types the strength variation may sometimes be fairly considerable. Manufacturers can usually supply the necessary data.

Methods of testing bricks for strength and the requirements for Engineering Bricks are given in British Standard 3921:1965.

Reference should be made to B.S. Code of Practice 111 (1964) for information on the design of calculated loadbearing brickwork. The appropriate Building Regulations should also be consulted since these do not in all cases follow precisely the Code of Practice. The various Building Regulations should also be referred to for the wall thickness required for uncalculated brickwork.

SOUND INSULATION

The sound insulation of brickwork, in decibels, for the frequency range 200—2,000 cycles per second is:
    4½in. brickwork = 45 decibels.
    9in. brickwork = 50 decibels.
    13½in. brickwork = 52 decibels.
    The use of an ordinary cavity wall in place of a 9in. solid wall is unlikely to provide any noticeable difference in sound insulation.

SOUND ABSORPTION

The sound absorption coefficient of untreated normal brickwork will be approximately:
    ·024 at 125 cycles per second.
    ·03 at 500 cycles per second.
    ·05 at 4,000 cycles per second.

MOISTURE MOVEMENT

A burnt clay product such as brick has very little movement itself but when combined with mortar some shrinkage of the brickwork can occur. The stronger the mortar the greater is the chance of such shrinkage becoming obvious.

In the past it has usually been assumed that moisture expansion did not occur. However, if bricks come straight from the kiln to site and are quickly used some expansion may occur.

The increasing tendency for brickwork to be built in close contact to concrete, and often restrained by that material, makes it necessary in such cases to consider the requirements arising from the differential movements of the materials.

For all those reasons the possibility of movement should be considered at the design stage and movement joints should be included accordingly.

FROST RESISTANCE

Well fired clay bricks can have an excellent frost resistance but where there is an unusual degree of exposure and where brickwork is liable to frequent saturation (e.g. in parapet walls, retaining walls or below D.P.C. on damp sites) special care may be required.

At present there is no entirely satisfactory laboratory test for frost resistance, although work in this direction continues. British Standard 3921: 1965 refers to bricks of three qualities: Internal; Ordinary; Special. Those of Ordinary quality should be durable in the external face of a building while those of Special quality should resist extreme exposure. On Frost Resistance the British Standard does not give any requirement except for bricks of Special quality and then it is that manufacturers shall produce evidence from exposure of at least three years under conditions similar to those in which the bricks will be used.

Water absorption tests in themselves are not a guide to frost resistance and are included in the British Standard only for Engineering Bricks and for bricks to be used for damp-proof courses.

FIRE RESISTANCE

The fire resistance of brickwork is very good. 8½in. of unplastered brickwork can give the highest rating of fire resistance, 6 hours, according to B.S. Code of Practice 111. Values for other thicknesses are tabulated in C.P. 111 and in the various Building Regulations. It should be noted that in some Regulations the thickness of brickwork required to achieve a given period of fire resistance varies according to whether the brickwork is loadbearing or non-loadbearing. For specific cases therefore reference should be made to the appropriate Building Regulations.

RECOMMENDED MORTAR MIXES

	TYPE OF CONSTRUCTION AND POSITION IN BUILDING	DEGREE OF EXPOSURE TO WIND AND RAIN	TIME OF CONSTRUCTION	RECOMMENDED MIXES (1) (PARTS BY VOLUME)	
External Walls Clay bricks Clay blocks	Normal construction not designed to withstand heavy loading Above damp-proof course	Sheltered and moderate conditions	Spring and Summer	1:2:8-9 1:8 1:3	Cement : lime : sand. Cement : sand with mortar plasticizer. Hydraulic lime : sand
		Sheltered and moderate	Autumn and Winter	1:1:5-6 1:5-6 1:2	Cement : lime : sand Cement : sand with mortar plasticizer. Hydraulic lime : sand
		Severe conditions	All seasons	1:1:5-6 1:5-6 1:2	Cement : lime : sand. Cement : sand with mortar plasticizer. Hydraulic lime : sand.
	Normal construction not designed to withstand heavy loading Parapets, free-standing walls or below damp-proof course	All conditions	All seasons	1:1:5-6 1:5-6 1:3	Cement : lime : sand Cement : sand with mortar plasticizer. Cement : sand.
Clay bricks over 5,000 lb/sq. in. crushing strength	Engineering construction All positions	All conditions	All seasons	1:3	Cement : sand.
Internal Walls (including partitions)	Normal	—	Spring and Summer	1:2:8-9 1:8 1:3:10-12 1:3	Cement : lime : sand. Cement : sand with mortar plasticizer. Cement : lime : sand. Hydraulic lime : sand.
			Autumn and Winter	1:2:8-9 1:8 1:1:5-6 1:5-6 1:2	Cement : lime : sand. Cement : sand with mortar plasticizer. Cement : lime : sand. Cement : sand with mortar plasticizer. Hydraulic lime : sand.
Tall chimneys	—	All conditions	All seasons	1:2-3 1:2:8-9	Hydraulic lime : sand. Cement : lime : sand.

(1) "Lime" refers to non-hydraulic or semi-hydraulic lime. The proportions given are for lime putty. If the lime is measured as the dry hydrate, the amount can be increased up to 1½ vols. for each vol. of lime putty; the hydrate should preferably be soaked at least overnight before use. Where a range of sand contents is given (e.g., 5-6, 8-9, or 10-12), the higher should be used for sand that is well graded, and the lower for coarse or uniformly fine sand.

METHOD OF MEASUREMENT

The method of measurement of brickwork for building work, excepting work measured in accordance with the Scottish Mode, is governed by the Standard Method of Measurement of Building Works, Fifth Edition, revised March 1964 and published by The Royal Institution of Chartered Surveyors of 12 Great George Street, Westminster, London, S.W.1. and The National Federation of Building Trades Employers of 82 New Cavendish Street, London, W.1.

The following is a brief summary of the basic principles involved in measurement in accordance with the foregoing method of measurement:

1. Brickwork of two brick thickness and over in each of the following classifications shall be reduced to one brick thick and billed separately in square yards.

Brickwork of under two brick thickness in each of the following classifications shall be billed separately in square yards stating the thickness.

The classifications are as follows:

Walls, filling old openings, skins of hollow walls, dwarf supports under fittings, tanks, etc., isolated piers and chimney stacks, battering walls, brickwork used as formwork, refractory brick linings as flues, brick damp proof courses and vaulting.

2. Faced brickwork shall be billed in square yards as extra over the brickwork on which it occurs and pointing shall be included with the item.

Half brick walls and one brick walls built fair both sides or entirely of facings shall be billed separately in square yards.

3. For such items as backing to masonry, thickening old walls, tapered walls, grooved bricks, eaves filling, rough cutting, rough chases, rough arches, bonding, facework to walls, piers, chimney stacks, returns, quoins, arches and tumblings, fair cutting, fair angles and chases, facework to plain and ornamental bands and cornices, tile creasings, sills, thresholds, copings and steps, key blocks, corbels, bases and cappings, pavings, boiler seatings and flues

reference should be made to the Standard Method of Measurement.

The method of measurement of brickwork for building work measured in accordance with the Scottish Mode is governed by the Scottish Mode of Measurement of Building Works revised and operative from 1st January, 1958, and with subsequent amendments to individual trades obtainable from the Scottish Branch of the Royal Institution of Chartered Surveyors, 7 Manor Place, Edinburgh 3, and 48 West Regent Street, Glasgow, C.2.

The method of measurement of brickwork for civil engineering work is governed by the Standard Method of Measurement of Civil Engineering Quantities reprinted 1964 and obtainable from The Institution of Civil Engineers, Great George Street, London, S.W.1.

EFFLORESCENCE AND SALTS

Efflorescence is the formation of crystals when water soluble salts are dried out to the surface. It usually shows up as white patches and is likely to be most prominent in the Springtime when dry weather follows a wet period. Normally efflorescence will disappear when the wall becomes wet again but may reappear in subsequent dry weather. Generally it will get less and less as time goes on. If it is desired to speed up the removal of efflorescence, dry brushing away of the salts may help. Washing is not recommended as this merely causes the salts to go into solution and be reabsorbed into the wall.

Salts may be introduced by the mortar or from other adjoining materials. Some salts may be present in the bricks and B.S. 3921:1965 describes methods of testing for efflorescence with results given in five gradings as 'nil', 'slight', 'moderate', 'heavy' or 'serious'. The B.S. permits a result of 'moderate'. The B.S. also sets out methods of determining the soluble salt content of bricks but lays down limiting standards only in the case of Special Quality bricks.

In most cases the best guide to probability of freedom from efflorescence can be obtained by inspecting buildings of similar brickwork but it must be realised that if building conditions or exposures vary these may cause different results.

Sulphate expansion in brickwork can cause damage. For it to occur there must be present soluble salts, tricalcium aluminate (from normal cement) plus water. It is impossible to lay down reasonable limits of sulphate content below which trouble will not occur as the result is so greatly influenced by the type of construction and the consequent access of water. Either the use of a very low sulphate content brick, or the use of a sulphate resisting cement or a method of construction ensuring reasonable water exclusion can be satisfactory.

BRITISH STANDARD 3921:1965 PROVIDES FOR ONE STANDARD SIZE OF BRICK, NOMINALLY 9 INS x 4½ INS x 3 INS. ACTUAL SIZE 8⅝ INS x 4⅛ INS x 2⅝ INS, WITH ALLOWANCE FOR TOLERANCES. THE STANDARD METHOD OF MEASUREMENT IS TO PLACE 24 BRICKS IN CONTACT, WHEN THE MEASUREMENTS SHOULD LIE WITHIN THE FOLLOWING LIMITS — LENGTH 204/210 INS, WIDTH 97½/100¾ INS, DEPTH 61¼/64¼ INS.

OTHER BRICK SIZES ARE ALSO AVAILABLE, INCLUDING BRICKS OF 2⅞ INS DEPTH, AND MODULAR BRICKS BOTH SMALLER AND LARGER THAN THE B.S. SIZE, AND ALSO A VARIETY OF LARGER CLAY BUILDING UNITS. TOLERANCE RANGES ON THESE ARE NOT INCLUDED IN THE BRITISH STANDARD AND INDIVIDUAL MANUFACTURERS SHOULD BE CONSULTED FOR INFORMATION ON THE LIMITS WITHIN WHICH THEIR PRODUCTS FALL.

SQUARES
and
CLOSERS

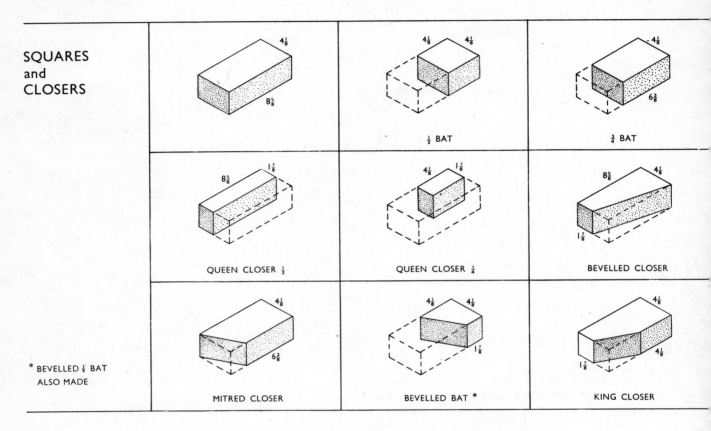

½ BAT

¾ BAT

QUEEN CLOSER ½

QUEEN CLOSER ¼

BEVELLED CLOSER

MITRED CLOSER

BEVELLED BAT *

KING CLOSER

* BEVELLED ¼ BAT
ALSO MADE

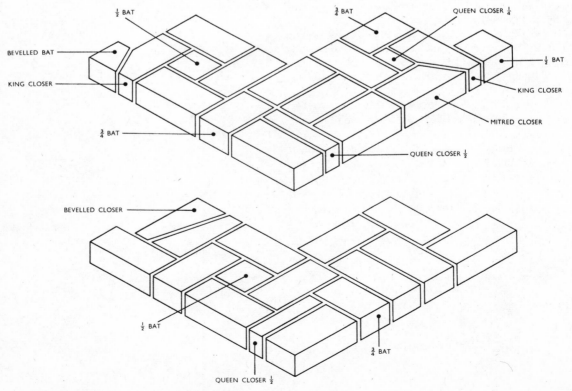

22

SQUINTS

SQUINTS ARE ALSO MADE
IN OTHER ANGLES

| 30° SQUINT | 45° SQUINT | 60° SQUINT |

RADIALS

Radius to specification

| RADIAL STRETCHER | RADIAL HEADER | CULVERT STRETCHER AND HEADER |

PLINTHS

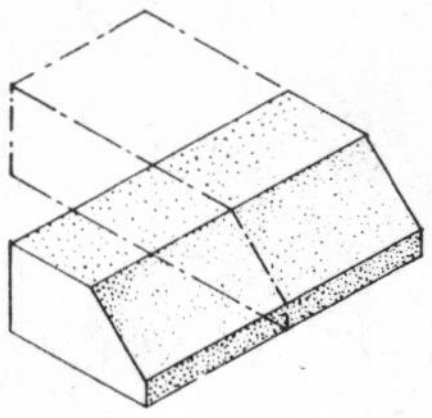

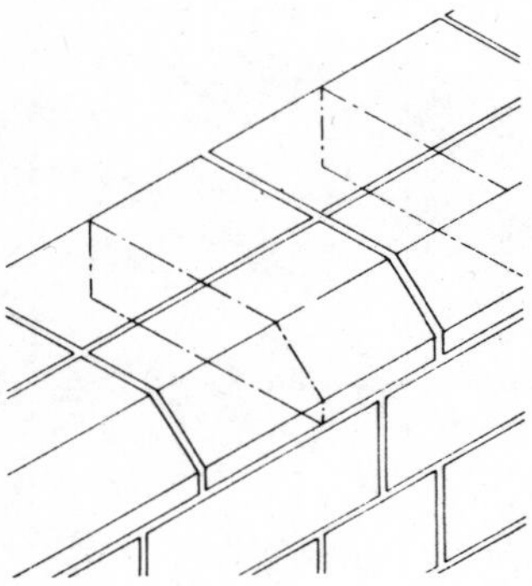

PLINTH INTERNAL RETURN
ALSO MADE AS 1½

| STRETCHER AND HEADER PLINTH | PLINTH EXTERNAL RETURN Handed | PLINTH INTERNAL RETURN 6½" Handed |

| PLINTH STOP Handed | PLINTH EXTERNAL ANGLE Handed | PLINTH INTERNAL ANGLE Handed |

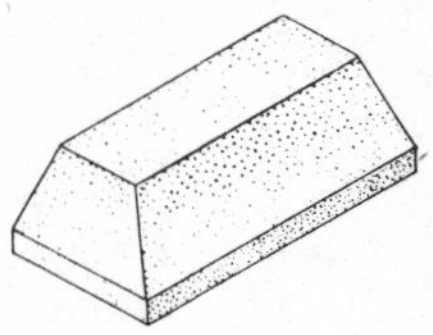

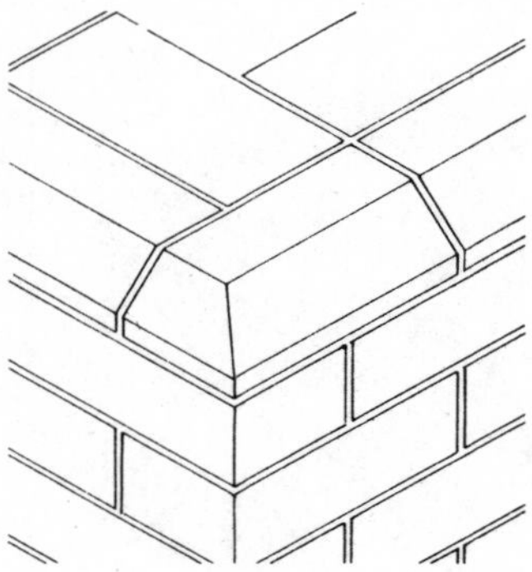

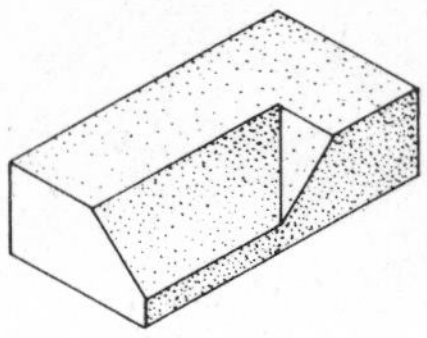

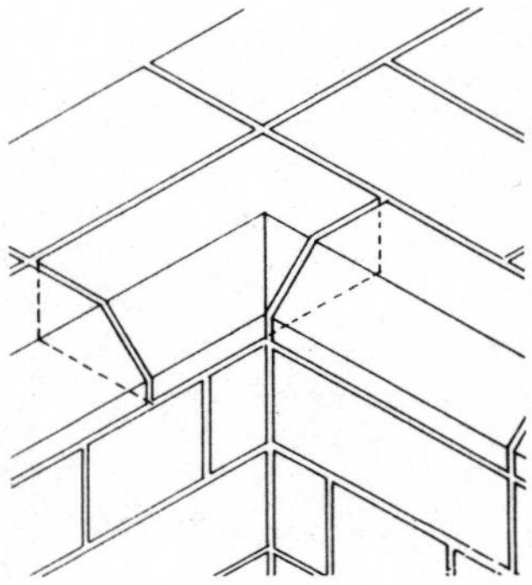

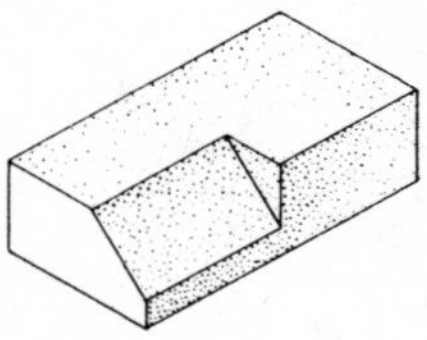

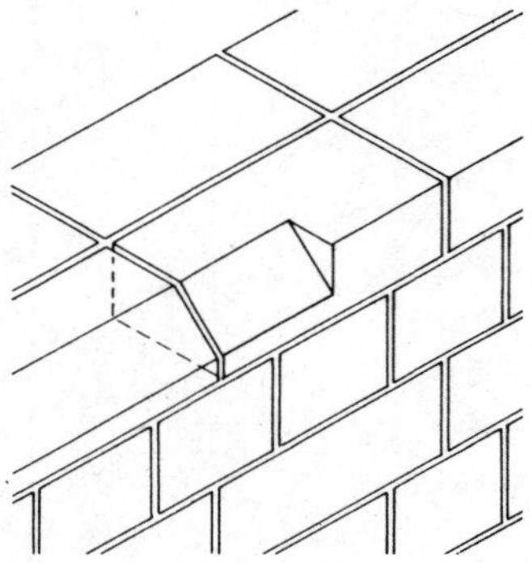

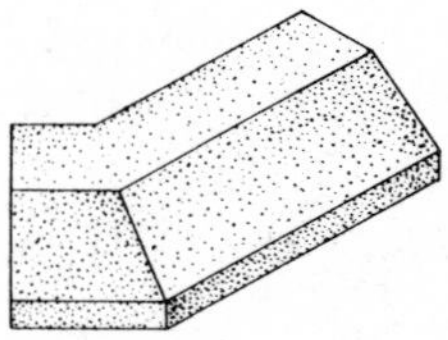

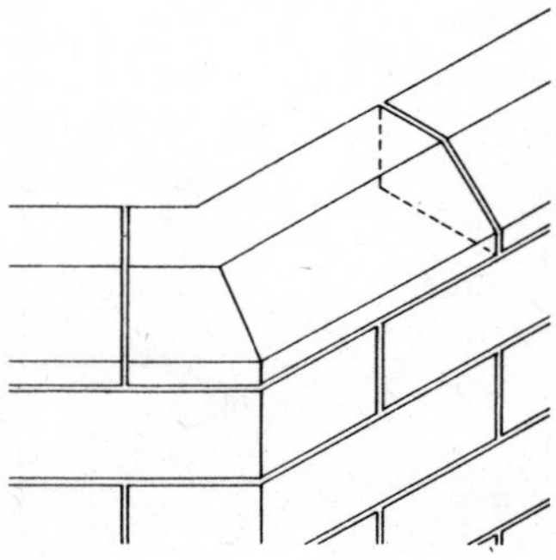

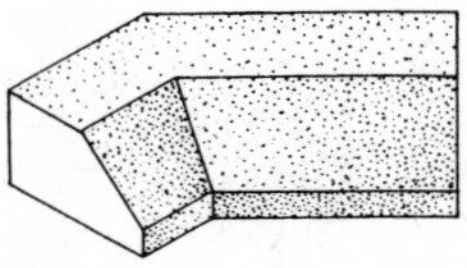

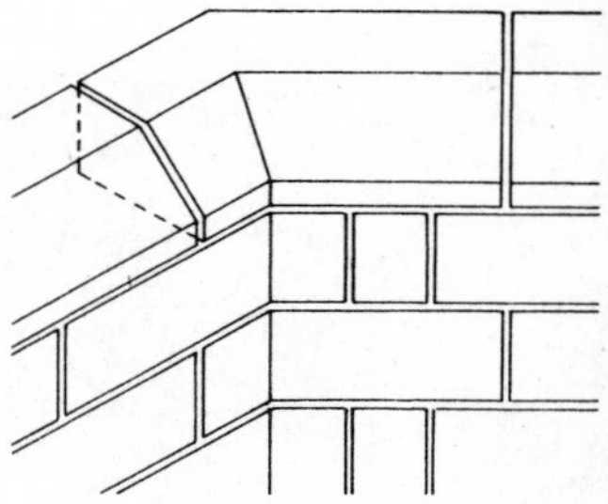

BRITISH STANDARD DOES
NOT DEFINE DIMENSIONS OR
ANGLES.
MANUFACTURERS WILL PRO-
VIDE THIS INFORMATION.

THE TOPS OF EXTERNAL WALLS ARE VERY EXPOSED TO RAIN AND FROST. WALLS SHOULD ALWAYS BE PROTECTED
BY A D.P.C. BENEATH COPINGS AND THE COPING BRICKS SHOULD BE OF A SUITABLE QUALITY. IN ADDITION TO THE
BRICKS ILLUSTRATED ABOVE AND ON SHEETS 5 & 6 AND COPINGS SHOWN ON SHEET 7. CLAY COPINGS OF OTHER
TYPES ARE DESCRIBED IN BRITISH STANDARD 3798: 1964.

SPLAYS
and
ANGLES

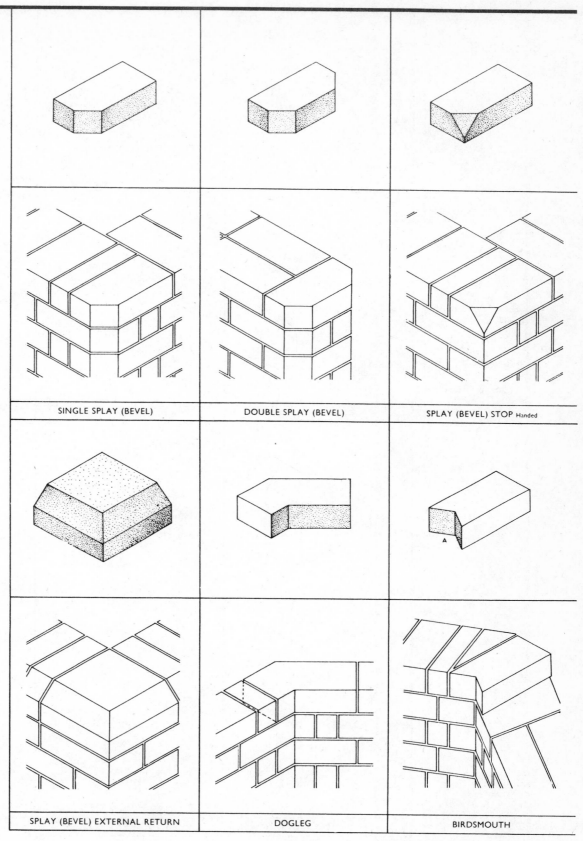

SINGLE SPLAY (BEVEL)

DOUBLE SPLAY (BEVEL)

SPLAY (BEVEL) STOP Handed

SPLAY (BEVEL) EXTERNAL RETURN

DOGLEG

BIRDSMOUTH

Dogleg and Birdsmouth are made 30°, 45° and 60° or to specification

BULLNOSE

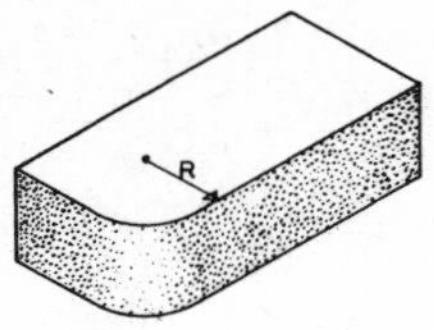

Radius R 2 1/16" or 1 1/8"

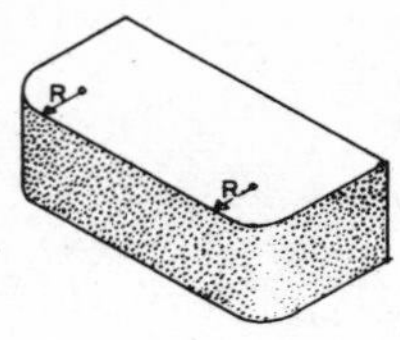

Radius R 2 1/16" or 1 1/8"

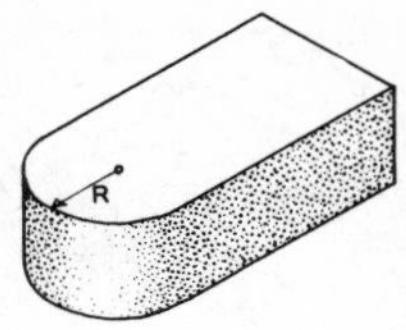

Radius R 2 1/16"

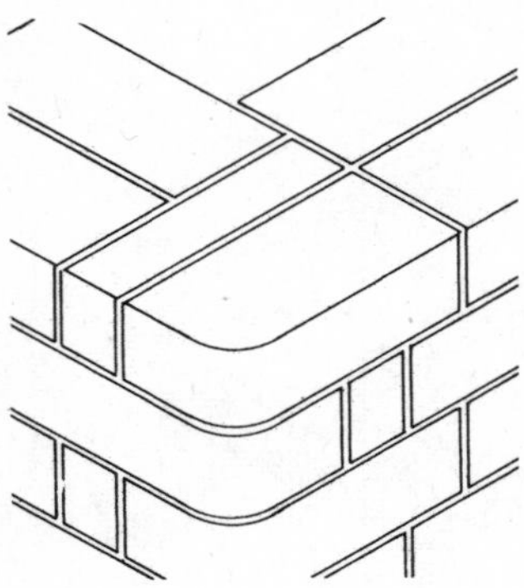

SINGLE BULLNOSE

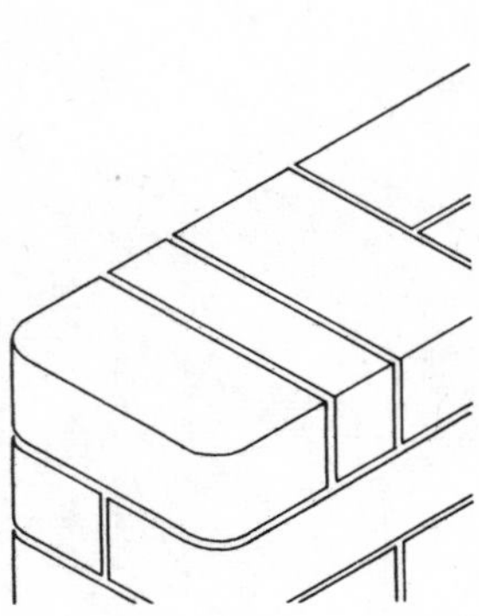

DOUBLE BULLNOSE

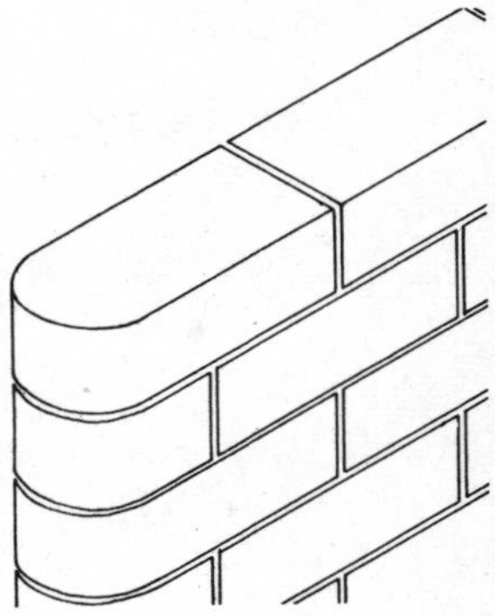

BULLHEAD (COWNOSE)

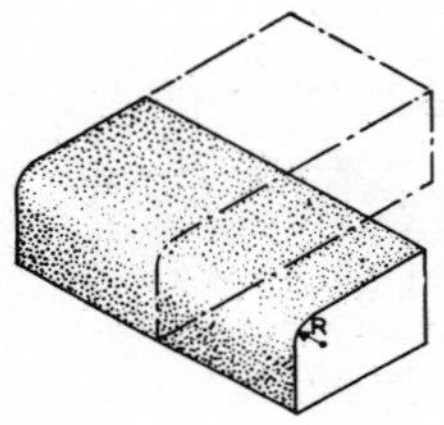

Radius R 2 1/16" or 1 1/8"

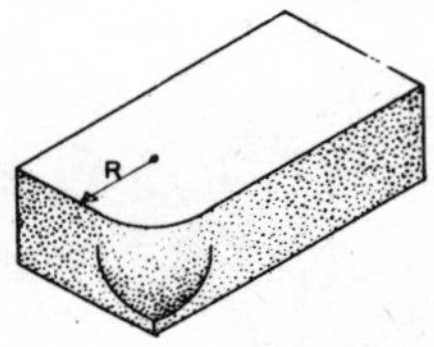

Also available double

Radius R 2 1/16" or 1 1/8"

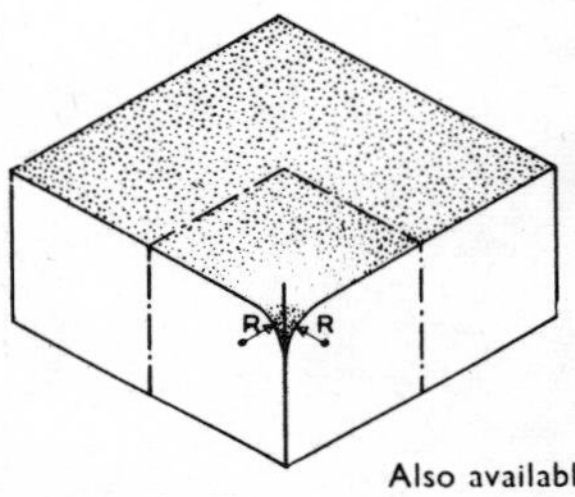

Radius R 2 1/16" or 1 1/8"

Also available for 2 5/8" depth

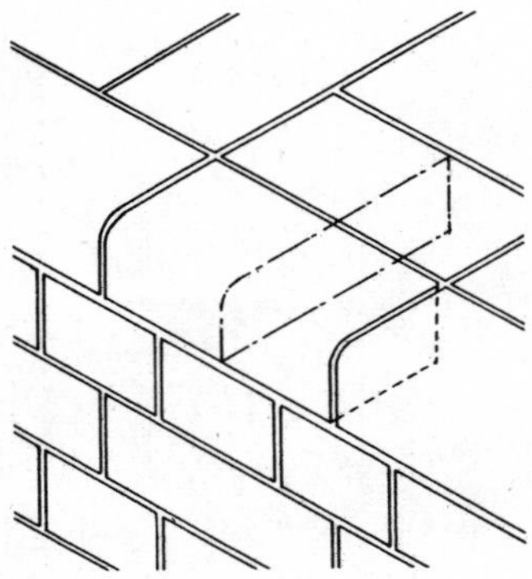

BULLNOSE ON FLAT (stretcher & header)

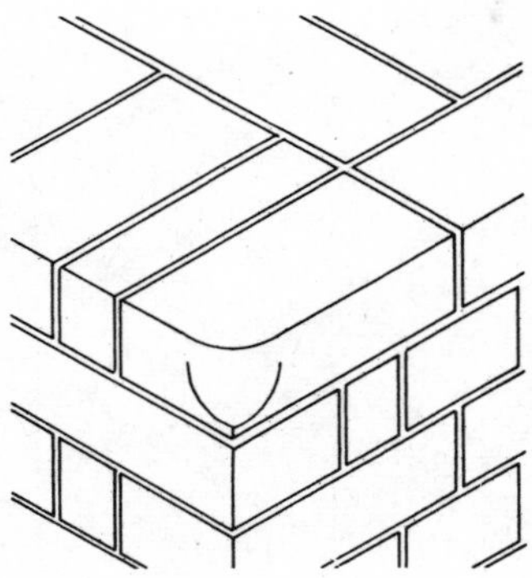

BULLNOSE STOP Handed

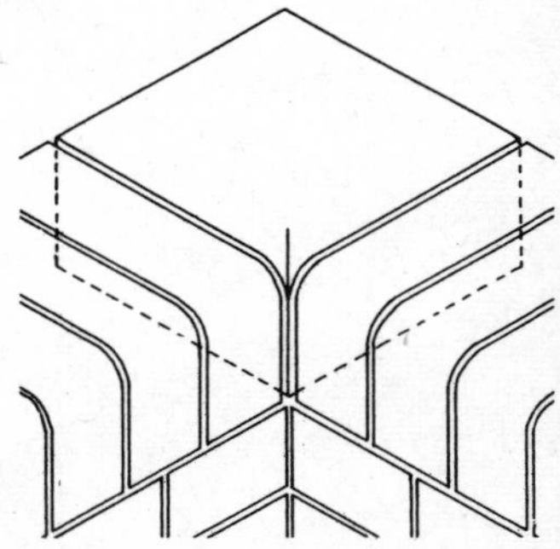

INTERNAL RETURN OR MITRE

BRITISH STANDARD DOES NOT DEFINE RADII BUT NORMALLY THEY ARE AS SHOWN.

THE TOPS OF EXTERNAL WALLS ARE VERY EXPOSED TO RAIN AND FROST. WALLS SHOULD ALWAYS BE PROTECTED BY A D.P.C. BENEATH COPINGS AND THE COPING BRICKS SHOULD BE OF A SUITABLE QUALITY. IN ADDITION TO THE BRICKS ILLUSTRATED ABOVE AND ON SHEETS 3 & 6 AND COPINGS SHOWN ON SHEET 7, CLAY COPINGS OF OTHER TYPES ARE DESCRIBED IN BRITISH STANDARD 3798: 1964.

BRICK SHAPES—BULLNOSE (see also Sheet 5)

BULLNOSE

Radius R $2\frac{1}{16}$" or $1\frac{1}{8}$"

Radius R $2\frac{1}{16}$" or $1\frac{1}{8}$"

Radius R $2\frac{1}{16}$" or $1\frac{1}{8}$"

EXTERNAL RETURN ON FLAT Handed

EXTERNAL RETURN ON EDGE Handed

INTERNAL RETURN ON EDGE Handed

Radius R $2\frac{1}{16}$" or $1\frac{1}{8}$"

Radius R $2\frac{1}{16}$" or $1\frac{1}{8}$"

Radius R $2\frac{1}{16}$" or $1\frac{1}{8}$"

INTERNAL RETURN ON FLAT Handed

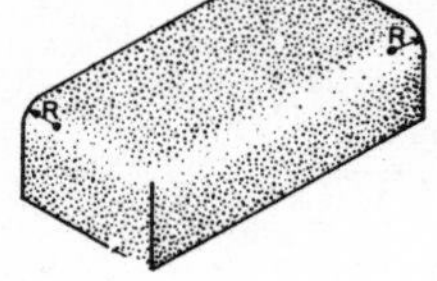

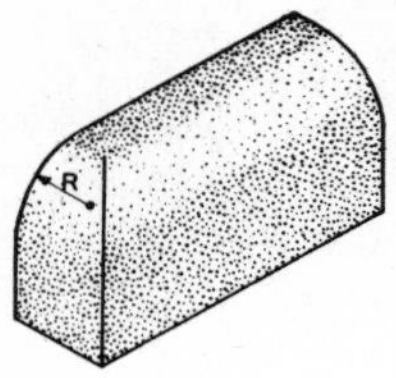

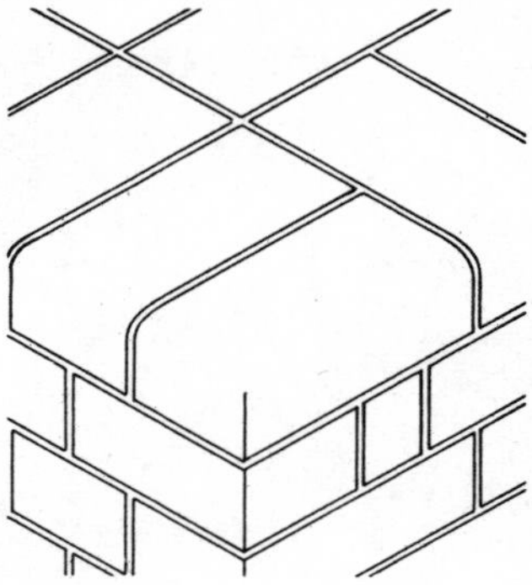

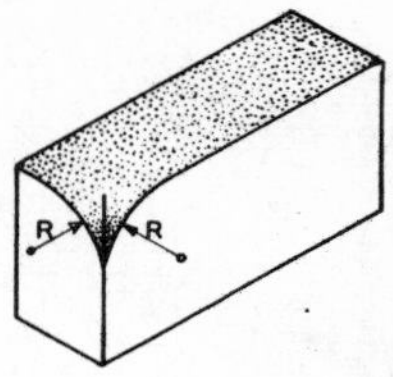

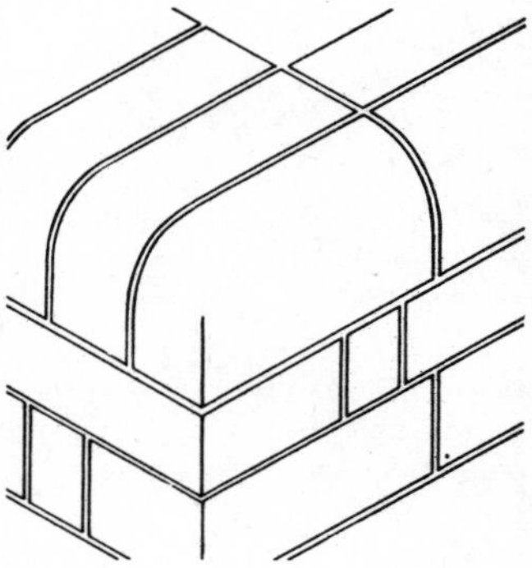

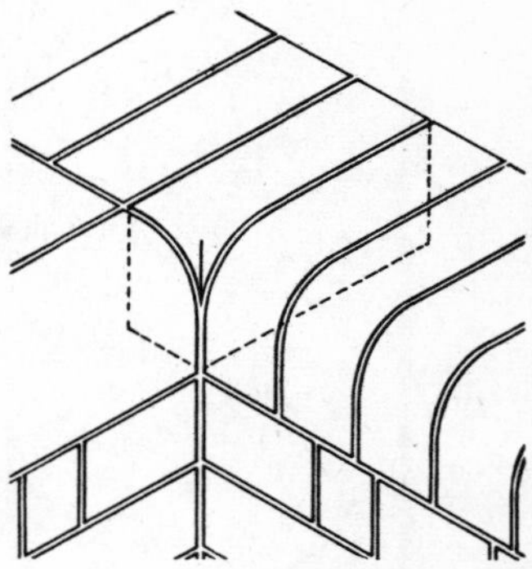

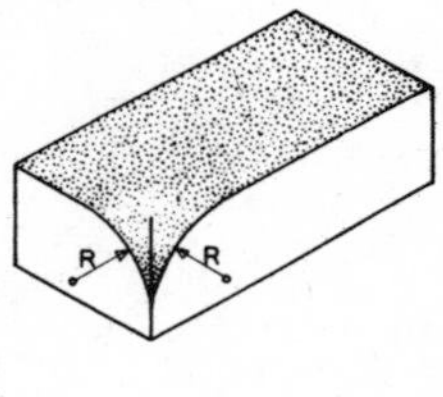

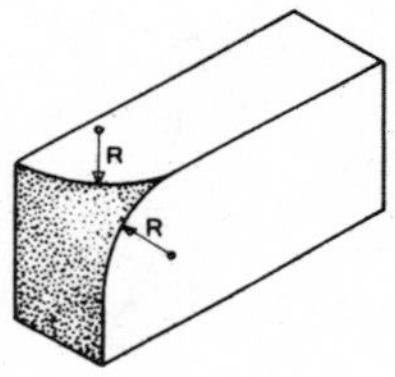

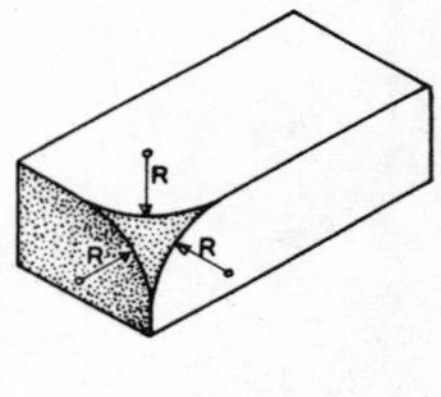

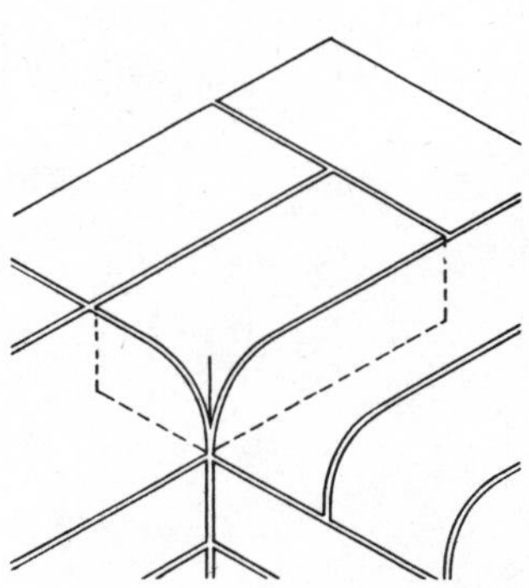

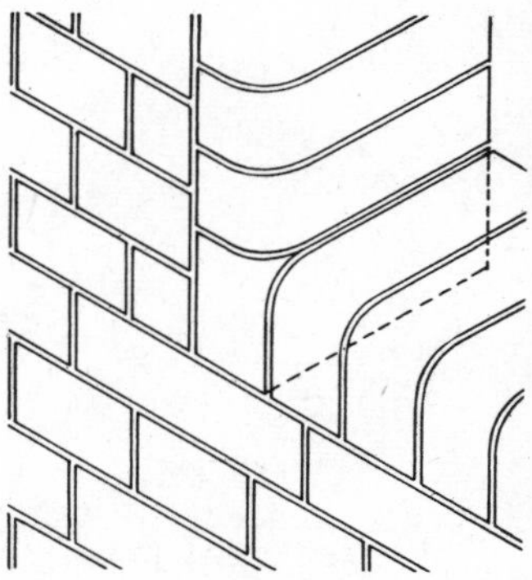

RETURN ON EDGE (CILL & JAMB)

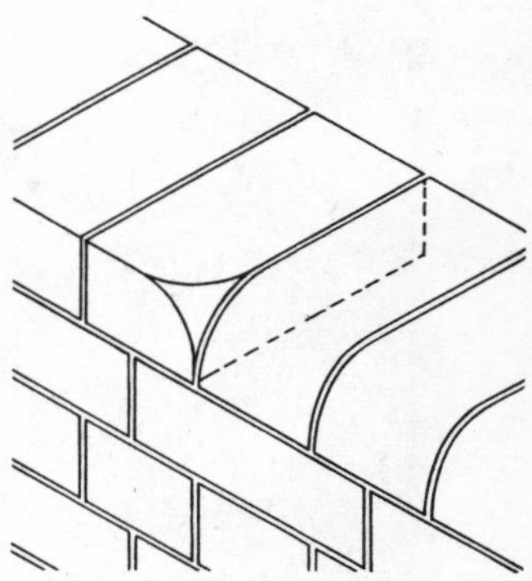

RETURN ON FLAT Handed

BRITISH STANDARD DOES NOT DEFINE RADII BUT NORMALLY THEY ARE AS SHOWN.

THE TOPS OF EXTERNAL WALLS ARE VERY EXPOSED TO RAIN AND FROST. WALLS SHOULD ALWAYS BE PROTECTED BY A D.P.C. BENEATH COPINGS AND THE COPING BRICKS SHOULD BE OF A SUITABLE QUALITY. IN ADDITION TO THE BRICKS ILLUSTRATED ABOVE AND ON SHEETS 3 & 5 AND COPINGS SHOWN ON SHEET 7, CLAY COPINGS OF OTHER TYPES ARE DESCRIBED IN BRITISH STANDARD 3798: 1964.

COPINGS

FOR 9" WALLS

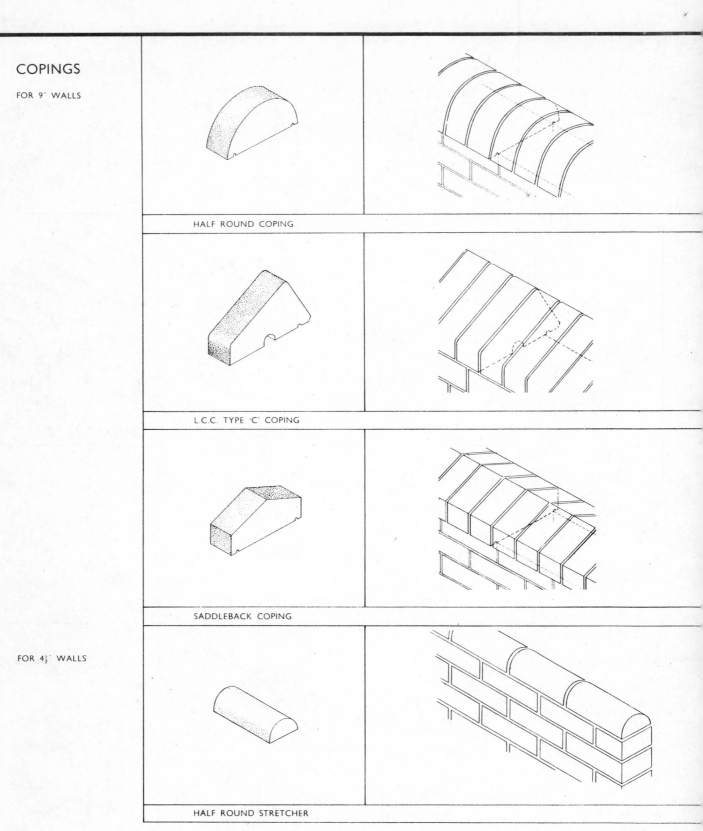

HALF ROUND COPING

L.C.C. TYPE 'C' COPING

SADDLEBACK COPING

FOR 4½" WALLS

HALF ROUND STRETCHER

THE TOPS OF EXTERNAL WALLS ARE VERY EXPOSED TO RAIN AND FROST. WALLS SHOULD ALWAYS BE PROTECTED BY A D.P.C. BENEATH COPINGS AND THE COPING BRICKS SHOULD BE OF A SUITABLE QUALITY. IN ADDITION TO THE COPINGS SHOWN ABOVE, AND THE BRICKS ILLUSTRATED ON SHEETS 3, 5 & 6, CLAY COPINGS OF OTHER TYPES ARE DESCRIBED IN BRITISH STANDARD 3798: 1964.

BRICK BONDS

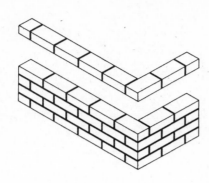

STRETCHER BOND

STRETCHER BOND

The normal bond for walls of half-brick thickness.

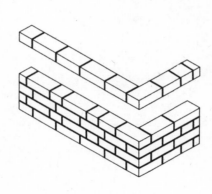

STRETCHER WITH SNAP HEADERS

STRETCHER WITH SNAP HEADERS

An easy way to improve the appearance of half-brick walling. A variety of patterns is possible.

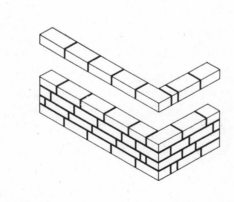

RAKING STRETCHER BOND

RAKING STRETCHER BOND

Economical and more interesting than normal Stretcher Bond.
Joints tend to become very prominent unless mortar colour is chosen with care.

BRICK BONDS

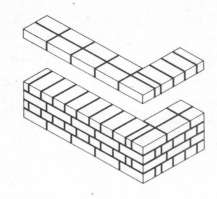

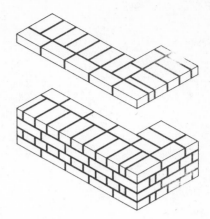

ENGLISH BOND
A strong bond and easy to lay, but has a certain monotony of appearance.

ENGLISH BOND

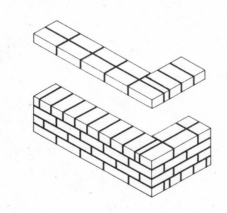

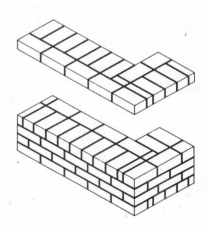

ENGLISH GARDEN WALL BOND
More economical in facing bricks than true English Bond. Frequency of header courses can vary.

ENGLISH GARDEN WALL BOND

BRICK BONDS

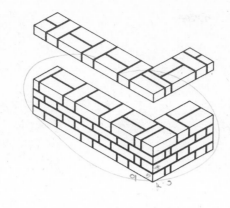

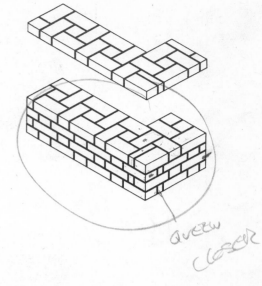

FLEMISH BOND
A simple pattern which is sometimes thought to give a more attractive appearance than English Bond.

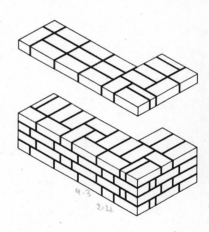

FLEMISH BOND

FLEMISH GARDEN WALL BOND

This requires a fair area of wall to show the pattern, and perpends need to be kept true, especially if headers differ from stretchers in colour.

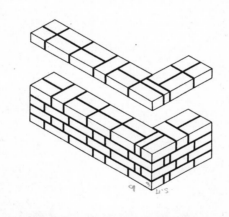

FLEMISH GARDEN WALL BOND

BRICK BONDS

left

MONK BOND

Has two stretchers to one header in each course with headers staggered. Complicated to lay, but gives an interesting appearance.

right

FLEMISH CROSS BOND

Similar to Flemish, but with two additional headers in place of a stretcher at intervals. Needs a large wall area to show well.

left

VERTICAL OR STACK BOND

Used mainly for panel infills as it is not a good bond for strength.

right

BASKET PATTERN

One of many possible decorative bonds.

left

PROJECTING BRICKS

Frequency and size of projections need relating to size of wall and distance from which it is seen.

right

HEADER BRICK ON EDGE

This unusual bond gives a very strong horizontal emphasis.

MONK BOND

FLEMISH CROSS BOND

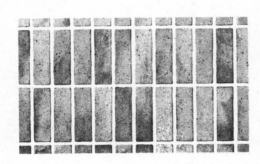

VERTICAL OR STACK BOND

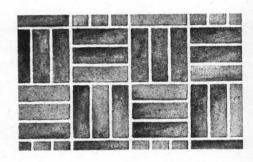

BASKET PATTERN

PROJECTING BRICKS

HEADER/BRICK ON EDGE

BRICK BONDS

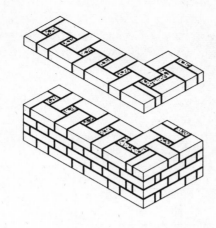

QUETTA BOND

QUETTA BOND
Vertical reinforcement in the voids, which are then filled with mortar.
The reinforcement is usually connected to foundations, floors and roof.

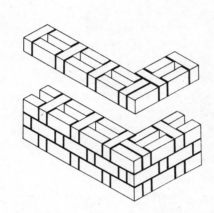

RAT-TRAP BOND

RAT-TRAP BOND
An economical wall of unusual appearance but with only partial cavities. Does not ensure resistance to rain.

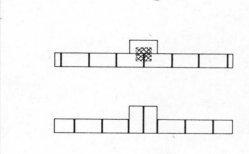

HALF BRICK WALL WITH PIER

HALF-BRICK WALL WITH PIER
Difficult to obtain a good appearance. This method results in a straight joint but bridges this with reinforcement.

THE AMOUNT OF HEAT TRANSMITTED THROUGH WALLS, FLOORS, ROOFS, ETC. IS PROPORTIONAL TO A COEFFICIENT 'U' WHICH IS DEFINED AS THE AMOUNT OF HEAT TRANSMITTED IN B.T.U.'S PER SQUARE FOOT PER HOUR PER DEGREE FAHRENHEIT DIFFERENCE BETWEEN INDOOR AND OUTDOOR TEMPERATURES. THIS COEFFICIENT IS MADE UP OF A VALUE FOR THE STRUCTURAL MATERIAL PLUS TWO VALUES FOR SURFACE RESISTANCE (INTERNAL AND EXTERNAL), THEREFORE AN INCREASE IN VALUE OF MATERIAL DOES NOT GIVE A PROPORTIONAL OVERALL INCREASE.

U-VALUES CAN BE CALCULATED QUITE SIMPLY. AN EXCELLENT GUIDE FOR THIS PURPOSE IS THE "COMPUTATION OF HEAT REQUIREMENTS FOR BUILDINGS" PUBLISHED BY THE INSTITUTE OF HEATING AND VENTILATING ENGINEERS.

TABLE OF U-VALUES FOR BRICK WALLS UNDER NORMAL EXPOSURE					
ALL WALLS PLASTERED INTERNALLY					
$4\frac{1}{2}''$ SOLID	9" SOLID	11" CAVITY VENTILATED	11" CAVITY UNVENTILATED	$13\frac{1}{2}''$ SOLID	18" SOLID
0.57	0.43	0.34	0.30	0.35	0.29

NOTES: THE CONDUCTIVITY (k) FOR BRICKWORK IS USUALLY ACCEPTED AS 8.0 B.T.U./SQ. FT./HR./°F/IN. THICKNESS BUT WILL VARY SOMEWHAT ACCORDING TO THE TYPE OF BRICK AND THE MOISTURE CONDITIONS OF THE WALL. IMMEDIATELY AFTER CONSTRUCTION WHEN THE BRICKWORK IS DRYING OUT INSULATION VALUES ARE LIKELY TO BE LOWER THAN WHEN THE WORK IS DRY.

BUILDING REGULATIONS NOW REQUIRE MINIMUM THERMAL INSULATION VALUES FOR CERTAIN CLASSES OF BUILDINGS. THE 1965 BUILDING REGULATIONS FOR ENGLAND AND WALES REQUIRE EXTERNAL WALLS OF DWELLINGS TO HAVE A U-VALUE OF 0.30 OR BETTER AND THIS IS DEEMED TO BE SATISFIED BY AN 11 IN. CAVITY BRICK WALL RENDERED OR PLASTERED ON ONE SIDE.

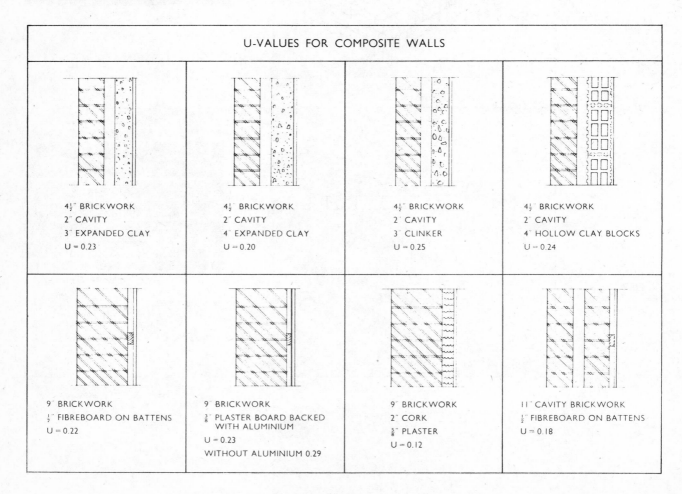

U-VALUES FOR COMPOSITE WALLS

$4\frac{1}{2}''$ BRICKWORK
2" CAVITY
3" EXPANDED CLAY
U = 0.23

$4\frac{1}{2}''$ BRICKWORK
2" CAVITY
4" EXPANDED CLAY
U = 0.20

$4\frac{1}{2}''$ BRICKWORK
2" CAVITY
3" CLINKER
U = 0.25

$4\frac{1}{2}''$ BRICKWORK
2" CAVITY
4" HOLLOW CLAY BLOCKS
U = 0.24

9" BRICKWORK
$\frac{1}{2}''$ FIBREBOARD ON BATTENS
U = 0.22

9" BRICKWORK
$\frac{3}{8}''$ PLASTER BOARD BACKED WITH ALUMINIUM
U = 0.23
WITHOUT ALUMINIUM 0.29

9" BRICKWORK
2" CORK
$\frac{5}{8}''$ PLASTER
U = 0.12

11" CAVITY BRICKWORK
$\frac{1}{2}''$ FIBREBOARD ON BATTENS
U = 0.18

MATERIALS

WATER

Must be clean and correct as to quantity. Drinking water supplies are almost always suitable. Quantity should be sufficient to make the mortar workable but no more.

SAND

Owing to heavy cost of transport a local sand is generally used. It should conform to B.S. 1200:1955 which states that the sand should be naturally occurring or of crushed rock or gravel: it should be hard, clean and free from adhering coating. The presence of clay will increase shrinkage of mortar. Grading is important and the B.S. gives appropriate limits. Good quality of sand is especially needed for work in cold weather.

CEMENT

Is usually either normal setting Portland (see B.S. 12:1958) or normal setting Portland Blast Furnace cement (see B.S. 146:1958).

LIMES

May be either eminently-hydraulic, semi-hydraulic or non-hydraulic and may be in the form of quicklime for site slaking or hydrated lime ready slaked. Non-hydraulic and semi-hydraulic are covered by B.S. 890:1940. Eminently-hydraulic is not covered by a B.S. but recommendations for quality are included in Code of Practice 121.101 (1951) (Brickwork).

TYPES OF MIX

In choosing the mix to be used there are several factors to consider. The strength required, the weather conditions at time of building, the degree of exposure after building, the importance of resisting rain penetration, the effect of the mix on ease of working and the appearance of facing brickwork.

The Table on the back of this sheet gives recommended mixes for general work. Unless high strength or exposure makes a strong mix essential it is better to use a relatively weak cement lime or a cement plasticizer mix. Shrinkage is likely to be less and the addition of lime together with a careful choice of colour of sand makes possible a wider range of colour of the mortar without adding colouring pigments. If colouring pigments are used they should be in the form of pre-mixed mortar as otherwise it is extremely difficult to keep the colour consistent. It is essential that the manufacturer's instructions are strictly followed.

For calculated load-bearing brickwork tables of basic compressive stresses are included in B.S. Code of Practice III and in the various Building Regulations and reference should be made to these for all such work. It should be noted that the hardening time for any particular type of mortar depends upon temperature conditions. In C.P. III it is stated that the hardening times quoted in the main Table should be increased by 50% when temperatures are between 40 and 50 Deg. F. and by 100% when temperatures are below 40 Deg. F.

JOINTING AND POINTING

The appearance of the brickwork will depend a good deal upon the shape of the joint. Five types of joint are illustrated. There is some argument as to the respective merits of finishing the joint as the work proceeds as against raking and pointing. Probably finishing as the work proceeds will give a joint rather more resistant to frost but it may be difficult to keep the colour consistent throughout a large area of wall. If raking and pointing is done then the raking must be very thorough and should be to a depth of $\frac{3}{4}$ inch. The type of joint chosen needs to be related to the kind of brick used. For example a flush joint might not give the desired appearance with bricks having a heavily textured finish.

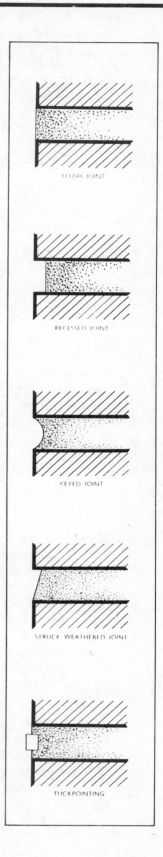

FLUSH JOINT

RECESSED JOINT

KEYED JOINT

STRUCK WEATHERED JOINT

TUCKPOINTING

BS 3921:1969 metric format and work sizes
BS 187:1970: metric units calcium silicate (sandlime and flintlime) bricks

These are fractionally smaller than the Imperial Standard Sizes of BS.3921:1965. (For a direct comparison of sizes see Metric Sheet 225.2.)

The **FORMAT SIZE is 225 x 112.5 x 75 mm** which includes the brick and its mortar joints.

To conform to this with 10 mm joints requires a **WORK SIZE of 215 x 102.5 x 65 mm** for the actual brick.

In brickwork

The format length (225) is the spacing of stretcher perpends.

The format width (112.5) is the spacing of header perpends.

The format height (75) is the coursing height.

It should also be noted that:

(a) the actual length of a brickwork panel is less by one joint, eg, 10 mm, than the overall distance between format lines;

(b) the actual widths of openings between brick reveals are greater by one joint, eg 10 mm, than the width between format lines;

(c) the height of brickwork, measured conveniently between the tops of the courses, is equal to the format height multiplied by the number of courses; and the clear height of an opening measured to the brickwork is therefore greater by one bed joint than the coursing height of the opening.

The above are illustrated overleaf.

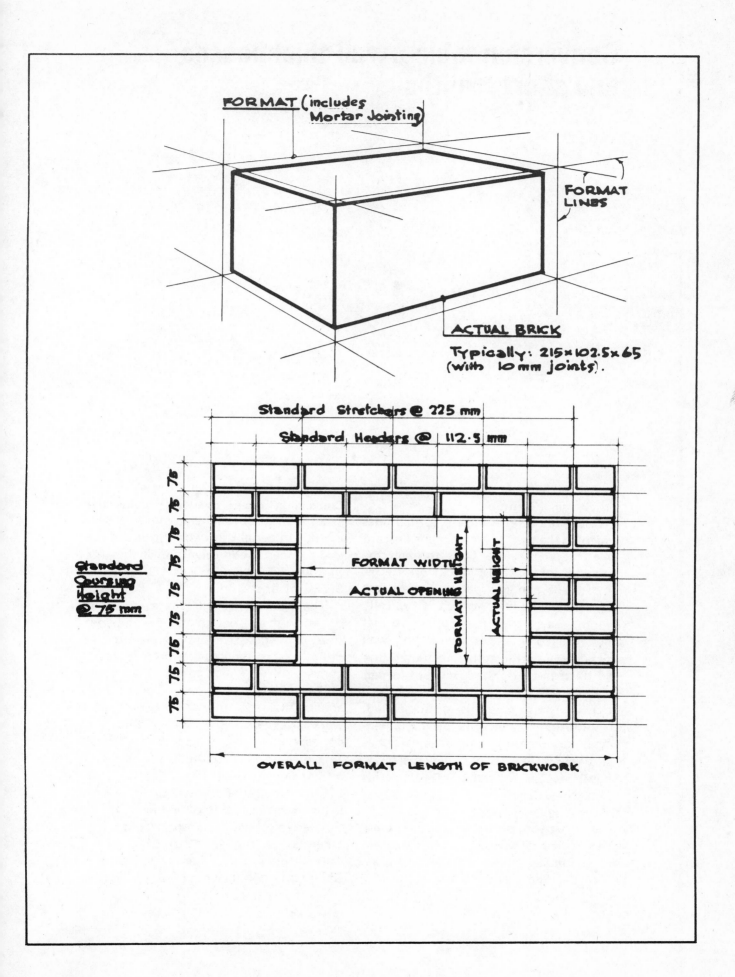

FORMAT (includes Mortar Jointing)

FORMAT LINES

ACTUAL BRICK

Typically: 215×102.5×65
(with 10 mm joints).

Standard Stretchers @ 225 mm

Standard Headers @ 112.5 mm

Standard Coursing Height @ 75 mm

FORMAT WIDTH

ACTUAL OPENING

FORMAT HEIGHT

ACTUAL HEIGHT

OVERALL FORMAT LENGTH OF BRICKWORK

Conversion tables: wall thicknesses and short lengths

visually significant.

Brickwork sizes are to the format lines which, in the case of plan dimensions, are taken as the centre lines through the perpend joints. Bare brickwork faces, ie of walls and at opening reveals, are set back half-a-joint, which is typically 5 mm, from their appropriate format lines. This practice of designating wall thicknesses and lengths, including the widths of piers, by the spacing of format lines; eg 112.5 mm (or 4½ in) for half-brick walls and so on; is recommended. It conforms to the principle of dimensional coordination and enables zones for brick walls and their lengths to be coordinated with modular planning grids in buildings that conform to BS.4330 for Controlling Dimensions.

Such brickwork sizes, ie to format lines, are the **Coordinating sizes** for the brickwork. Further notes on these are given on Metric Sheet 225.3.

Conversion tables overleaf provide the exact metric equivalents in mm for existing BS. Imperial Brickwork Sizes and the exact imperial equivalents in inches for the new BS.3921:1969 Part 2 METRIC SIZES of brickwork.

The new metric sizes are 'rounded down' with respect to the old imperial sizes; ie the new sizes approximate to the old sizes according to the rounded down conversion of : 1 inch equals 25 mm.

Because the exact conversion is 1 inch equals 25.4 mm the new metric brickwork sizes are 1.6% smaller than the corresponding imperial brickwork sizes. This difference is measurable for sizes of one stretcher length and over; eg for 225 mm the reduction is 3.6mm which is just over one-eighth of an inch.

The typical width of mortar jointing is taken as 10 mm in metric brickwork. This is fractionally wider than 3/8 inch (9.525 mm); but the difference is not

EXISTING IMPERIAL		FORMAT DIMENSION	NEW METRIC	
INCHES	MM		MM	INCHES
3/8	9·525	TYPICAL JOINT	10	0·394
3/4	19·05			
1	25·4		25	0·984
1½	38·1		37·5	1·476
2	50·8		50	1·969
2¼	57·15	QUEEN CLOSER	56·25	2·215
3	76·2	COURSING HEIGHT	75	2·953
4	101·6	I.S.O. MODULE	100	3·937
4½	114·3	HALF BRICK	112·5	4·429
6¾	171·45	3/4 "	168·75	6·644
7	177·8		175	6·89
8	203·2		200	7·874
9	228·6	BS.3921 STRETCHER	225	8·858
10½	266·7	CAVITY WALL	275	10·827
11	279·4			
12	304·8	PLANNING MODULE	300	11·81
13½	342·9	1-½ BRICKS	337·5	13·288
15½	393·7	CAVITY WALL	387·5	15·256
18	457·2	2 BRICKS	450	17·717
22½	571·5	2½ "	562·5	22·146
27	685·8	3 "	675	26·575
31½	800·1	3½ "	787·5	31·004
36	914·4	4 "	900	35·433
39·37	1000	ONE METRE	1000	39·37
40½	1028·7	4½ BRICKS	1012·5	39·862
45	1143	5 "	1125	44·291
49½	1257·3	5½ "	1237·5	48·72
54	1371·6	6 "	1350	53·15
58½	1485·9	6½ "	1462·5	57·579
63	1600·2	7 "	1575	62·008
67½	1714·5	7½ "	1687·5	66·437
72	1828·8	8 "	1800	70·866

CONVERSION FACTOR
1 INCH = 25·4 MM

CONTINUED ON METRIC NO. 3

Conversion tables: wall lengths

Stretcher bond

the coordinating format line.

Elsewhere, for example at joints between brickwork and other components, such as at door and window jambs, the format line for the brickwork may not coincide with the coordinating line for the other component. A similar lack of coincidence may occur in respect of the format lines, for the widths of piers and for wall thicknesses, and the planning grid lines that define their zones; it is not uncommon for fair brickwork faces, where this is feasible, to be erected on the zone lines.

For buildings that are designed with planning grids the brickwork should therefore be set out and erected only after careful reference to the designers' working details, ie of junctions with other components and of the relation of the bare faces of the brickwork to the planning grid.

See also typical jointing and bonding details in this metric sheet series.

Conversion tables overleaf for standard brickwork lengths are continued from Metric Sheet 225.2 (Wall thicknesses; piers; and short lengths); note that the imperial sizes for these longer lengths are given in feet and inches.

As defined on Metric Sheet 225.2 the sizes given are the coordinating sizes for standard brickwork, ie to format lines at the centres of perpend joints (see also Metric 225.1).

In standard brickwork the average width of perpend joints is 10 mm and acceptable variations in this will generally provide for manufacturing variations as well as in some cases, for minor variations in setting out and erection.

Where one brickwork component is joined into another, eg at a pier in a wall or at a corner, their corresponding format lines coincide to become co ordinating lines appropriate to both. At each perpend joint between the two brickwork components the actual bricks are set back half-a-joint on either side of

EXISTING IMPERIAL		FORMAT DIMENSION	NEW METRIC	
FT - IN	MM		MM	FT - IN
6 - 4½	1943·1	8½ BRICKS	1912·5	6 - 3·3
6 - 9	2057·4	9 "	2025	6 - 7·72
7 - 1½	2171·7	9½ "	2137·5	7 - 0·15
7 - 6	2286	10 "	2250	7 - 4·58
7 - 10½	2400·3	10½ "	2362·5	7 - 9·01
8 - 3	2514·6	11 "	2475	8 - 1·44
8 - 7½	2628·9	11½ "	2587·5	8 - 5·87
9 - 0	2743·2	12 "	2700	8 - 10·3
9 - 4½	2857·5	12½ "	2812·5	9 - 2·73
9 - 9	2971·8	13 "	2925	9 - 7·16
10 - 1½	3086·1	13½ "	3037·5	9 - 11·59
10 - 6	3200·4	14 "	3150	10 - 4·02
10 - 10½	3314·7	14½ "	3262·5	10 - 8·44
11 - 3	3428	15 "	3375	11 - 0·87
11 - 7½	3543·3	15½ "	3487·5	11 - 5·3
12 - 0	3657·6	16 "	3600	11 - 9·73
12 - 4½	3771·9	16½ "	3712·5	12 - 2·16
12 - 9	3886·2	17 "	3825	12 - 6·59
13 - 1½	4000·5	17½ "	3937·5	12 - 11·02
13 - 6	4114·8	18 "	4050	13 - 3·45
13 - 10½	4228·1	18½ "	4162·5	13 - 7·88
14 - 3	4343·4	19 "	4275	14 - 0·31
14 - 7½	4457·7	19½ "	4387·5	14 - 4·74
15 - 0	4572	20 "	4500	14 - 9·17
15 - 4½	4686·3	20½ "	4612·5	15 - 1·59
15 - 9	4800·6	21 "	4725	15 - 6·02
16 - 1½	4914·9	21½ "	4837·5	15 - 10·45
16 - 6	5028·2	22 "	4950	16 - 2·88
16 - 10½	5143·5	22½ "	5062·5	16 - 7·31
17 - 3	5257·8	23 "	5175	16 - 11·74
17 - 7½	5372·1	23½ "	5287·5	17 - 4·17
18 - 0	5486·4	24 "	5400	17 - 8·6
18 - 4½	5800·7	24½ "	5512·5	18 - 1·03

Planning grids, wall zones and controlling lines

Planning grids of reference lines are used by designers to facilitate dimensional coordination of components in a building; ie to enable the use of pre-sized standard components and to reduce the need for special site-fitted components. The appropriate size for planning grids is 300 x 300 mm.

Wall zones are the spaces allocated for walls and which are defined on plans by reference lines that may be directly related to the planning grid. Preferred widths of zones for loadbearing walls include : 100; 200; 300; 400 mm.

Controlling lines represent key reference planes in a building and include those which on plans determine the spacing apart of loadbearing walls and columns and the widths of their zones. Preferred spacings are multiples of 300 mm either to zone reference lines, ie zone boundaries, or to zone axes, eg (but not necessarily) zone centrelines.

When planning grids are used it is obviously necessary to relate these to the controlling lines and generally these coincide; otherwise their displacements should be selected from the preferred increments of BS.4011 in the same order.

Designers may exercise certain options with respect to walls and their finishes in relation to their zones; actual walls need not always fill their zones and finishes may be added either within or without the zone boundary; non-standard, ie neutral, zones may be used in exceptional circumstances and, similarly, walls may extend beyond the zone boundaries. In all cases it is important that the use of dimensionally coordinated standard components should not be inhibited.

Some typical examples of metric brick walls and their zones for use with planning grids are provided overleaf; and corresponding working detail examples are the subject of Metric Sheets 225.5 and 225.6.

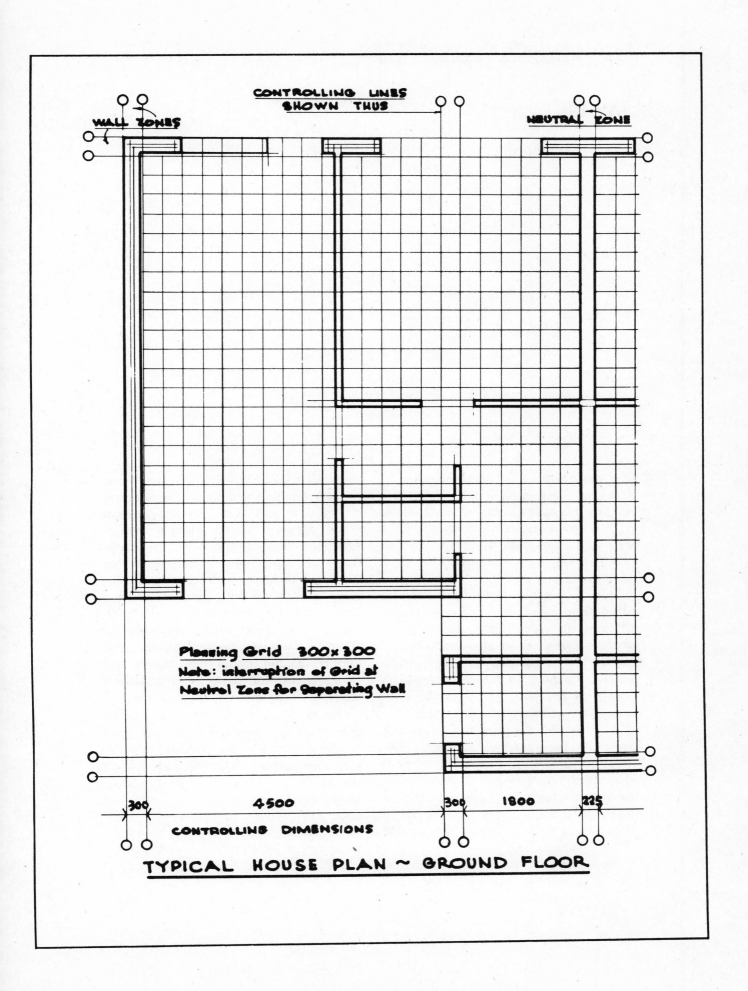

CONTROLLING LINES
SHOWN THUS

WALL ZONES

NEUTRAL ZONE

Planning Grid   300 x 300
Note: interruption of Grid at
Neutral Zone for Separating Wall

300   4500   300   1800   225

CONTROLLING DIMENSIONS

TYPICAL HOUSE PLAN ~ GROUND FLOOR

Typical working details 1(A)

Plan dimensions

For the dimensional coordination of brickwork with other components in a building it is appropriate, when detailing, to distinguish between brickwork built prior to the erection of components to which it will be jointed and that built subsequently.

This is because the effects of the unavoidable inexactitude of building are different in each case and because erection clearances are also dissimilar.

Prior-built brickwork, for which typical examples are detailed overleaf, is both pre-sized and pre-positioned with respect to adjacent building spaces for which it provides a solid and unyielding boundary. Erection clearances that may be required at junctions/joints with adjacent components as well as provision for size variations will in general require wider, ie coarser, joints than is usual with components that are built into brickwork.

The setting out of prior-built brick-

work in relation to site reference lines for the building and its erection should therefore be as accurate as possible; and it is appropriate to include design reference lines on the working details. Site reference lines will however be displaced to a varying extent from their true, ie design, position; and in working details the spacing of actual brickwork faces from these lines will only be typical. Moreover, any sizes, eg joint widths, should be average values and taken to include permissible variations whether these are specified or not.

Where permissible variations smaller than those normally associated with good building practice are required, special measures to assist the bricklayer will be needed.

Other notes on prior-built brickwork are also appropriate to brickwork built subsequently to the erection of components for which refer to Metric Sheet 225.6.

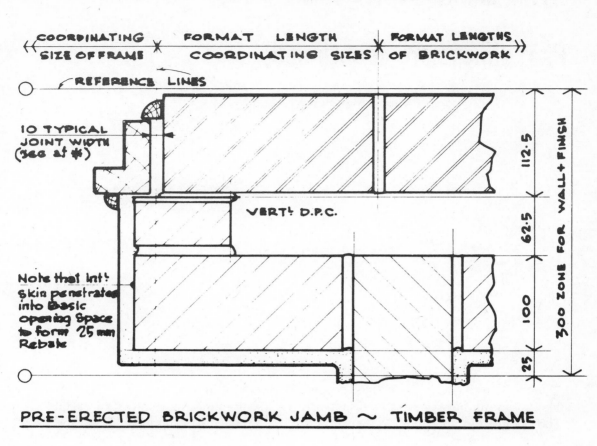

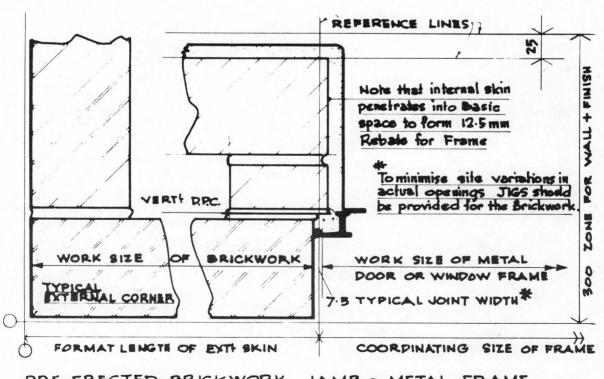

PRE-ERECTED BRICKWORK JAMB ~ TIMBER FRAME

PRE-ERECTED BRICKWORK JAMB ~ METAL FRAME

Typical working details 1(B)

Plan dimensions (continued)

Typical working details are given overleaf of joints between insitu brickwork and pre-erected other components.

For purposes of dimensional coordination these other components should be pre-erected accurately in relation to the appropriate reference lines for the building; and the design reference lines are included on the working details for this reason. Except where such lines may be related to adjacent prior-built brickwork, eg in traditional 'hole-in-the-wall' construction, they can have little or no meaning in relation to the insitu brickwork for which the jointing face of the pre-erected component provides adequate coordination.

For such working details little or no erection clearance is required for the adjoining insitu brickwork and very fine unfilled joints become feasible; the brickwork itself acts as a jointing component and is required to incorporate all size variations that may occur as a result of using such joints.

To minimise inaccuracies in establishing the site reference lines for the building these should be set out with survey instruments and care should be exercised in preserving these on site, eg at pegs and profiles. Some variation from the design reference lines is unavoidable and it follows that brickwork on site whether prior-built or insitu may be required to occupy non-standard sizes of building space.

Minor variations in the widths of perpends may be used to incorporate variations of the order of 1% or less; but, in general, non-standard brickwork lengths will require special formats. Reference should be made to Metric Sheet 225.7: BONDING. In this context it should be noted that for external skins of facing brickwork at door and window jambs a degree of dimensional flexibility may be obtained by building the frames into rebates behind such skins.

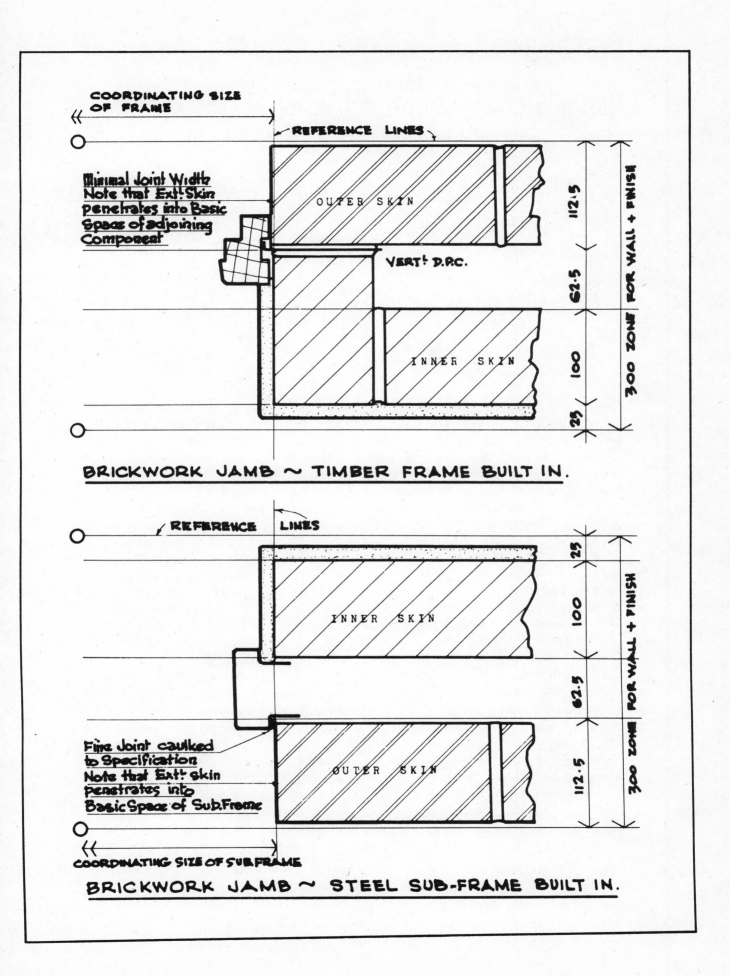

COORDINATING SIZE OF FRAME

REFERENCE LINES

OUTER SKIN

Minimal Joint Width
Note that Ext! Skin penetrates into Basic Space of adjoining Component

VERT! D.P.C.

INNER SKIN

112·5
62·5
100
25

300 ZONE FOR WALL + FINISH

BRICKWORK JAMB ~ TIMBER FRAME BUILT IN.

REFERENCE LINES

INNER SKIN

Fire Joint caulked to Specification
Note that Ext! Skin penetrates into Basic Space of Sub.Frame

OUTER SKIN

25
100
62·5
112·5

300 ZONE FOR WALL + FINISH

COORDINATING SIZE OF SUB-FRAME

BRICKWORK JAMB ~ STEEL SUB-FRAME BUILT IN.

Bonding - BS 3921:1969 metric bricks

⅓rd and modified ½ stretcher bonds

Details of bonding BS.3921:1969 Metric Format bricks into walls and piers are similar to those for BS.3921 : 1965 Imperial Format bricks; and geometrically similar stretcher, header and standard special formats are required for the various types of bonding and junctions. The use of a conversion scale of 25 mm to 1 inch for the format sizes, ie of the actual brick shapes *plus* joints, enables the same details to be used for both metric and imperial.

It should be noted that a separate British Standard is to be issued for the metric special formats corresponding to those previously included in Appendix D of BS.3921:1965; and their relation to the standard format is to be included.

As may be seen from the conversion tables on Metric Sheets 225.2 and 225.3 the actual sizes of metric bricks and brickwork are fractionally smaller than corresponding imperial sizes.

For the design of other than standard brickwork lengths, special formats and bonding may be used, notably in facing leaves for which alternative methods of producing every whole multiple of 300 mm are illustrated overleaf as examples. These alternatives are: Third Stretcher Bonding and Modified Half Stretcher Bonding and the following notes are relevant.

In Third Stretcher Bonding the lap is one-third of the stretcher. When used with standard 225 mm stretchers the special formats at stopped ends are 75 and 150 mm; and 187.5 mm formats are required in addition at returned ends. Such bonding is plainly suited to 300 x 100 format bricks for which the additional format length at ends is 200 mm.

In Modified Half Stretcher Bonding a special format length of 187.5 mm is used in place of a standard stretcher of 225 mm, generally at ends, as required to achieve coordination.

Because of site variations in building spaces for brickwork the supply to order of special formats will not always ensure dimensional coordination; and such should only follow verification of design lengths on site.

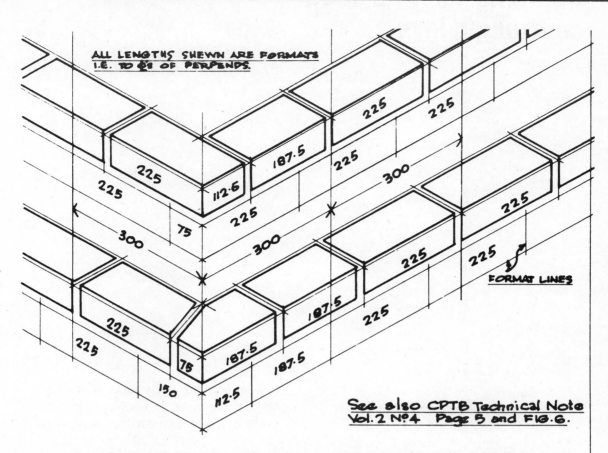

ALL LENGTHS SHEWN ARE FORMATS I.E. TO ¢ OF PERPENDS.

FORMAT LINES

See also CPTB Technical Note Vol.2 Nº4 Page 5 and FIG.6.

TYPICAL QUOINS ~ THIRD STRETCHER BONDING

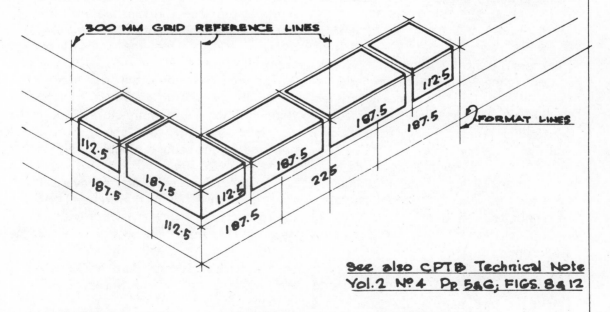

300 MM GRID REFERENCE LINES

FORMAT LINES

See also CPTB Technical Note Vol.2 Nº4 Pp. 5&6; FIGS. 8&12

MODIFIED HALF STRETCHER BONDING

Coursing heights and controlling levels

The coursing height of brickwork includes the mortar bed joint and is conveniently taken as the vertical distance between the tops of bricks in successive courses. In standard brickwork the coursing height is the format height (see Metric Sheet 225.1); for BS.3921:1969 Metric Bricks this height is 75 mm, ie 4 courses rise 300 mm. The bed joint thickness is typically 10 mm.

Controlling levels are used in design to determine key building planes at finished floor and ceiling levels. Intermediate controlling levels are also used at sills and heads of openings in walls. The purpose of such controlling levels is to assist in the sizing of the coordinating heights of building components; notably of windows and doors as well as of walling elements; staircases; plumbing stacks and so on.

Coordinating heights of brickwork are necessarily multiples of its coursing height and in dimensionally coordinated building these are directly related to the coordinating heights of adjoining other components and or to their appropriate controlling levels. Standard brickwork is readily coordinated to heights that are multiples of 75 and 300 mm; but special provisions need to be made where multiples only of 100 mm are used for the coordinating heights of other components and or for the spacing of controlling levels.

One such provision is the use either of 34 or 35 courses for 2600 mm floor to floor heights, ie in multi-storey Local Authority housing; the corresponding coursing height being respectively either 76.47 or 74.29 mm.

The foregoing are illustrated overleaf for typical two storey housing and the discrepancies for which special provisions need to be made are also indicated. Typical working details that embody such provisions for standard brickwork are included on Metric Sheets 225.9 et seq.

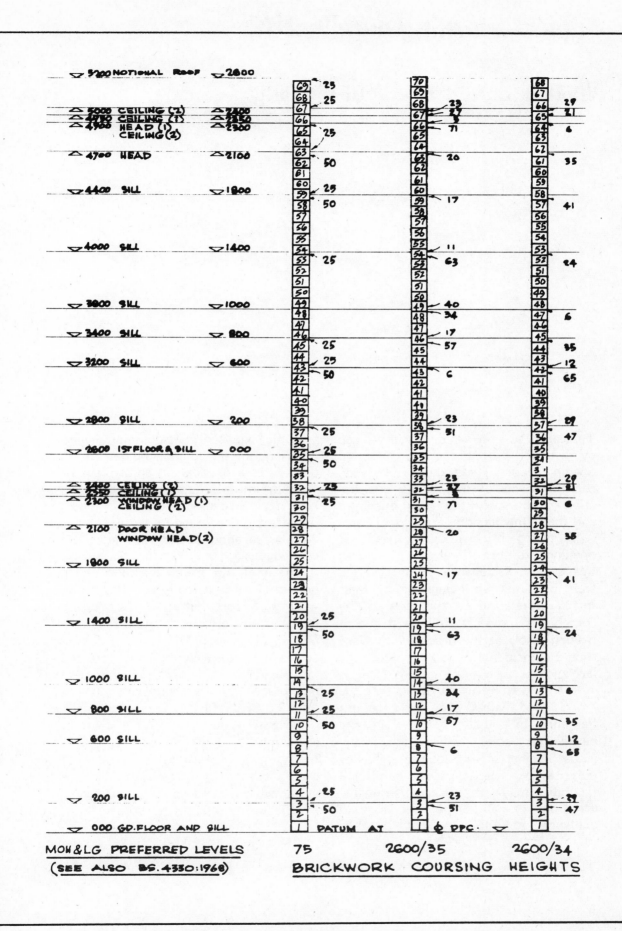

MOH&LG PREFERRED LEVELS   75   2600/35   2600/34

(SEE ALSO BS.4330:1968)   BRICKWORK COURSING HEIGHTS

Typical working details 2(A)

Vertical dimension - door opening

Typical head and sill details are given overleaf for a timber framed external opening door. These details illustrate certain aspects of the vertical coordination of brickwork at such openings; and the following notes are also relevant.

In facing brickwork the heads of openings are normally at coursing levels; this avoids discrepancies either at the arch or lintel springings or at their soffits in relation to the tops of the component frames. It follows that controlling levels at the heads of openings should coincide with brickwork coursing levels.

It should however be noted that the mortar bed joints provide a greater flexibility in coursing height than may be necessary solely for variations in the actual heights of individual bricks. For purposes of design such additional flexibility may either imply permissible variations on site or it may be used explicitly for minor adjustments, for example; between controlling levels and otherwise unvarying coursing heights. In the example overleaf an adjustment of 10 mm in the coursing height of the head of the opening has been so made; the effect on each bed joint thickness is on average about ½% of the individual coursing height, ie 0.36 mm.

The level and overall thickness of the horizontal DPC at the top of the footing walls are factors that affect the coursing levels in the brickwork over. (See also Metric Sheet 225.10.) It is not uncommon to use a double thickness bed joint and for the level of the DPC membrane to be approximately at Finished Ground Floor Level (FFL).

Account is taken of these factors to accommodate a 25 mm penetration of the hardwood threshold sill member of the door frame below FFL; this member is shown also with a DPC bedded in mastic in a reduced thickness joint.

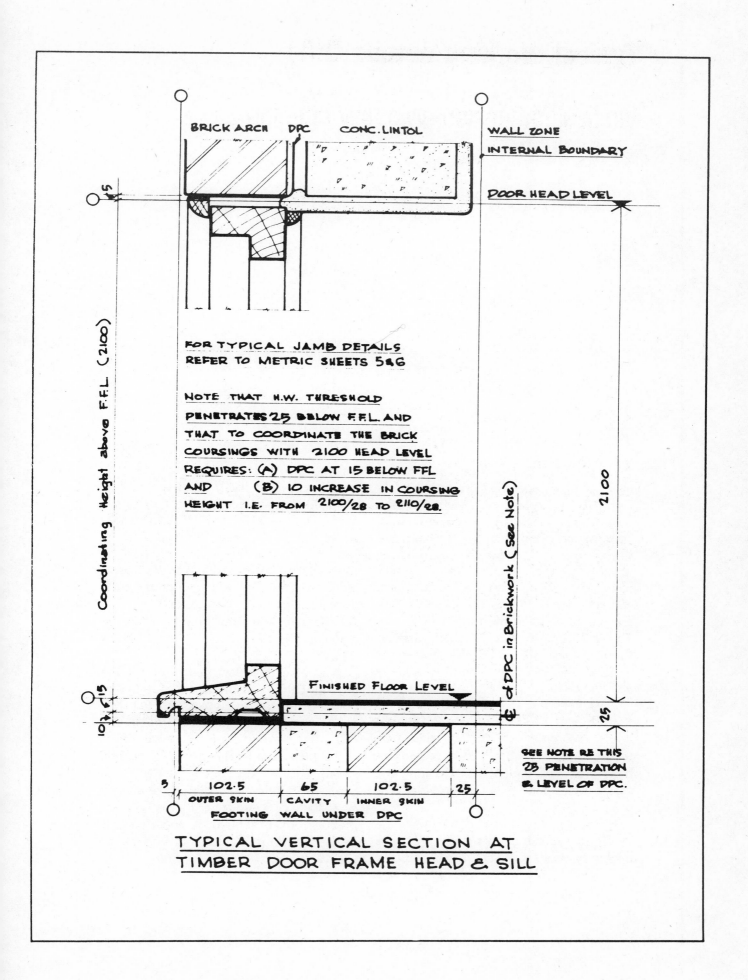

BRICK ARCH   DPC   CONC. LINTOL

WALL ZONE
INTERNAL BOUNDARY

DOOR HEAD LEVEL

5

Coordinating Height above F.F.L. (2100)

FOR TYPICAL JAMB DETAILS
REFER TO METRIC SHEETS 5&6

NOTE THAT H.W. THRESHOLD
PENETRATES 25 BELOW F.F.L. AND
THAT TO COORDINATE THE BRICK
COURSINGS WITH 2100 HEAD LEVEL
REQUIRES: (A) DPC AT 15 BELOW FFL
AND   (B) 10 INCREASE IN COURSING
HEIGHT I.E. FROM 2100/28 TO 2110/28.

2100

℄ of DPC in Brickwork (See Note)

15

Finished Floor Level

℄

25

10

SEE NOTE RE THIS
25 PENETRATION
& LEVEL OF DPC.

5   102.5   65   102.5   25
     OUTER SKIN   CAVITY   INNER SKIN
     FOOTING WALL UNDER DPC

TYPICAL VERTICAL SECTION AT
TIMBER DOOR FRAME HEAD & SILL

53

Typical working details 3(A)

Vertical dimension - window opening

A typical vertical section is provided overleaf of an opening in standard brickwork for a ground floor window, the heights for each of which are fully coordinated. That is to say the head and sill controlling lines and the coordinating planes for the window are mutually coincident and also coincide with coursing levels in the brickwork which are spaced apart at multiples of 75 mm. The relation to the Finished Ground Floor Level (FFL) of the horizontal DPC at the top of the footing walls is also shown.

This example is appropriate only to the 600 and 1800 sill and the 2100 head levels recommended for housing (see Metric Sheet 225.8); but it provides a basis for subsequent details of the adjustments that may be provided for other heights of window.

Moreover the head detail may be taken as typical on the basis that it is necessary only for the coordination plane for the top of the window to coincide with a coursing level. Thus where two controlling levels for heads of openings occur, such as 2100 and 2300, the brickwork coursings may be displaced in relation to either one or possibly both of these and the corresponding coordination plane of the window or door; such displacement need not exceed 25 mm. All further adjustments may then be made at the sill line; see Metric Sheets 225.11 and 225.12.

A double thickness bed joint for the ground DPC is assumed and for design purposes (in the detail overleaf) the DPC membrane is ideally laid at 5 mm below FFL. For reasons given on Metric Sheet 225.9 the actual level of the DPC may be varied accidentally or by design; the precision implied by a difference in level of as little as 5 mm is hardly realisable in practice. In other words a permissible deviation from this level is implicit in such a detail.

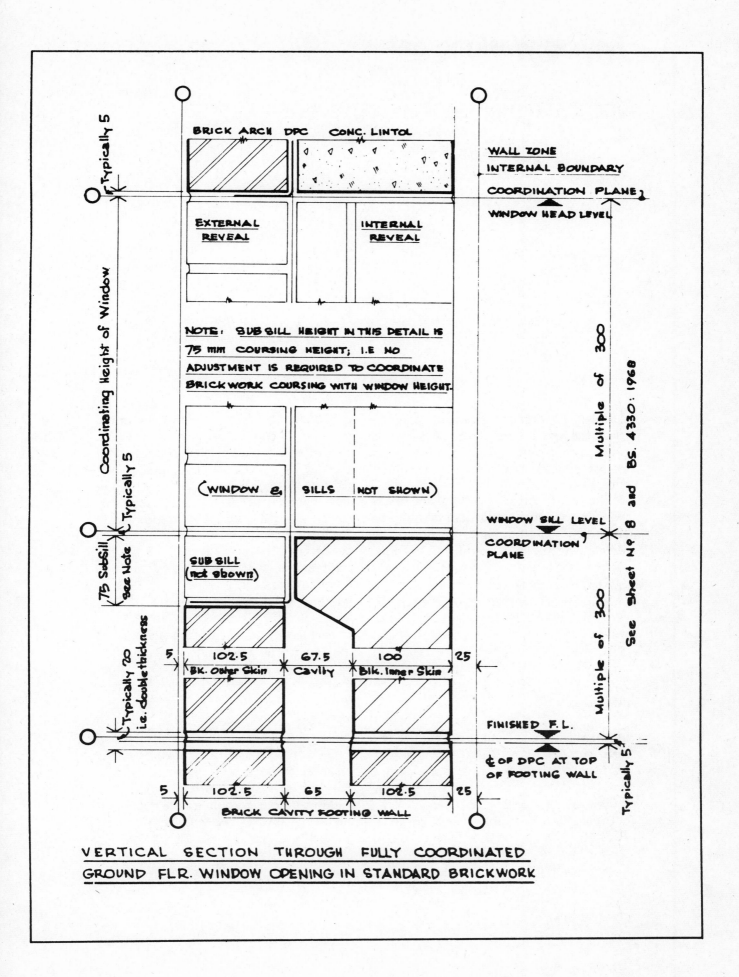

VERTICAL SECTION THROUGH FULLY COORDINATED
GROUND FLR. WINDOW OPENING IN STANDARD BRICKWORK

Typical working details 3(B)

Vertical dimension - window sills 1

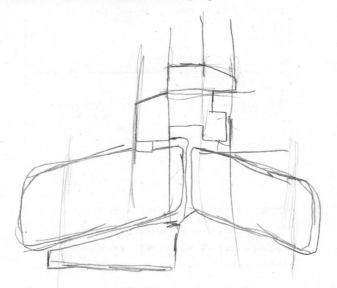

Subsills, as detailed overleaf, may be used for the coordination of standard brickwork, coursing at 75 mm, with standard metric steel windows in timber subframes.

The coordination planes for the brickwork and for the window are taken to be coincident at the head as noted on Metric Sheet 225.10; and adjustments, such as those indicated on Metric Sheet 225.8, are made at the sill line.

Adjustments respectively of plus 25, zero, and minus 25 penetration by the window into the coursing height of the brickwork opening are indicated. In other words subsill zone heights of 50; 75; and 100 mm are used for design purposes.

For the first case a reduced height subsill member is used. For the second case a standard coursing height subsill member suffices. For the third case a standard coursing height subsill member is used together with a deep bed joint.

Conventional timber subframes ex 75 x 50 (3 in x 2 in) have been provided

the bedjoints (from 10 mm). As noted for; but it should be noted that these add, at each member, slightly more than 25 mm to the coordinating (and work) sizes of the steel windows. The excess at each member needs to be provided for. At the sill this is shown as a reduction in the bedjoints (from 10 mm). As noted on Metric Sheets 225. 5 – 9 similar adjustments, at head and jambs, to those required for comparable variations in building exactitude may be made. Subframes ex 75 x 75 add very nearly 50 mm at each member and do not give rise to the same need for minor adjustment. It is observed that, by using the larger frames at head and cill in the first case shown overleaf and at the cill only in the third case, the need for other than the standard coursing height subsill is eliminated. In other words the adjustments may be made wholly in the subframe thicknesses.

Similar details apply to clayware and concrete subsills. Alternative sills and undersills are shown on Metric 225.12.

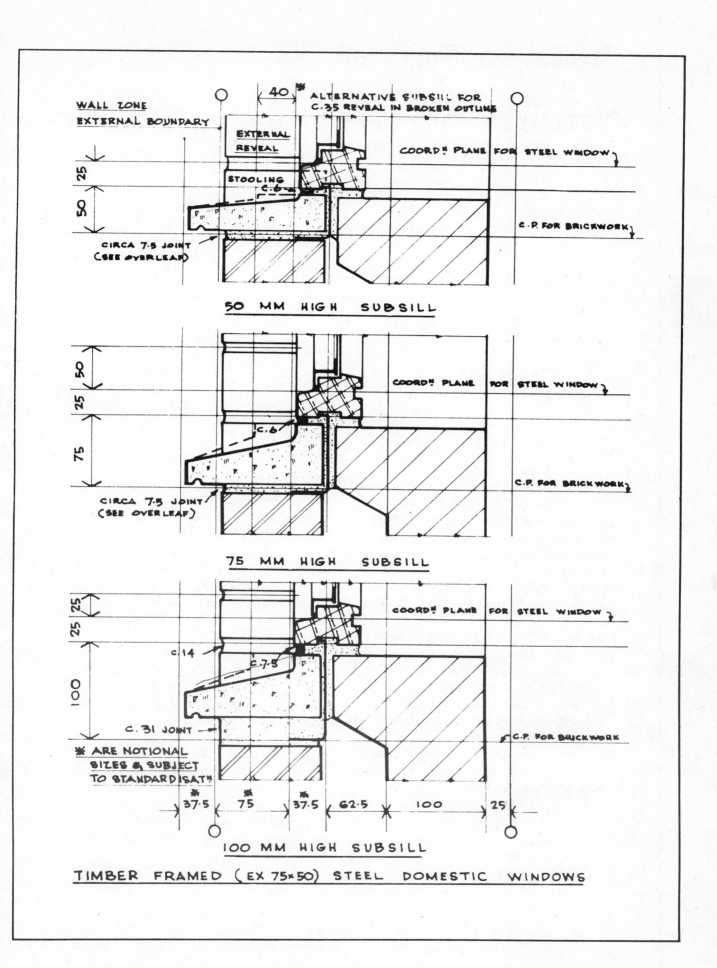

WALL ZONE
EXTERNAL BOUNDARY

40 ※ ALTERNATIVE SUBSILL FOR
C.35 REVEAL IN BROKEN OUTLINE

EXTERNAL
REVEAL

COORD⁵ PLANE FOR STEEL WINDOW

STOOLING
C.6

25

50

C.P. FOR BRICKWORK

CIRCA 7·5 JOINT
(SEE OVERLEAF)

50 MM HIGH SUBSILL

50

25

COORD⁵ PLANE FOR STEEL WINDOW

C.6

75

C.P. FOR BRICKWORK

CIRCA 7·5 JOINT
(SEE OVERLEAF)

75 MM HIGH SUBSILL

25

25

COORD⁵ PLANE FOR STEEL WINDOW

C.14

C.7·5

100

C.31 JOINT

C.P. FOR BRICKWORK

※ ARE NOTIONAL
SIZES & SUBJECT
TO STANDARDISAT⁹

37·5 75 37·5 62·5 100 25

100 MM HIGH SUBSILL

TIMBER FRAMED (EX 75×50) STEEL DOMESTIC WINDOWS

Typical working details 3(C)

Vertical dimension – window sills 2

Typical sill details are provided overleaf for windows in the outer leaf of cavity brick walls coursing at 75 mm; these are for steel windows and sills and also for timber windows with conventional projecting sills.

The coordinating heights of the windows, ie to the coordinating plane at the underside of steel frame or timber sill, are shown respectively penetrating plus 25, zero and minus 25 below the corresponding coursing level of the brickwork. (See also Metric Sheet 225.11.)

Vertical coordination in each case is effected by the use of an appropriate undersill arrangement.

In the first type the undersill coordinating height is 50; for the steel windows and sills this is provided by insitu fine concrete or mortar filling beneath the sill. For the timber windows a cut course of 50 height is used; and the alternative of a brick-on-edge course with 22.5 bed-joint is indicated, in broken outline, as a second preference.

In the second type the undersill coordinating height is 75; for the steel windows this is provided by a cut course of 50 plus 25 mortar filling beneath the sill. As second preference a brick-on-edge course with mortar filling is indicated in broken outline. For timber windows a standard brick course suffices.

In the third type the undersill coordinating height is 100; for steel windows this is provided by a standard brick course of 75 plus mortar filling beneath the sill. For timber windows a 25 creasing course is shown. Two possible alternatives are: (1) a 35 overall mortar bed joint; (2) a brick-on-edge course together with a reduction in associated bedjoints (the average reduction for two joints being 12.5). These alternatives may be considered as second and third preferences respectively.

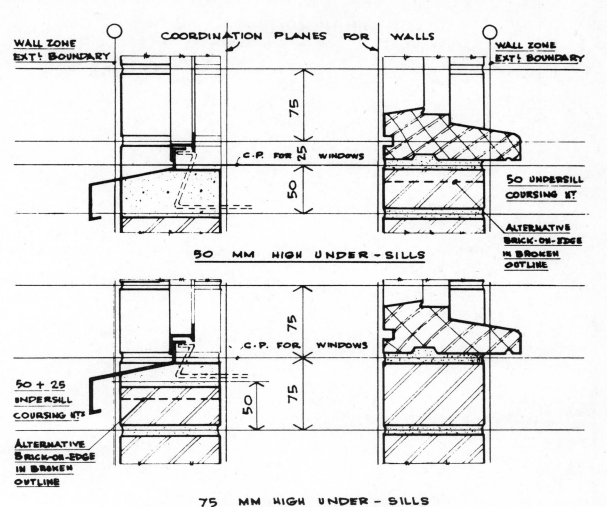

COORDINATION PLANES FOR WALLS

WALL ZONE EXT! BOUNDARY

75

25

C.P. FOR WINDOWS

50

WALL ZONE EXT! BOUNDARY

50 UNDERSILL COURSING HT

ALTERNATIVE BRICK-ON-EDGE IN BROKEN OUTLINE

50 MM HIGH UNDER-SILLS

75

C.P. FOR WINDOWS

75

50

50 + 25 UNDERSILL COURSING HT?

ALTERNATIVE BRICK-ON-EDGE IN BROKEN OUTLINE

75 MM HIGH UNDER-SILLS

NOTE: ALL BED JOINTS TYPICALLY 10 MM; COURSING HT 75 UNLESS NOTED.

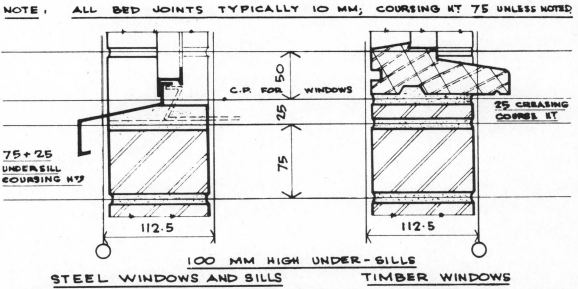

50

C.P. FOR WINDOWS

25

75

75 + 25 UNDERSILL COURSING HT?

112.5

25 CREASING COURSE HT

112.5

100 MM HIGH UNDER-SILLS

STEEL WINDOWS AND SILLS   TIMBER WINDOWS

TYPICAL WINDOW SILLS IN EXTERNAL HALF-BRICK WALLS

SECTION 2
BRICKS IN USE

Bricklaying under winter conditions

K. Thomas, MSc CEng MIStructE FIOB ARTC
Chief Technical Officer–The Brick Development Association

INTRODUCTION

The hazards of bricklaying under winter conditions are well known and it is all too easy to halt the building process and await more favourable conditions. In the United Kingdom this is rarely necessary and is most certainly against the National interest, not only in lack of turnover and reduced profits but in human terms this means that building trade employees suffer a reduction in wages and are exposed to undesirable working conditions.

Bad weather creates management problems but with efficient site pre-planning and a realistic understanding of the problem, winter working is a challenge to the enthusiastic building technologist to be prepared for adverse weather conditions having taken the relevant precautions.

The problems of winter building are basically two-fold, i) working under wet conditions and ii) working under freezing conditions, perhaps the most uncomfortable for the building worker is a combination of the two. Wet weather is a hazard which can occur in any season but during the winter months (November to March) up to between 25% and 50% of the total working hours can be affected by wet weather depending upon location. Indeed, in Great Britain throughout the year an average of one working hour in five is affected by rainfall.

The purpose of this article is basically to recommend a procedure for winter working, with emphasis on the precautions to be taken during frosty weather.

In cold weather mortar sets and hardens much more slowly than at normal temperatures. Even when the air temperature drops to 10°C (50°F) the rate of setting and hardening is noticeably reduced. When the temperature falls still further to freezing point, the setting and hardening processes cease. If newly laid mortar is allowed to freeze before it has time to harden, the expansion of the water as it turns to ice disrupts the mortar joints and, even though setting and hardening will start again when the temperature rises and the ice thaws, the jointing material will be weak and porous and may need to be discarded.

Delays and difficulties due to cold weather can be substantially reduced if the following precautions are taken:

(a) Weather forecasts

Brickwork is particularly vulnerable to freezing during the first few days after laying. Even though the brickwork may have been laid during mild weather a sudden frost can cause extensive damage. Early warning of low temperatures is therefore essential if the work is to be protected in time. Forecasts for any area of the United Kingdom are obtainable free of charge from the Weather Services and general weather forecasts for the next 24 hours, with an outlook for a further 24 to 48 hours are given on radio, television and in the press. The forecasts are for large areas; forecasts for smaller areas are available on the GPO automatic telephone weather service. The Meteorological Office can prepare special forecasts for a particular building site and comprehensive details are readily available elsewhere.[1] In addition a climatology service known as CLIMEST is available to the builder and provides information on average variations in the weather as distinct from the actual weather at a specified time. Charges are made for this service but are of a modest nature.

(b) Working temperatures

Temperatures should be checked regularly adjacent to new brickwork and the work protected to prevent freezing for at least 3 days after laying. When night temperatures do not fall below – 4°C (25°F) and day-time temperatures rise above freezing point, insulating material such as sacking, fibreglass or straw quilts, may be sufficient without additional heat; the insulating material must of course be covered with polythene or similar impervious sheeting materials to maintain it in a dry condition. For more severe conditions it will be necessary to supply heat in addition to coverings and insulating quilts.

It is important that sturdy and easily read thermometers be used to measure the temperature of the air and materials. In addition to the normal maximum/minimum type of thermometers for measuring air temperatures, sites should also have robust soil-type thermometers so that temperatures can be measured and *not guessed*. These thermometers should be used for measuring the temperatures of sand stockpiles, mixing water, and mixed mortar. A careful record of all such temperature measurements should be kept, thus ensuring a regular site procedure.

(c) Storage of materials

Bricks and blocks should be stacked clear of the ground and completely covered with tarpaulins or polythene sheeting, bearing in mind that the drier the the materials, the less susceptible they are to frost attack. The protective sheeting should completely cover the stacks on all sides and should be weighted down to stop the wind blowing them off, Fig 1. Where bricks and blocks are used under conditions of artificial heating they should be stored under such conditions for at least 24 hours before use.

Bricks should not be wetted to reduce their suction rate during frosty weather.

If sand must be stored out of doors, it should be covered immediately after delivery and stockpiled in

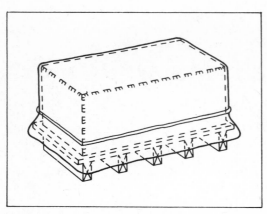

Figure 1 *Bricks stacked clear of the ground and covered with waterproof sheeting.*

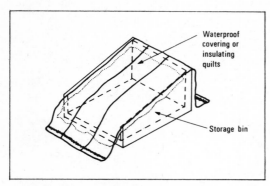

Waterproof covering or insulating quilts

Storage bin

Figure 2 *Sand protected from frost.*

storage bins or on a firm dry base laid to falls to provide drainage, Fig 2. Frozen sand must not be used in mortar.

Cement should be stored well above ground level on a timber floor or platform, preferably in a dry structure, but if this is impracticable it should be completely covered with weighted tarpaulins or polythene sheeting. Consignments should be so placed as to permit inspection and used in the order of delivery. Cement affected by dampness should never be used.

Hydrated lime should be stored in the same manner as cement.

Ready-mixed lime-sand for mortar should be stored on a timber platform or steel sheeting and covered with weighted tarpaulins or polythene sheeting immediately after delivery.

Loose plank covering of materials is not advisable. Careless storage of materials increases the cost of building, because the removal of ice and snow and the thawing of materials used for bricklaying are absolutely necessary before construction may be commenced. The cost of material protection is small relative to the total contract cost and this fact should not be ignored.

(d) Mortar

Mortar mixes weaker than 1:1:6 cement:lime:sand should not generally be used externally in cold weather and it is recommended that unless a stronger mortar is specified for structural brickwork or work below ground level damp-proof courses, a 1:6 cement-sand mortar with an air-entraining plasticizer be used. This will have the advantage of earlier setting than a mix containing lime, and overcoming the problem of maintaining lime-putty in a frost free condition.

Mortar plasticizers, which entrain air in the mix, now provide an alternative to or may be used in conjunction with lime. The air bubbles introduced into the mortar by the plasticizing agent serve to increase the volume of the binder paste, filling the voids in the sand, and this correspondingly improves the working qualities with less water to expand on freezing without disrupting the motar.

The writer is not aware of any documented evidence that commercially available plasticizers cause efflorescence, but some claims have been made to this effect. Certainly, domestic detergents should not be used as mortar plasticizers, as many of them contain sodium sulphate which could contribute to efflorescence.

The use of aerated mortars with dry bricks can in some instances affect the bond, and some authorities[2] recommend additions of water-retaining additives, such as cellulose ethers to counteract this.

Stockpiles of sand should be protected as discussed in (c) and this protection should only be removed or partly removed when sand is being taken from storage. If the stockpile becomes partially frozen, a simple polythene or tarpaulin tent with a low-output air heater or a coke brazier used for 24 hours will generally be sufficient to thaw out and partially warm the sand.

If the temperature is expected to fall below minus 4°C (25°F) the mixing water should be heated using a thermostatically controlled bottled gas burner, an electric immersion heater or even simple braziers. It is important to exercise control over the temperatures and an appropriate range would be 50–65°C (120–150°F). To save loss of heat and to avoid flash setting when using hot water, the materials should be gauged dry and the hot water added last.*

To ensure a uniform and intimate mix of the cementing materials and sand, a mechanical mixer should be used where possible.

When mixing plasticized cement-sand mortars or masonry cement mortars, care should be taken not to add too much water at the start, as these mortars become more fluid as air is entrained. The old-fashioned roller type mortar mill is unsuitable for aerated mortars because it tends to 'roll out' the entrained air. Prolonged mixing of these mortars in other types of mixer can lead to excessive air entrainment and subsequent weak mortars.

In the UK the major suppliers of ready-mixed lime-sand for mortar normally include an air-entraining agent in their mixes during cold spells. It is imperative that users of this excellent quality-controlled product should not gauge admixtures to improve frost resistance as over-aeration can only be harmful to the mortar and finished brickwork.

Additions of frost inhibitors based on calcium chloride should never be used in mortar joints as, apart from being ineffective (ie there is no evidence as far as the writer is aware that sufficient heat can be generated in a normal mortar joint to depress the freezing point), they cause deliquescence with the subsequent danger of corrosion of embedded steel and an increased possibility of efflorescence.

Small quantities of sand can also be heated by spreading it on corrugated metal sheets over a heater or by banking it around pipes in which fires have been lit. Having produced a satisfactory mortar for winter working it is important to keep it warm and this can be achieved by gently heating it before use on a metal sheet over a heater, remembering to turn it over once or twice to ensure even warmth. The mortar should then be used as soon as possible if it is not to cool too quickly, especially in windy weather.

(e) Extended curing time

As mortars take longer to gain strength at low temperatures every care should be taken not to load the brickwork or blockwork too soon. When the air temperature is 5°C (41°F) for example, mortar may take two to three times longer to reach the required strength than it does under normal weather conditions. In addition, an extra day should be added for each day on which the temperature falls below freezing point.

If super-sulphate cement is used advice on mix proportions should be sought from the manufacturers if those using the cement are unfamiliar with its properties. It is susceptible to extremes of temperature and great care should be taken in the UK in frosty weather, as low temperatures tend to delay its setting time and serious difficulties may result.

(f) Rendering and plastering

Internal rendering and plastering does not usually suffer damage from frost action provided that cold winds are kept out and the walls themselves are not extremely cold when the rendering or plastering is applied. Whenever possible all windows should be glazed or covered with polythene, and during cold spells warm air heaters should be placed in the room the day before rendering or plastering so that the walls and materials may be brought to a reasonable temperature. Heating should continue for at least 48 hours after completion of the work. Large temperature differentials should be avoided as this tends to cause crazing in the finished rendering or plasterwork. Temperatures should also be kept above freezing point for lightweight plasters during the application and hardening period. Dry lining will, of course, avoid all the difficulties inherent in cold weather rendering and plastering.

External rendering should not be carried out during frosty weather.

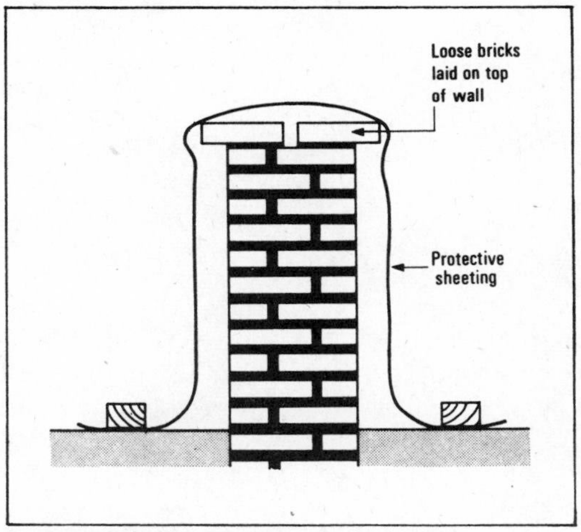

(g) Protecting completed work

Brickwork and blockwork should be covered as the work proceeds to protect it against freezing for from three to seven days, depending upon conditions; cold winds and draughts can be very damaging to new mortar. Special care should be taken with single-leaf walls as they are more readily attacked by frost than thicker walls, especially when exposed on both sides. One simple method giving nominal protection is to put a close course of bricks on top of the wall at the end of each day, letting them project about 2 in on either side; then cover at least the work carried out during the past 24 hours with polythene sheeting or similar waterproof covering, Fig 3. Additional insulation may be provided under the covering in very cold weather.

During exceptionally cold weather it is advisable to use heating to ensure that brickwork is unaffected. Small protective enclosures with forced air heaters may be suitable and strong, windproof enclosures can be formed using scaffold tubes and polythene sheeting, Figs 4, 5, 6, 7. More examples are illustrated elsewhere[3]. When bricklaying inside a building all openings should be sealed and where heat is needed portable space heaters are convenient and a guide to the output required to heat various types of structure is given in Standard Practice for Winter Working, published by NFBTE[4].

Walls should not be heated on one side with no protection on the other. Enclosures should be arranged to allow a circulation of warm air on both sides of the wall.

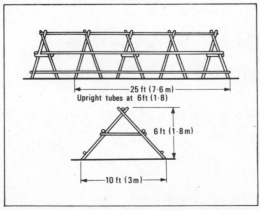

Figure 3 *Brickwork protected after laying.*

Figure 4 *Example of typical scaffold tube frame for bricklayers tent.*

Figure 5 *Underbuilding for cottages in freezing temperatures at Linwood, nr. Paisley.*
Figure 6 *Shelter for brick cladding at the 11th storey level. Scottish Special Housing Association site at Clydebank.*
Figure 7 *Internal view of shelter for brick cladding – SSHA Clydebank site.*

5   6 7

Bricklaying under winter conditions

(h) Protection against rain and snow

Individual materials and recently completed brickwork should always be protected against rainwater. The precautions for individual materials are similar to those described above for frosty weather and elaboration is considered unnecessary. It is essential that newly completed brickwork is protected at the end of each day's work against rainwater or in any period of interruption through rain. It is particularly important that brickwork constructed of bricks containing perforations should be covered during intervals of bricklaying. In multi-storey construction or where scaffolding is used, it is important that the plank adjacent to the brickwork and any mortar boards are turned back at the completion of each day's work to avoid splashing of the finished surface.

Rain soaked brickwork takes a long time to dry, especially in the winter months, and this often means delay in plastering and decorating. Care should be taken therefore to ensure that any brickwork known to be saturated at the time of frost should also be properly protected to prevent damage by freezing.

A simple test to determine if new brickwork is frozen is to apply a blow lamp to the mortar joints. Any apparent softening of mortar is a certain indication that the material is frozen. Under these conditions work must stop until such time as normal drying can continue.

Any work affected by snow or ice due to the coverings being displaced, should be thawed with live steam, a blow lamp or blow torch carefullly applied. The heat should be sustained long enough to thoroughly dry out the brickwork. If it is frozen or damaged, defective parts should be replaced before bricklaying continues.

INDIVIDUAL MATERIALS
Bricks

Bricks should be stacked and protected as described in (c). It is unlikely that many bricks used in the UK will have an excessive suction rate during the winter months. If clay bricks have a suction rate in excess of 20gm/dm²/min or the bricklayer considers they will be difficult to lay because of the suction rate, the bricks must not be wetted during cold or frosty weather. To overcome the problem of high brick suction a 'water-retentive' mortar should be used, ie a mortar containing a proportion of lime will improve its water retentivity and resistance to suction. Calcium silicate bricks should never be wetted to reduce their suction rate and it is desirable that a highly water-retentive mortar be used as an alternative to wetting the bricks.

It is not usually necessary to heat bricks before laying under winter conditions in Great Britain but in exceptionally cold weather there may be some advantage in doing so. Recommendations have already been made regarding bricklaying indoors in (f).

In the unlikely event of bricks being delivered 'hot from the kiln' they should not be built into the work until at least 2 days have elapsed from the time of drawing from the kiln.*

Cements

All cements should be stored as described in (c) above. Cement should always be kept dry prior to use but should *never* be heated before gauging with mixing water.

All building materials undergo size changes due to wetting and drying and for further information on this subject the reader is referred elsewhere. [5,6]

Special cements
Super-sulphate cement

This cement is not a Portland cement and the precautions listed under (e) should be noted. Cold and frosty weather may delay its setting time.

Masonry cement

Masonry cements marketed in the United Kingdom all contain approximately 75% Portland cement and 25% of a fine inert filler such as a ground limestone and an air-entraining agent. British masonry cements contain a higher proportion of Portland cement than those of many other countries, and for this reason it is unnecessary to gauge additions of Portland cement to obtain mortar of relatively high strength. The strength of masonry cement mortars in the UK is in fact only controlled by altering the proportion of sand, the normal range of volume proportions being from 1:3 to 1:7 masonry cement:sand. It is very important that the manufacturer's recommendations on mix proportions should be strictly followed. As already stated masonry cement contains a plasticizer and additional air-entraining agents *must not* be incorporated as over air-entrainment could cause serious difficulties.

The author has visited numerous sites where troubles have occurred due to the misuse of masonry cement. These are usually where the operative on site has assumed masonry cement to be Portland cement. The results of such misuse can be responsible for frost attack of the mortar; frost/sulphate attack of the mortar and sub-standard mortar strengths. In each case such malpractices have necessitated the pulling down of work and rebuilding, a tragedy which should never have occurred had the site staff known the properties of the material they were handling.

Masonry cement is an excellent material for normal two-storey domestic construction and in some instances for, perhaps, more adventurous building, but unfortunately it is grossly misused and often confused with ordinary Portland cement.

Sand

Poor sand should never be used, particularly in cold weather. Very fine sand has a particularly high surface area and if used in weak mortar mixes or in conjunction with masonry cement it has the effect of diluting the ratio of cement to sand making the resulting mortar more vulnerable to frost and sulphate attack. Excess of pigment can have a similar effect, particularly those based on carbon black.[7]

Dirt or loam in the sand will make the mortar weak and may also slow down the rate of hardening of the cement. It is therefore essential that only clean sand is used.

During periods of continuous heavy frost stock piles of sand should be maintained at an approximate temperature of 24°C (75°F) using one of the following methods:

a) Passing steam through lances or coils inserted in the sand.
b) Passing heat from bottled gas burners through pipes laid in ground beneath stock piles or bins.
c) Erecting a protective enclosure over the sand and heating the enclosed area with suitable space heaters.
d) Using electric surface heaters or electric blankets (ie reinforced PVC insulating quilts which include heating elements).

Small quantities can be heated by spreading the sand on corrugated metal sheets over a heater or by banking it around pipes in which fires have been lit.

Damp proof courses and flashings

Some damp proof course materials (particularly those composed of bituminous materials) tend to harden and become unworkable in cold weather. To overcome such difficulties these materials should if possible be stored in a warm dry atmosphere prior to use. Alternatively, it may be necessary to heat the materials, but a blowlamp unless recommended by the manufacturers should never be used. Where flashings require to be shaped (eg to form trays etc), great care should be taken during cold weather to ensure that they are not damaged, as slight tears tend to extend with some materials due to creep (not only lead) and extensive damage can occur requiring extremely expensive remedial action later.

Recommended mortar mixes

Tables 1 & 2 are included to give information on the selection of mortar mixes for different constructions under varying degrees of exposure and frost hazard. The guiding principle in using the tables is to prescribe a mortar that contains no more cement than is necessary to give adequate strength in the brickwork, unless there is good reason for choosing a richer mix □

SUMMARY

1 Anticipate bad weather by using the meteorological services.
2 Protect brick stacks against wetting.
3 Use 1:1:6 mortar or 1:6 cement:sand mortar with plasticizer but, do not use a stronger mortar unless necessary for structural brickwork or brickwork below dpc.
4 Heat the sand and mortar in extremely cold weather.
5 Keep the cement and hydrated lime dry.
6 Do not use very fine or dirty sand.
7 Do not use calcium chloride.
8 Do not wet the bricks prior to laying.
9 Never use frozen materials.
10 Keep the finished brickwork above freezing point for at least 3 days after laying.

ACKNOWLEDGEMENTS

The author is indebted to the Department of the Environment for permission to reproduce figures 1 to 4 and Tables 1 and 2. Particular thanks are due to Mr J. R. Smith, Advisor on Winter Building to the DOE for his co-operation and advice.

Figures 5, 6 and 7 are reproduced by kind permission of Scottish Special Housing Association (Building Department).

REFERENCES

1 Anon. Advisory leaflet No 40. 'Weather and the Builder' DOE HMSO.
2 'Model Specification for Loadbearing Clay Brickwork' Special Publication No 56. Drafted by the Structural Ceramics Advisory Group of the Building Science Committee B Ceram RA 1967.
3 Anon. Advisory leaflet No 74. 'Protective Screens and Enclosures' DOE HMSO
4 Anon. 'Standard Practice for Winter Working' NFBTE
5 Thomas. K., 'Movement Joints in Brickwork' CPTB Technical Note Vol 1. No 10 July 1966.
6 Foster. D. 'Some Observations on the Design of Brickwork Cladding to Multi-Storey RC Framed Structures'. The Brick Development Association Technical Note Vol 1 No 4. September 1971.

7 Thomas. K., Coutie. M.G., & Pateman. J., 'The Effect of Pigment on some Properties of Mortar for Brickwork' To be published. SIBMAC Proceedings 1971.

Table 1 Selection of mortar types

Type of construction	Exposure conditions	Early frost hazard	Type of bricks or blocks	Mortar designation
EXTERNAL WALLS				
Retaining wall	Any	Yes or no	Clay, concrete or calcium silicate	1
Parapet, free standing wall or below damp-proof course	Any	Yes or no	Clay	1, 2 or 3
			Concrete or calcium silicate	3
Between eaves and damp-proof course	Severe	Yes or no	Clay, concrete or calcium silicate	3
	Sheltered or moderate	Yes	Clay, concrete or calcium silicate	3
		No	Clay	3 or 4
			Concrete or calcium silicate	4
INTERNAL WALLS AND PARTITIONS		Yes	Clay	3
			Concrete or calcium silicate	4
		No	Clay, concrete or calcium silicate	4 or 5

Table 2 Equivalent mortar mixes (proportions by volume)

Mortar designation (from Table 1)	Hydraulic –lime: sand	Cement: lime: sand	Masonry –cement: sand	Cement: sand with plasticizer
1	—	$1:0-\frac{1}{4}:3$	—	—
2	—	$1:\frac{1}{2}:4-4\frac{1}{2}$	—	—
3	—	$1:1:5-6$	$1:4\frac{1}{2}$	$1:5-6$
4	$1:2-3$	$1:2:8-9$	$1:6$	$1:7-8$
5	$1:3$	$1:3:10-12$	$1:7$	$1:8$

∧ Increasing strength but decreasing ability to accommodate movements caused by settlement, shrinkage, etc.

Direction of change in properties (within any one mortar designation)

————> Increasing resistance to ————> damage by freezing

<———— Improving bond and consequent <———— resistance to rain penetration

Tables 1 & 2 are reproduced from Building Research Station Digest No. 58 (Second Series) – Mortars for Jointing.

Interim notes on reinforced brickwork

INTRODUCTION

Reinforced brickwork was first used in this country almost 150 years ago, but the greatest subsequent development of this form of construction has taken place abroad, for example, in the United States and India. Comparatively little work in reinforced brickwork has been carried out in this country, and this may be due in part to the lack of design codes. The Structural Ceramics Advisory Group (SCAG) of the British Ceramic Research Association has set up a Working Party to prepare a design guide on reinforced and prestressed brickwork. This interim note is published to bridge the gap until that document is available.

The recommendations at present contained in CP.111: Part 2: 1970 relate only to walls and columns, not to beams or laterally loaded panels or slabs, but, in the absence of any other Code, it seems appropriate to turn to this one for some guidance on permissible stresses.

Test have shown that the structural performance of reinforced brickwork is similar to that of reinforced concrete, so that the established formulae for the latter material can quickly be adapted. It is assumed that all behaviour is linearly elastic, Hooke's Law being obeyed, and the tensile strength of brickwork is neglected, so it is assumed that all tensile forces are resisted by the reinforcement. The modular ratio is taken to be constant for any particular form of construction, dependent only upon the strength of the brick used, because different bricks have different elastic modulii.

At present all design is based on the elastic theory, although the SCAG document when produced will be in Limit State terms.

1 Horizontal reinforcement Normally, this takes the form of light reinforcement in the bed joints, and is used in non-loadbearing panels to increase the resistance to lateral wind loading, and in all forms of construction to resist the stresses resulting from moisture and thermal movements. Similar horizontal reinforcement is used in the vicinity of concentrated loads, to increase the wall strength locally, and in cases where minor differential foundation settlement may occur, to limit cracking.

2 Vertical reinforcement Normally, this is placed in a cavity between two leaves of brickwork, which is afterwards grouted, or is run up through the spaces in a special bond, for example Quetta Bond, where, again, it is finally grouted into place. But, in early applications, specially shaped bricks were used, which facilitated threading the reinforcement through the actual brickwork. Vertical reinforcement is used to turn brickwork into a member capable of acting as a vertical cantilever or beam, in such applications as retaining walls and tanks.

3 Full flexural reinforcement This consists of fairly heavy steel bars, both horizontal and vertical, usually grouted in the cavity, to produce a brickwork beam capable of spanning horizontally across openings. To date, the indications are that reinforced brickwork is cheaper than reinforced concrete in most situations.

MATERIALS

As far as possible the materials should comply with the recommendations given in the Model Specification for Loadbearing Clay Brickwork, issued by the British Ceramic Research Association and available free from the Brick Development Association. Calcium silicate bricks should comply with BS 187: Part 2: 1970. In addition, all bricks used should have a crushing strength greater than 15 MN/m², and the mortar should not be weaker than $1 : \frac{1}{2} : 4\frac{1}{2}$ (cement: lime: sand) or equivalent, to provide adequate protection to the reinforcement. If used, admixtures and masonry cement must be carefully selected, because some types can give rather poor bond between the mortar and both the bricks and the steel.

MODULAR RATIOS

Adopt the values given in Table 6A of CP 111: Part 2: 1970, as revised June 1971, for the ratios of the modulus of elasticity of steel reinforcement to that of the brickwork.

PERMISSIBLE STRESSES

In general, the permissible stresses given in Clause 321 of CP 111: Part 2: 1970, as revised, can be used, but it should be remembered that these are really for walls and other vertical load-carrying members, and so need modifying for other types of structures. Some suggestions are given below:

(i) Direct and flexural compression. The Code reduction factors for slenderness are to guard against instability, and additionally for eccentricity of loading, to cover the magnification of the resulting moment, at mid-height of the member, due to lateral deflection. Consequently, both factors apply only to members carrying substantial in-plane forces.

In simple in-fill panels, free-standing walls and similar structures, which are loaded laterally and carry only self weight vertically, the behaviour is flexural, and the extreme fibre stress in compression in the brickwork can exceed the appropriate basic stress by $33\frac{1}{3}\%$. A further 25% overstress, both in brickwork and steel, may be permitted when due only to the action of wind.

In members subjected simultaneously to lateral and in-plane forces, it is suggested that these be proportioned to satisfy the expression

$$\frac{v}{V} + \frac{l}{L} = 1$$

where
v = design in-plane loading
V = in-plane loading, acting alone, which will produce the maximum permissible direct compressive stress, making due allowance for all reduction factors.
l = design lateral loading
L = lateral loading, acting alone, which will produce

Figure 1 *(front cover) Dorset Water Board offices, Poole. The regularly spaced structural piers around the perimeter of the building are in loadbearing brickwork with lightly reinforced concrete cores. Architects: Farmer & Dark.*

Figure 2 *(page 2) Anglia Square car park, Norwich. Parapet walls have light reinforcement in the bed joints. Architects: Alan Cooke & Partners.*

Figure 3 *Prestressed brickwork water tank (120,000 gal capacity) designed by Donald Foster RIBA of Structural Clay Products Limited.*

the maximum permissible flexural compressive stress, that is basic stress + $33\frac{1}{3}$ %.

(ii) Shear The values quoted in the Code are for stresses resulting from racking in walls, and can be used for all in-plane shear stresses in that situation.

In all other cases, including transverse shear, the capacity of the brickwork alone should be taken to be 0.10 MN/m² , and any excess stress should be carried on correctly positioned shear reinforcement. The brickwork should be assumed to be inoperative in shear and reinforced accordingly, following reinforced concrete practice, when the calculated stress exceeds 0.50 MN/m² , or in positions where the brickwork may be cracked.

In no case should the calculated stress exceed 1 MN/m² .

DEFLECTION
The calculated deflection, based upon the gross uncracked brickwork section, and neglecting the reinforcement, should be limited to 1/500th of the span or 8 mm, whichever is the smaller.

WORKMANSHIP
As far as possible the requirements of Model Specification for Loadbearing Clay Brickwork should be followed.

The construction of brickwork with horizontal reinforcement in the bed joints presents no particularly original problems, although the thickness of these joints should be limited to 10 mm, and the reinforcement should be carefully centred.

The construction of grouted cavity work takes two forms known as low lift and high lift respectively. In low lift work, the brickwork is carried up about 200 mm, with a narrow 20 mm cavity, and the grouting is performed by hand using a fairly stiff mix, which is tamped into place. In high lift construction, the whole wall is built to a storey height, leaving a cavity about 75 mm in width. The cavity is cleaned out, using a pressure jet of air or water, stop ends are inserted at about 8 m. intervals and grouting is then carried out, using a fluid mix suitable for pumping, in lifts not exceeding 1 m. The grout is consolidated during placing by vibration or tamping.

PRESTRESSED BRICKWORK
The design of prestressed brickwork should follow the general principles of prestressed concrete but using the permissible stresses appropriate to reinforced brickwork. When considering loss of prestress, it should be noted that creep in brickwork is very small and clay brickwork tends to expand early in its life, whilst calcium silicate brickwork tends to shrink. The numerical values for these effects should be ascertained for the type of brick it is proposed to use, but the losses are normally much smaller than in concrete, so there is little justification for permitting increased stresses at transfer, except that the reduction factors do not apply at this stage, if stressing internally with tendons. Because losses are only indirectly related to brick strength, it is possible to economically prestress quite low strength brickwork, in direct contrast to concrete, where high strength material is essential to limit losses.

MORTARS FOR BRICKWORK

INTRODUCTION

Mortars influence the compressive strength, durability and resistance to rain penetration of brickwork; consequently, it is important to select carefully, and then correctly use, the mortar for any particular application.

TYPES OF MORTAR

The normal grades of mortar are given in the table below. Straight cement/sand or lime/sand mixes are now rather unusual; the former tend to be harsh and the latter, although very workable, are weak and slow setting. The strength and set conferred by cement, and the workability conferred by lime, can be obtained in 'gauged' mixes of cement/lime/sand, although mortars of roughly equivalent properties are said to be produced using plasticizers or masonry cements.

MORTAR SELECTION

General

The various mixes given against any mortar grade have roughly equivalent strength and workability, but the other characteristics differ. The presence of lime improves the water retentivity, while retarding the set, possibly rendering the mortar more liable to frost damage. Plasticizers and masonry cement introduce air bubbles into the mortar and these adversely affect both the bond and the resistance to rain penetration. Therefore, mortars in columns 5 and 6 of the table below have the greater frost resistance, and those in columns 2 and 3 have the greater bond.

Normally, the mortar should be weaker than the brick, so that any minor cracking will be confined to the mortar, where it is less obvious. However, strong dense mortar is essential in positions of extreme exposure, particularly with clay bricks, where there is a danger of sulphation.

In calculated loadbearing brickwork, reinforced brickwork or prestressed brickwork, the mortar is selected by the engineer to suit the particular strength requirements and must not be altered without permission. Masonry cements, plasticizers and additives of any sort should not be used unless approved by the engineer.

Recommended mixes

The mortar given in the recommendations on page 3 is the weakest that is likely to be suitable for the location, assuming that the correct grade of brick is being used. Mixes containing lime are always to be preferred for use with calcium silicate (CS) bricks.

MATERIALS

Mortar materials should comply with the relevant British Standards: Cement – BS 12: Portland cement (ordinary and rapid hardening)
Sulphate-resisting cement – BS 4027: Sulphate-resisting Portland cement
Lime – BS 890: Building limes
Sand – BS 1200: Building sands from natural sources
Water – must be clean and free from impurities; tap-water is normally satisfactory. Where the water supply is of doubtful quality, refer to BS 3148: Tests for water for making concrete.
Pigments – BS 1014: Pigments for cement, magnesium oxychloride and concrete.
Plasticizers – BS 4887: Mortar plasticizers.

BATCHING AND MIXING

Shovel batching is totally inadequate; cement is dry and will not stand up on the shovel, whereas sand is frequently wet and will stand up. Consequently, a shovelful of sand has a much greater volume than a shovelful of cement, so that the resulting mix will be lean and weak.

The materials must be gauged using a bucket or box, struck off level each time. The box can actually be bottomless so that the materials can be shovelled into the mixer, after gauging.

Enough water should be added to give the correct workability.

Clay bricks with excessive suction should be wetted to make them more manageable, but this is not usually done with calcium silicate bricks so, within reason, more water should be added to the mix, or a water-retaining additive should be employed. When mixing aerated mortars, care should be taken not to add too much water at the beginning, because these mortars become progressively more fluid as air is entrained

1 Mortar grade	2 Lime : sand	3 Cement : lime : sand	4 Cement : sand	5 Cement : sand & plasticizer	6 Masonry cement : sand	7 Typical compressive strength, N/mm²	
						7 days	28 days
1		1 : ¼ : 3	1 : 3			7.0	11.0
2		1 : ½ : 4½		1 : 4	1 : 3	3.5	5.5
3		1 : 1 : 5–6		1 : 5–6	1 : 4½	1.0	2.5
4	1 : 2	1 : 2 : 8–9		1 : 7–8	1 : 6	0.7	1.0
5	1 : 3	1 : 3 : 10–12		1 : 8	1 : 7		

Notes

1 Proportions are by volume. Where a range of sand content is given (eg 5–6), the larger quantity should be used for well graded sands, and the smaller for sands that are uniformly coarse or uniformly fine. Due to bulking, the volume of damp sand may need to be increased.

2 'Cement' means ordinary Portland cement, rapid hardening Portland cement or sulphate resisting Portland cement. Other cements may be employed in special circumstances, but the maker's instructions should be followed meticulously.

3 'Lime' means non-hydraulic or semi-hydraulic lime, and the proportions given are for lime putty. If the lime is measured as dry hydrate, the amount can be increased by up to 50%.

4 Calcium chloride should not be used as an additive.

LOCATION	Type of brick	Cold*** weather working	Warm weather working
Internal walls	All	4 (aerated) or 3	4 (5 if not loadbearing)
Inner leaf of cavity walls	Clay	3	3
	CS	4 (aerated) or 3	4 (5 if not loadbearing)
Backing to external solid walls	Clay	3	3
	CS	4 (aerated) or 3	4
External walls between eaves and dpc sheltered or moderate exposure, rendered or tile hung	Clay	3	3
	CS	4 (aerated) or 3	4
External walls between eaves and dpc severe exposure	All	3**	3**
Below dpc but 150 mm above ground	Clay	3	3
	CS	3	4
Below ground or within 150 mm of ground	All	2*	3*
Parapets (unrendered)	Clay	2**	2**
	CS	3	3
Parapets (rendered one side)	Clay	3**	3**
	CS	4 (aerated) or 3	4
Sills and copings (using standard bricks, not purpose made units)	All	2**	2**
Freestanding walls	All	2**	2**
Earth retaining walls	All	2*	2*
Tall chimneys	All	3**	4**
Brickwork dpc's	Clay	1	1

Notes
Sulphate-resisting cement will be required where there are sulphates present in the groundwater.
** *Sulphate-resisting cement will be required if the clay bricks are not 'special quality'.*
*** *'Cold weather' is roughly December to March in Great Britain.*

during the mixing process.

Every batch of mortar should be thoroughly mixed mechanically for at least three minutes, and then used within two hours of mixing.

WINTER WORKING

In general, mortar mixes weaker than Grade 3 should not be used externally in cold weather. Stronger mixes, mixes that do not contain lime and aerated mixes have greater frost resistance.

Sand stored out of doors should be covered immediately after delivery, and in periods of prolonged frost will need to be heated before use. Small quantities can be warmed on metal sheets over a heater, stacks can be warmed by passing steam through coils embedded in them or by erecting shelters around them and then using space heaters.

In very cold weather – below —4°C – the mixing water also should be heated to about 40°C, but to minimise heat loss and to avoid flash setting the materials should be gauged dry and the hot water added last. Cement must not be heated, but the mortar itself can be gently warmed on metal sheets until it is needed.

Additives based on calcium chloride should not be used. They impart little or no frost resistance, but encourage dampness in the hardened mortar which may lead to corrosion of the wall ties.

No bricklaying should be carried out when the temperature is below 3°C, unless adequate precautions have been taken to ensure a minimum temperature of 4°C in the work when laid, for example, by covering and space heating as necessary. The brickwork must be maintained at a temperature above freezing point, until the mortar has hardened, which may take up to one week.

READY-MIXED MORTARS

Ready-mixed lime/sand, complying with BS 4721 and delivered wet to site, and dry cement/sand mixes both may be used for mortar, and both can be supplied ready pigmented.

Ready-mixed lime/sand is supplied in several categories, generally intended to be used only after gauging on site with ordinary Portland or sulphate-resisting cement in the correct proportions, and with sufficient extra water to produce the necessary workability; neither lime nor sand should be added. The supplier will be able to advise on suitable batching procedures but, for most work, it will be adequate if both materials are batched by volume, using gauge boxes. Batching by the shovelful should not be done, because it is wildly inaccurate.

The supplier should be consulted before ordering ready-mixed lime/sand, to ensure it is of the correct category for the grade of mortar required, although the Table below may be used for guidance.

The wet ready-mixed lime/sand should be stored under cover on site, on a hard standing, and be protected from freezing during the winter. Heaps of different strength should be clearly identified.

Dry cement/sand mixes are usually based upon masonry cement, or Portland cement with a plasticizer, and should be mixed with water only and not lime. It may be possible to alter the grade of the mortar, by the addition of accurately gauged quantities of cement or sand to the dry mix on site, but this should be confirmed with the supplier. It is safest to assume that these mixes are not sulphate resisting.

Dry cement/sand mixes should be stored on site in the same manner as cement.

PIGMENTS

Pigments are always used in small quantities and should be weighed carefully; excessive use, particularly of carbon black, can affect the strength of the resulting mix.

Once a mix has been established producing an acceptable colour, nothing should be changed. Altering the proportions or the source of the supply of the sand or cement would almost certainly alter the colour of the mortar. Remember that the colour of a mortar changes while it is drying and setting □

Grade of mortar required	Cement : lime : sand	Ready mixed lime: sand by volume	Site mixing cement : ready-mix
1	1 : ¼ : 3	1 : 12	1 : 3
2	1 : ½ : 4½	1 : 9	1 : 4½
3	1 : 1 : 6	1 : 6	1 : 6
4	1 : 2 : 9	1 : 4½	1 : 9
5	1 : 3 : 12	1 : 4	1 : 12

THE SELECTION OF BRICKS FOR PERFORMANCE

INTRODUCTION

Clay bricks to BS3921 are manufactured, basically by firing naturally occurring deposits of clay or shale, to a single standard work size of 215 × 102.5 × 65 mm, which gives a format size of 225 × 112.5 × 75 mm, when used with a 10 mm joint. Many brickmakers are able to produce special bricks to BS4729, which are compatible with standard bricks, and some manufacture bricks to a metric module, in accordance with the recent BS Draft for Development.

Calcium silicate bricks to BS187 are manufactured, by autoclaving lime with sand and sometimes other aggregates, to a single standard work size of 215 × 103 × 65 mm, which gives the same format size as clay bricks. Some metric modular bricks are also produced.

CLAY BRICKS

Nomenclature

No systematic nomenclature has been evolved, and the designations in use today are based on at least six different factors. These are listed below, with some examples, but the latter are neither exhaustive nor necessarily important.

1. Place of origin – eg Leicester Red, London Stock
2. Raw materials – eg marl, gault
3. Method of manufacture – eg hand-made, wire-cut, pressed
4. Use – eg facing, engineering, common
5. Colour – eg yellow, multi, brindle
6. Surface texture – eg sandfaced, rustic.

Formerly, when transport was restricted, the description of a place of origin or type of clay was usually sufficient to specify a brick quite closely. Now bricks are likely to be transported long distances, and more fundamental technical descriptions are to be preferred.

Two designations in general use should be noted: 'Fletton' is the name widely applied to the pressed bricks made from the Lower Oxford Clay, which occurs in the Peterborough and Bedford regions. Bricks of this kind form almost one-half of the total British output. 'Stock' is the term often loosely used to refer to the brick manufactured in a particular district, notably South-East England; one well-known variety is the yellow coloured 'London Stock'.

Classification

Clay bricks are classified according to variety, quality and class.

1. Varieties

1.1 *Common* Common bricks are suitable for general building work but have no special claim to give an attractive appearance. The name Common is applied to the many varieties of clay bricks, which fall outside the categories of Facing and Engineering bricks.

1.2 *Facing* Facing bricks are specially made or selected to give an attractive appearance when used without rendering or plaster or other surface treatment of the wall.

1.3 *Loadbearing* Loadbearing bricks, which may be either Facings or Commons, according to their appearance, conform to specified average compressive strength limits depending upon their class (see below).

1.4 *Engineering* Engineering bricks are dense and strong, and also conform to specified limits for absorption and strength. This is the most precisely defined variety, since BS3921 lays down performance characteristics as shown below:

It must be stressed that for bricks to qualify as 'Engineering', they must possess *both* the required strength and absorption properties. There are many excellent Loadbearing, Common and Facing bricks which exceed the compressive strength requirements for Engineering Class B, and some even exceed those for Class A, but they cannot be so classified, because their absorption characteristics exceed the specified limit.

Note that the term 'Semi-Engineering' has no defined meaning and should not be used.

1.5 *DPC* Clay bricks with an average absorption of 4.5% or less are specially made by some suppliers for use as a damp proof course, and these are called DPC bricks.

2. Quality There are three qualities of bricks:

2.1 *Internal* Bricks described as Internal Quality are suitable for internal use only, and may need protection on site during winter.

2.2 *Ordinary* Bricks of Ordinary Quality are less durable than Special Quality, but normally durable in the external face of a building. Generally, they do not need protection on site when stacked during one winter.

2.3 *Special* Special Quality bricks are durable even when used in situations of extreme exposure where the structure may become saturated and be frozen, for example, parapets and retaining walls.

Engineering bricks normally attain this standard of durability. Some Facing and Common bricks do also, but this should not be assumed in any particular instance unless it is claimed by the manufacturer.

Special Quality bricks also have clearly defined limits for soluble salts content, to minimise the risk of sulphation in the mortar.

3. Class

Loadbearing bricks, whether Facing or Common, are divided into eight classes, graded according to compressive strength:

Class	Average compressive strength N/mm²
1	7.0
2	14.0
3	20.5
4	27.5
5	34.5
7	48.5
10	69.0
15	103.5

Intermediate classes may be interpolated.

The strength of a clay brick is not a complete guide to its durability, so that selection is normally based on quality.

Engineering	Average compressive strength N/mm² not less than	Average absorption % weight not greater than
A	69.0	4.5
B	48.5	7.0

CALCIUM SILICATE BRICKS

Nomenclature

The generic name is 'calcium silicate' bricks. Depending on the materials used, however, the more specific term 'sandlime' and 'flintlime' are acceptable designations. Other features, such as the place of manufacture or the brick colour, are often used in naming a brick.

Classification

Eight classes of calcium silicate bricks are recognised in BS187, graded largely according to compressive strength (see table below).

The strength of a calcium silicate brick is a much closer guide to its durability than that of a clay brick, so that selection is normally based on class.

PERFORATED BRICKS

Perforating an extruded clay brick generally confers distinct advantages, in the form of improved compaction and more uniform firing. BS3921 states that such a brick may be regarded as solid, if the perforations consist of small holes which do not exceed 25% of its volume. Testing has shown that walls built of such bricks are as strong and resistant to rain penetration as walls constructed of similar truly solid bricks, and no attempt should be made to fill the perforations during construction. Nevertheless, the presence of even this moderate amount of perforation reduces the fire resistance of the wall (1).

FROGGED BRICKS

A proportion of both clay and calcium silicate bricks are manufactured with depressions in the bed faces known as frogs. In calculated loadbearing work, the frogs should be laid uppermost and filled with mortar as the work proceeds.

BS3921 states that a clay brick may be regarded as solid if the frog does not exceed 20% of its volume. Frogs of this size do not influence the notional period of fire resistance of walls built from either clay or calcium silicate bricks (1).

SELECTION OF BRICKS

Bricks are frequently selected for appearance only, although there are many other important factors. In loadbearing or reinforced brickwork, the class of brick will be selected by the engineer, to suit the particular stress involved, and this must not be changed without permission. The quality of clay bricks, however, may be changed provided that the same class is used.

The recommendations given overleaf list the minimum types of brick likely to be durable in the given locations, assuming that the appropriate grades of mortar are used (1, 2), and that all other constructional details are correct.

The following general points should be noted:

Clay bricks

There is no substitute for experience, in assessing the frost resistance of clay bricks, and no satisfactory accelerated test has yet been devised. Bricks stronger than 48 N/mm² are normally durable, although this is not always true, and there are also many weak bricks that are completely frost resistant. Bricks of Ordinary Quality are normally durable when used between eaves and dpc, but this should not be taken for granted in the very exposed parts of Scotland, for example, and care should always be exercised when using any brick in a new geographical location. Bricks of Special Quality are always frost resistant.

Clay bricks tend to expand after firing and the provision of movement control joints is advisable (1).

Calcium silicate bricks

Although these bricks can be tinted during manufacture, their natural colour is 'off-white', which is particularly suitable for light-wells and for internal use, where the need for a plaster finish is sometimes eliminated. Calcium silicate bricks are free from soluble salts, and so do not cause efflorescence or sulphation in the mortar.

They should not be used where they will be exposed to acids, acid fumes, sea-water or sewage, although sulphates, naturally occurring in some soils and groundwaters, do not affect them.

Although they should not be exposed to high temperatures, they may be used in domestic chimneys.

Calcium silicate bricks tend to shrink after manufacture, and the provision of movement control joints is advisable (1) □

REFERENCES

1. CP121 : 1973, Walling, Part 1, **Brick and Block Masonry**.
2. 'Mortars for Brickwork', BDA Practical Note No 2, September 1973 (available free from BDA).

CALCIUM SILICATE BRICKS

Class	Average Compressive Strength N/mm²	Drying Shrinkage*
1	7.0	—
2A	14.0	0.025
2B	14.0	0.035
3A	20.5	0.025
3B	20.5	0.035
4	27.5	0.025
5	34.5	0.025
7	48.5	0.025

* As a percentage of the original wet length

Location	Rendered	Constructional weather conditions	Clay brick quality	Class of CS Brick
Backing to external solid walls	Yes	Cold*	Ordinary	1
Inner leaves of cavity walls	Yes	Warm	Internal	1
Internal walls and	No	Cold*	Ordinary	2
Partitions	No	Warm	Internal	2
Outer leaves of cavity walls above dpc below ground or below dpc but 150 mm above ground		All	Ordinary	2
External facing to solid construction. Walls within 150 mm of ground level		Cold*	Special	3
		Warm	Ordinary***	3
External free standing walls excluding capping and coping, walls below ground		All	Ordinary	3
Parapets	Yes (one side only)	All	Ordinary	3
	No	Cold*	Special	3
		Warm	Ordinary***	3
Sills, Copings**	—	All	Special	4
Earth Retaining Walls	—	All	Special	4
As Damp Proof Course	—	All	DPC or Engineering 'A'	Not suitable
Sewerage Work	—	All	Engineering 'A' or 'B'	Not suitable

Note:
  * Cold conditions are roughly – December to March in Great Britain.
  ** Using standard bricks, not purpose made units.
  *** Special quality bricks are always to be preferred.

CLEANING OF BRICKWORK

GENERAL

Staining on brickwork is generally due to external causes but sometimes may result from salts in the bricks themselves. Staining can mar the appearance of brickwork but incorrect cleaning techniques can cause permanent damage, consequently, any proposed method of cleaning should be tried out in a small unobtrusive area and left for as long as possible, but at least a week or so, to judge the results, before the whole job is tackled.

The techniques given below are intended for 'do-it-yourself' work in removing relatively small areas of staining. A specialist contractor should be engaged for cleaning large areas of brick, for example, general cleaning of industrial grime from a building.

Where chemicals are to be used, the brickwork should always first be saturated with clean water, to prevent it absorbing the chemicals, and washed down thoroughly with clean water afterwards. Normally, it is preferable to employ wooden scrapers and stiff fibre brushes to avoid damaging the bricks. Adjacent features, such as metal windows, and the area at the foot of the wall should be protected from splashing with the chemicals. In places where it is not possible to make a mess, or where the stain is very localised, most of the cleaning liquids can be applied as a poultice by thickening them with an inert filler such as talc, bentonite or powdered chalk. Many of the chemicals recommended below are caustic or poisonous, so that care should be taken and protective clothing and

goggles should be worn. Volatile solvents should only be used indoors under conditions of good ventilation.

Cleaning techniques differ in some cases for clay and calcium silicate bricks, so it is necessary to identify the brick in question, and the following notes should help, although it is always best to consult the brickmaker, when known. Calcium silicate bricks, which are relatively modern, have only been used within about the last 40 years. In their natural state, they are white or off-white and smooth surfaced, although they can be tinted to almost any colour and roughened by sand-blasting or acid-etching. Clay bricks have a much longer history, can be produced with heavy surface rustications and sand coatings, and are still by far the commonest bricks in general use.

Persistent efflorescence, extensive salt staining from within the bricks themselves and vegetable growth on the brickwork are due to excessively wet conditions, and, unless the basic problem of water penetration is cured, the staining will repeatedly reappear. In such cases, it is necessary to overhaul faulty flashings and damp proof courses, repair leaking rainwater downpipes, renew copings, etc.

Where chemical treatments are advocated in the recommendations which follow, readers are asked to note that it is essential for both clay and calcium silicate brickwork to be washed down afterwards, using copious quantities of water to clear all traces of chemicals from the surface, adjacent areas and drains.

Front cover and this page
Georgian brickwork in Bedford Row, London, restored to something closely approaching its original colour and condition, despite centuries of exposure to a polluted urban atmosphere.

Opposite page
Left Cleaned and uncleaned brickwork on neighbouring properties.
Right Large areas of brickwork cleaning and restoration usually necessitate the services of a specialist contractor.

CLAY BRICKWORK

Remember to saturate the brickwork with clean water, before applying any chemical, and wash down with clean water afterwards.

Stain	Method
Oil	Sponge or poultice with white spirit, carbon tetrachloride or trichlorethylene. Good ventilation is essential indoors.
Paint	Apply commercial paint remover or a solution of trisodium phosphate (1 part to 5 parts of water by weight), allow the paint to soften and remove with a scraper. Wash the wall with soapy water and finally rinse with clean water.
Efflorescence (white crystals or white furry deposit)	This usually disappears rapidly from new brickwork due to the action of wind and rain but the process can be accelerated by repeated washings, allowing the wall to dry out at intervals. Brushing down the wall at times of maximum efflorescence will help, although this should not be done on sand-faced bricks, to avoid removing the face. The salts brushed off should not be allowed to accumulate at the base of the wall, otherwise they may be carried back into the brickwork by subsequent rain.
Lichens and mosses	These can be killed with a solution of zinc or magnesium silicofluoride (1 part to 40 parts of water by weight) or, alternatively, a proprietary weed killer. Vegetable growth is generally indicative of damp brickwork and will usually reappear if this basic cause is not cured. (Green staining which does not respond to this treatment is probably due to vanadium salts from within the bricks.).
Vanadium (green)	Wash down with a solution of ethylene diamine tetra acetic acid (1 part to 10 parts of water). (Hydrochloric acid should never be used on vanadium stains since it 'fixes' them and turns them brown. Such brown stains can sometimes be removed using a strong caustic soda solution, but there is a risk of damaging the bricks.).
Rust or iron	Wash down with a solution of oxalic acid (1 part to 10 parts of water by weight), or make a mixture of glycerine, sodium citrate and warm water in the proportions of 7 : 1 : 6, then add an inert filler to make a paste; apply this to the stain and leave for several days to dry. (Brown staining which does not respond to this treatment, particularly at the junction of the brick and mortar, is probably due to manganese.).
Manganese (dark brown)	Brush the stain with a solution of 1 part acetic acid and 1 part hydrogen peroxide in 6 parts of water.
Timber (brown or grey)	These stains are due to water spreading tannin or resin from the timber across the bricks and mortar, and can normally be removed by scrubbing with a 1 : 40 solution of oxalic acid in hot water.
Water	Water running regularly down the surface of brickwork produces pattern staining and this can frequently be removed by scrubbing following wetting with a high pressure mist spray of cold water. If this is not effective, the treatment recommended for mortar should be followed.
Mortar	Where possible, remove larger pieces with a scraper, then wash down with a dilute solution of hydrochloric acid (1 : 10 by volume).
Lime	Follow treatment recommended for 'Mortar'.
Smoke and soot	Scrub with a household detergent. The more stubborn patches can be pulled from the brick pores using a poultice based on trichlorethylene, although good ventilation is needed indoors.
Tar	Except where bricks are liable to surface damage, remove excess with a scraper, then scrub with water and an emulsifying detergent. If necessary, finally sponge or poultice with paraffin.

CALCIUM SILICATE BRICKWORK

It is always advisable to consult the brick manufacturer before embarking upon any cleaning of calcium silicate brickwork, particularly where it is intended to use an acid. Remember to saturate the brickwork with clean water before applying any chemical and wash down with clean water afterwards.

Badly stained brickwork of natural (white) calcium silicate bricks cannot always be satisfactorily cleaned to give a uniform appearance. The appropriate stain removing treatments should be employed, and then it may be desirable to abrade the whole surface, to give a new clean face, or to apply a decorative coating, for example, of alkali-resisting paint.

Stain	Method
Oil	Scrub with an oil emulsifying detergent in water. Allow to dry thoroughly, then, if necessary, poultice with white spirit, carbon tetrachloride or trichlorethylene. Good ventilation is essential indoors.
Paint	Apply commercial paint remover or a solution of trisodium phosphate (1 part to 5 parts of water by weight), allow the paint to soften and remove with a scraper. Wash the wall with soapy water and finally rinse with clean water. In very bad cases it may be necessary to grind off the face of the brickwork in the affected area.
Lichens and mosses	These can be killed with a 1 % solution of sodium pentachlorophenate or, alternatively a proprietary weed killer, recommended by the makers. There will usually be an obvious black residue of dead material which should be removed by scrubbing with water. Vegetable growth is generally indicative of damp brickwork and may reappear if this basic cause is not cured.
Rust	Wash down with a solution of oxalic acid (1 part to 10 parts of water by weight).
Timber (brown or grey)	These stains are due to water spreading tannin or resin from the timber across the bricks and mortar, and they can normally be removed by scrubbing with a 1 : 40 solution of oxalic acid in hot water.
Water	Water running regularly down the surface of brickwork produces pattern staining which frequently can be removed by scrubbing following wetting with a high pressure mist spray of cold water. If this is not effective, the treatment recommended for mortar should be followed.
Mortar	To remove the larger pieces of mortar, lightly abrade the surface using a brick of the same colour, then wash down with a very dilute solution of hydrochloric acid (1 : 20 by volume). The surface of some calcium silicate bricks may be damaged by acid, so this process should be used cautiously.
Tar	As much as possible should be removed, first by scraping, then by scrubbing with water and an emulsifying detergent. Allow to dry, then, if necessary, poultice with paraffin.
Smoke and soot	Scrub with a household detergent. The more stubborn patches can be pulled from the brick pores using a poultice based on trichlorethylene, although good ventilation is essential indoors.
Lime bloom	This is a thin white film over the surface of the bricks which disappears with the effects of rain and wind – so that no treatment is necessary. Lime staining from an external source should be treated in the manner recommended for mortar.

COMMERCIAL CLEANING

A relatively large area is required, before cleaning by a commercial company becomes an economical proposition, and the complete side of an average house would be a typical minimum size.

Commercial companies now generally use water or chemicals. Steam is little used. It is always advisable to have a trial section cleaned, because the change in colour of the brickwork when the dirt is removed can be surprising, and this trial application will ensure that the cleaning technique is suitable.

Saturating large areas of clay brickwork with water, which is necessary with this process, may lead to a fresh outburst of efflorescence, but this will be only temporary and will disappear with the effects of rain and water. The possibility of this water penetrating through older solid brickwork should not be overlooked.

Some stains, particularly paints, may penetrate deep into the pores of the bricks and surface cleaning will not remove them. In these cases commercial companies will undertake sand- or gritblasting, but because this will affect the surface texture and colour of the bricks, it should only be undertaken as a last resort, after judging the effects on a trial area.

Most commercial companies will also undertake repairs and remedial work concurrently with cleaning □

FURTHER READING (Obtainable from HMSO)

1. Building Research Station Digest No 46: 'Design and Appearance – 2'.
2. Building Research Station Digest No 113: 'Cleaning external surfaces of buildings'.
3. Department of the Environment Advisory Leaflet No 75: 'Efflorescence and stains on brickwork'.

Prefabricated brickwork
An outline of international developments

INTRODUCTION

Throughout the early and middle years of the past decade, many brickmakers in the United Kingdom and in countries abroad carried out research into prefabricated brick or ceramic construction. For a variety of reasons, among which may be noted the decline in industrialised building and in construction generally towards the end of the sixties – and the continuing low cost of traditional in situ brickwork – few of the developments that emerged in this country found their way into commercial application. In various European countries, however, prefabricated brickwork made steady progress and is now being used for a continually expanding range of applications. Recent progress has been such, in fact, that many European brickmakers are of the opinion that sooner or later a very large proportion of their output will be supplied to the construction industry in prefabricated panel form.

For the reasons outlined above, most of the illustrations in this short survey are of overseas developments and include several examples drawn from Holland where a consortium of brickmakers has evolved an effective system for the production of very large prefabricated brick, or brick and concrete, panels. The current situation in Holland is of additional interest in that the Dutch have an equally strong tradition of facing brickwork, broadly similar material/site labour costs to those obtaining in the UK, and a climate which – like our own – has led to very unsatisfactory results with other claddings and finishes on industrialised buildings and other types of construction.

Horizontal casting (Delta system). Bricks being laid face down in a bond pattern.

METHODS OF PANEL ASSEMBLY

There are three basic methods of panel assembly: jig laying, horizontal casting and vertical casting. Numerous variations of technique exist within these generic categories.

Jig laying

Many early prefabrication systems were based on jig laying – ordinary bricklaying using jigs to improve productivity and facilitate quality control. This method requires relatively simple equipment and production facilities. Thus, the 'factory' can be established either at the brickworks, the contractors' yard, or on the site itself.

Horizontal casting (Delta system). Thin grout being pumped into the joint spaces to bond the bricks into a single panel of brickwork.

Horizontal casting

Although quite large single-leaf and cavity wall panels have been produced by jig laying methods, storey height panels, for example, require either some means of raising the bricklayer and his materials to the optimum laying height (scaffolding) or of progressively lowering the work. Horizontal casting techniques were evolved to overcome this limitation. In broad principle, all of them involve the placing of bricks in a bond pattern face down in a horizontal tray or mould. A thin grout is then poured into the

joint spaces so as to bond the bricks into a single panel of brickwork. One British development of this method is almost entirely automated.

Vertical casting

First developed in the United States, this method was imported into this country by a British manufacturer who has since made considerable improvements to the process. In this form of prefabrication, bricks arranged in a bond pattern are tightly clamped between two vertical faces in a special machine. Grout is then fed in at the top of the machine until all the joint spaces are filled. Although it requires much more complicated equipment than most horizontal casting methods, this is a very efficient technique which enables large panels (10ft x 10ft) to be produced very quickly, absolutely true and plumb, and with perfectly filled mortar joints.

Two horizontally cast single-leaf brick panels joined by a special tie to form a cavity wall.

Detail of toothed joint between vertically cast panels.

UK DEVELOPMENTS

As noted earlier, although several British brickmakers have evolved efficient production techniques, there has, to date, been no significant demand for prefabricated brick panels in the construction industry. Nevertheless, development work has enabled substantial technical progress to be made in three potential areas of application:

Loadbearing panels

Studies to date indicate that there is no insurmountable technical reason why prefabricated loadbearing brickwork should not be used for a number of low-rise and high-rise applications – particularly housing. An interesting early British development in this direction was a jig-laying system used to construct a two-storey office building out of 12in thick cavity wall elements with maximum dimensions of 5ft 3in x 3ft. This system was subsequently progressed to a Mark 2 version, based on storey-height panels produced by horizontal casting methods, and this was used to construct a mock-up of a two-storey house.

Mock-up of a two-storey house produced by a British system based on horizontal casting techniques.

Multi-storey car park, Maidenhead. Vertically cast panels used to considerable environmental advantage.

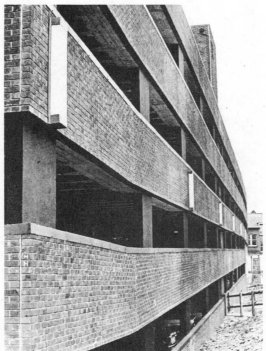

Detail of Maidenhead car park.

Non-loadbearing panels

Non-loadbearing panels probably have a more extensive potential field of application – particularly as a cladding to framed structures such as offices, hotels, schools, factories and hospitals. As compared with loadbearing panels, these pose fewer manu-facturing and structural problems and could be made widely available, using any of the three basic manu-facturing techniques. A recent and very effective use of prefabricated non-loadbearing brick panels is to be seen at Maidenhead where panels produced by the vertical casting method have been used to great environmental advantage as the parapet walls of a multi-storey car park in the town centre.

Ceramic veneer panels

Prefabricated concrete panels finished with thin clay facings in brick and other shapes have been widely available for some time and, unlike brick panels, have been used for a reasonably wide variety of building applications. Of particular interest in this context is a building of outstanding architectural merit at Poole, Dorset, where prefabricated brick slip panels were used as permanent shuttering for the in situ concrete edge and fascia beams.

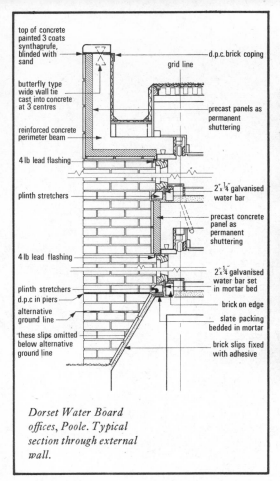

top of concrete painted 3 coats synthaprufe, blinded with sand

d.p.c. brick coping

grid line

butterfly type wide wall tie cast into concrete at 3 centres

precast panels as permanent shuttering

reinforced concrete perimeter beam

4 lb lead flashing

plinth stretchers

$2'' \times \frac{1}{4}''$ galvanised water bar

precast concrete panel as permanent shuttering

4 lb lead flashing

plinth stretchers
d.p.c in piers
alternative ground line

these slips omitted below alternative ground line

$2'' \times \frac{1}{4}''$ galvanised water bar set in mortar bed

brick on edge

slate packing bedded in mortar

brick slips fixed with adhesive

Dorset Water Board offices, Poole. Typical section through external wall.

Dorset Water Board offices, Poole. Architects: Farmer & Dark. Prefabricated brick slip panels were used as permanent shuttering for the concrete edge and fascia beams.

OVERSEAS DEVELOPMENTS

USA

The use of prefabricated brickwork is increasing in several parts of the country and recent developments include the evolution of a low-rise housing system. At the present time (August 1972) Chicago is the scene of an important international conference solely concerned with prefabricated brick construction. Perhaps one of the most interesting advances to be made in the USA has been the development of high bond mortars which overcome the one weakness of brickwork as a structural material – its lack of tensile strength. As far as is known, high bond mortars have not as yet been produced with the high water/cement ratios required for the pourable grouts used in the horizontal and vertical casting methods. Nevertheless, excellent use of their superior adhesive properties has been made in jig laid panels, of which the University Centre building in Austin, Texas, affords a most impressive example. For this very large multi-storey structure, storey-height panels were constructed in an on site 'factory'. The bond between bricks and mortar was such that the very large and heavy panels could be lifted by rings fixed in the top course of the brickwork – thus obviating the need either for specialised panel lifting equipment or of lifting bars being incorporated into the panels during manufacture. High bond mortars are now available in the UK.

South Africa

Prefabricated cladding panels have been used on numerous high-rise structures in Johannesburg. Recent developments include a low-rise housing system based on horizontally cast panels 4m x 2.6m.

Scandinavia

Prefabricated brickwork has made substantial progress in Scandinavia, particularly in Denmark and

University Centre building, Austin, Texas. Entirely clad with jig laid panels made with high bond mortar.

Private housing, Denmark, during construction.

Completed Danish housing. As in other countries, early applications of prefabricated brickwork involved the use of narrow storey or cill height panels.

Sweden. In Denmark alone, at least six companies are supplying loadbearing and non-loadbearing panels for applications such as low and high-rise housing (public and private sectors), offices, hotels, department stores, schools and factories. Early applications mostly involved the use of relatively narrow storey or cill height panels for cladding or infill functions. Over the past two or three years, however, very large load-bearing panels have been supplied to the construction industry. Similar trends are observable in Sweden. Low-rise housing systems have been successfully developed and marketed, and numerous multi-storey structures in the new city centre of Stockholm are entirely clad in prefabricated brickwork.

France/Switzerland

Developments in these countries have been principally directed towards clay and concrete building systems, of which the best known and most widely used are Costamagna, Preton and Fiorio. In that none of these is a clay facing method – two (Costamagna & Fiorio) being based on heavily fluted clay blocks – they are not of particular interest as far as UK requirements are concerned. Nevertheless, all three merit attention as notable developments in structural ceramics. By the beginning of 1970, their use had spread to the following countries:

Fiorio (France): Austria, Portugal, Spain and Italy.

Costamagna (France): Belgium, Spain, Venezuela, Austria, West Germany.

Preton (Switzerland): Australia, Belgium, Finland, Peru, South Africa, USA, West Germany.

Fiorio system. 20 storey block of flats in Toulouse. Unreinforced clay and concrete panel construction with prestressed clay block floors.

Fiorio system. Hostel block for Toulouse University.

Fiorio system. Detail of Toulouse University hostel block showing floor panels.

Typical Preton panel. A non-facing jig laid system using perforated bricks.

Holland

Ceramic veneer panels and a complete facing brick panel system have been successfully used in Holland for a number of years. Over the past two years, however, the appearance of the horizontally cast Delta panel system has indicated a marked improvement in production techniques and has considerably extended the potential range of applications. At the present stage of development, panels can be produced in any combination of sizes up to 9.5m long and 3.1m high and are suitable for a variety of industrial, commercial and housing applications. A complete low-rise housing system has also been developed.

Four types of panel are available:

Facing elements: (half-brick).

Industrial elements: (brick/concrete).

Sandwich elements: (brick/expanded polystyrene/concrete).

Cavity elements: (brick/cavity/expanded polystyrene/concrete).

For housing and other appropriate applications panels can be supplied complete with doors and windows.

In addition to dimensional flexibility and large panel size, the system also offers the advantage that facing brickwork can be in any bond, or combinations of bonds, and may incorporate recessed areas of decorative brickwork. Transported and erected, the cost of the panels is comparable with that of Dutch in situ brickwork.

The largest single application of the system to date is a housing development of 80 two-storey houses at Capelle on the outskirts of Rotterdam. However, panels for a further 200 houses are currently in production and several other housing projects have reached the planning stage. Among other notable applications of the system are a large factory faced with sandwich elements at Uithoon, near Amsterdam, and a very large prison complex at Maastricht which will be entirely clad with facing elements.

continued → 12

Delta system. Typical cavity elements.

Delta system. Very large storey-height panels can be supplied complete with doors and windows.

88

Housing, Capelle, near Rotterdam. Note variations of bond and recessed decorative areas.

Capelle. Typical rear elevation. Gable wall is in traditional brickwork.

*Delta system. Housing,
Eerbeek. The latest
development of the system
includes prefabricated
gable ends.*

*Factory at Uithoorn.
Entirely clad with Delta
sandwich elements.*

*Uithoorn. Detail of factory
interior showing bolted
connections between panels
and in situ rc frame.*

Maastricht, prison complex. The first of several hundred prefabricated brick panels being unloaded.

Flats, de Bilt, near Utrecht. Prefabricated brick slip panels used to replace failed timber cladding.

CONCLUSION

The general impression to be derived from current international trends is that following a fairly tentative start with relatively small panels, the evolution of better production techniques and the consequent availability of large, dimensionally flexible, panels is bringing about a considerable increase in demand. It may also be assumed that the general international trend of site labour costs rising much faster than material costs is having an equally significant effect.

Although, as far as the immediately forseeable future in the UK is concerned, it would appear unlikely that a lower unit cost can be achieved with prefabricated brickwork than with traditional in situ brickwork, there are clearly many applications where the use of prefabricated panels can have economically advantageous effects on the construction process as a whole. Considered as a replacement for other forms of cladding – as distinct from traditional brickwork – and bearing in mind the current emphasis on the quality of the built environment, it may also be reasonably argued that the superior functional and appearance properties of ceramic surfaces should alone be sufficient to ensure a future demand. It cannot be emphasised too strongly that, should such a situation arise, several member companies of the Brick Development Association have both the technical knowledge and the facilities to achieve quantity production of prefabricated brick panels □

Detail of slip panels at de Bilt.

Maastricht. Brick drying chambers faced with Delta system prefabricated brick panels in contrasting bonds, colours and textures.

Section 3
TECHNICAL INFORMATION

Loading Tests on Brick Walls built in Stretcher Bond*

Summary

LOADING TESTS have been carried out on five 9 in. thick storey height walls built in stretcher bond (without headers) to give a fair face on each side of the wall. Metal ties connected adjacent half-brick leaves and all joints were completely filled with mortar. Two types of mortar and a Class B engineering brick (crushing strength 8290 lb/in²) were used to build the test walls. Load was applied to the top of the wall either axially or with an eccentricity of load top and bottom equal to one-sixth of the thickness of the wall. The maximum loads carried by the walls are given below. They apply strictly to walls built with bricks and mortar similar to that used and to walls constructed and tested as described in this report.

Wall No.	Mortar Mix	Type of Load	Max. Load Tons Per Ft. Run
1	1:1:6 cement:lime:sand	Eccentric	54·5
2	do.	Axial	67·4
3	do.	Axial	72·7
4	1:¼:3 cement:lime:sand	Axial	87·4
5	do.	Axial	91·0

* *Department of Scientific and Industrial Research Building Research Station. Report of Special Investigation No. 2132.*

Fig. 1: View of 9 ft. x 4 ft. 6 in. x 9 in. thick wall in testing machine.

General

Loading tests have been carried out during the period 9th to 28th November, 1962, to examine the behaviour under vertical load of five 9 in. thick brick walls. The wall construction comprised two half-brick leaves built in stretcher bond and connected with metal ties to give a 9 in. thick wall without headers, and a fair face on both sides of the wall. All joints including the vertical through joint required by this construction were completely filled with mortar.

Materials used in walls

Brick:

Double-frog facing bricks were supplied for the construction. The crushing strength of these bricks was obtained from tests carried out by the method specified in B.S. 1257:1945 for bricks with frogs. The results are given in Table 1. The mean strength was 8290 lb/in² and the bricks would therefore satisfy the class B requirement for compressive strength for clay engineering bricks to B.S. 1301:1946.

TABLE 1

Compressive strength of double-frog bricks

No.	Area sq. in.	Failing Load Tons	Compressive Strength lb/in²
1	36·1	151·0	9360
2	38·7	129·0	7500
3	36·1	121·0	7500
4	35·3	146·5	9310
5	37·6	130·7	7780
6	36·1	115·5	7160
7	36·1	149·0	9470
8	35·3	118·0	7500
9	35·3	136·0	8640
10	36·5	132·0	8090
11	36·5	142·2	8720
12	35·3	132·7	8430

Mean Strength = 8290 lb/in²
Standard Deviation = 820 lb/in²
Coefficient of Variation = 9·9 per cent

$$\text{Range} = \frac{\text{Max}^m - \text{Min}^m}{\text{Mean}} = 27 \cdot 9 \text{ per cent}$$

Strength of mortar used to fill the frogs as required by B.S.1257 = 4800 lb/in²

Mortar:

A 1:1:6 (by vol.) cement:lime:sand mortar mix was used in building three of the five test walls and a 1:¼:3 (by vol.) mortar mix was employed for the remaining two walls. The strength of the mortars actually used in the wall construction was obtained from mortar cubes which were tested at the time the walls were loaded, i.e. approximately 5 weeks after construction. The results are included in Table 2.

Wall ties:

Metal ties were used to tie together the two 4½ in. thick stretcher courses of the 9 in. thick test walls. Standard strip wall ties were flattened and the modified tie was placed in the horizontal bed courses of the wall. Fifty of these ties were used in each wall and their location is shown in Fig. 4. The number of ties was about five times that normally called for in 11 in. cavity construction of similar height and width.

Description of test walls

The test walls were all 9 ft. high, 4 ft. 6 in. wide and 9 in. thick. Specific instructions given to the bricklayer called for all horizontal bed joints and vertical perpend joints to be completely filled and flushed up with mortar. A vertical through joint approximately ½ in. wide is left by the requirement for an overall thickness of 9 in. with this stretcher course construction. The bricklayer was instructed to completely fill this central through joint with mortar. For both 1:1:6 and 1:¼:3 mortar mixes the bricklayer was allowed to use the water he considered necessary to give the workability required to comply with the above instructions.

Test arrangements

One of the walls built with the weaker mortar mix was subjected to eccentric loading. The wall was positioned so that its vertical axis was 1½ in. to one side of the centre of the testing machine. This gave an eccentricity of load, in the same direction at the top and bottom of the wall, equal to one-sixth of the thickness of the wall. The load was applied through steel plattens fitted with 'knife-edge' assemblies. The height of the test wall measured between top and bottom knife edges and including a concrete base on which the wall was constructed was 10 ft. 8 in. (See Fig. 5(a).)

The remaining four test walls, comprising two walls with 1:1:6 mortar mix and two with 1:¼:3 mortar mix, were all subjected firstly to an eccentric load as described above,

TABLE 2

Details of test walls and summary of results

Wall No.	Age at Test Days	Mortar Mix:— cement:lime :sand	Mortar Compressive Strength lb/in²	Mortar Transverse Strength* lb/in²	Main Test Details	Maximum Load Tons
1	34	1:1:6	600	205	Eccentric loading test to failure. Eccentricity of 1½ in. at top and bottom of wall with end conditions as Fig. 5(a).	245
2	37	1:1:6	600	205	Eccentric loading (e = 1½ in.) up to a proof load of 130 tons; followed by axial loading to failure. End conditions as Fig. 5.	303
3	34	1:1:6	530	185	As wall No. 2	327
4	36	1:¼:3	2010	540	Eccentric loading (e = 1½ in.) up to a proof load of 140 tons; followed by axial loading to failure. End conditions as Fig. 5.	394
5	43	1:¼:3	2160	535	As wall No. 4	409

* Modulus of Rupture

(Eccentricity t/6)

Fig. 2: Wall after eccentric loading test.

View of vertical cracks and local cracking of brickwork on "Compression" side of wall.

loaded to a specified proof load and then unloaded. The test wall was then repositioned so that it was central in the testing machine and reloaded under axial loading condition (see Fig. 5(b)) to failure. The main details are summarised in Table 2. The height of all the test walls measured between the 'knife-edges' was 10 ft. 8 in. (see Fig. 5) and for consideration involving slenderness ratio of the test walls the effective height between lateral supports was taken as three-quarters of the overall height, i.e. an effective height of 8 ft.

The proof loads for walls 2, 3, 4 and 5 in the initial eccentric loading tests were specified as 130 tons for the walls of 1:1:6 mortar mix (walls 2 and 3) and 140 tons for walls with the stronger mortar (walls 4 and 5).

Measurements were made of the overall change in length of front and back surfaces of the test wall under applied load, and of local surface strains at mid-height and at each side of the wall. Gauges also recorded changes in the thickness at each end of the test wall. The positions of the various gauges used to record these measurements are shown in Fig. 5.

A general view of one of the walls in the testing machine prior to test is given in Fig. 1.

Results

Wall No. 1

The wall was subjected to eccentric loading and the load was applied in increments of 20 tons. The relationship between load and change in height of wall, and between load and surface strain is indicated in Figs. 6 and 7 respectively. It will be seen that the relationship for the surface nearer the line of application of the load, i.e. the compression side, was sensibly linear up to a load of 80 tons, but at the opposite side of the wall values of overall changes in height and in strain measured at mid-height under applied load were negligible.

Assuming a linear distribution of strain over the wall thickness, the average shortening of the wall at 80 tons was 0·017 in., corresponding to a mean strain of 246×10^{-6}. The average stress at this load was 370 lb/in², and the modulus of elasticity based on this assumption was therefore $1·51 \times 10^{6}$ lb/in². A special preliminary axial loading test was carried out on duplicate wall No. 2, and the value of the modulus of elasticity of the brickwork, calculated from the results of this loading, was $1·53 \times 10^{6}$ lb/in².

It would seem therefore that in an initial stage of loading, the wall was behaving as a 'solid' wall with surface stresses and a distribution of stress corresponding to that usually accepted theoretically for an eccentricity of load of one-sixth of the thickness of the wall.

The average strain at 80 tons measured over a short gauge length at mid-height of the wall was 300×10^{-6} (see Fig. 7), which was some 22 per cent greater than the average value of 246×10^{-6}, obtained from overall strains given by dial gauges positioned at each corner of the wall.

The load was increased up to a load of 130 tons at which stage the mid-height surface strains were 1200×10^{-6} compression, and 85×10^{-6} tension respectively at the two surfaces of the wall. The tension could be the result of a small lateral deflection under the load of 130 tons augmenting the initial test eccentricity of 1½ in. The load was then reduced to 20 tons and the sets measured at this load are indicated in Figs. 6 and 7. There had been no measurable change in the thickness of the wall throughout the loading and unloading cycle as measured by end dial gauges, and no visible damage of any kind was observed during this early load stage.

The wall was then reloaded to 130 tons and the four corner dial gauges were removed. The relationship between the strains measured at mid-height of wall and applied load as this load was steadily increased is given in Fig. 7.

A faint sound suggesting an initial partial breakdown in the bond between mortar and brick was heard at a load of 195 tons, but no cracking was noticed until a slightly higher load of 200 tons was attained. At this load a fine crack was observed by the naked eye at the interface between the mortar of the vertical through joint and the bricks at each end of the wall. The cracks extended downwards from the top of the wall for a distance of between 9 and 12 in., the width of the crack being between 0·002 and 0·003 in. at the widest point. This failure in bond tension at the top of the wall was not reflected in the dial gauge readings recording changes in thickness, and in subsequent tests more sensitive gauges were in fact used for this purpose.

The maximum load attained by the wall under the eccentric loading was 245 tons or 54·5 tons per ft. run of wall. The damage sustained by the wall near and at the maximum load, was in the form of local crushing of the brickwork and several discontinuous vertical cracks. This damage is shown in Fig. 2 and was confined to the compression side of the eccentrically loaded wall. The local crushing occurred in the top course and in a single brick at mid-height and at one side of the wall, while the vertical cracks passed along perpend joints and also through the bricks themselves (see Fig. 2). The maximum width of the crack at the vertical through joint at mid-thickness of the wall measured after the test was of the order of 1/100 in.

The mean failing load from later tests on similar walls No. 2 and No. 3 both axially loaded was 315 tons, and the ratio of eccentric load to axial load from the three tests was therefore 0·78. This ratio is greater than the ratio of permissible loads for eccentric loading of t/6 and axial loading usually accepted in design in this country, i.e. 0·625.

Wall No. 2

The wall was subjected to an axial load up to 80 tons in a preliminary test in order to obtain the modulus of elasticity of the brickwork, and a value of 1.53×10^6 lb/in² was calculated from the average shortening of the wall.

The wall was then repositioned in the machine to give an eccentricity of load top and bottom of $1\frac{1}{2}$ in., i.e. one-sixth of the wall thickness, and the change in height and surface strains under applied load are given in Fig. 8 and Fig. 9. The load-shortening and the load-compressive strain relationships were, as in Wall No. 1, linear up to a load of 80 tons. The modulus of elasticity calculated from the average shortening and the average stress, assuming linear distribution of stress, was 1.57×10^6 lb/in² and the close agreement with the value obtained from the preliminary axial loading again suggests an approach to a triangular distribution of stress as generally accepted for a 'solid' wall under an eccentricity of load of one-sixth of the wall thickness. The load was increased to a specified proof load of 130 tons and there was no noticeable sign of any damage or any indication of cracks in the central through joint. It will be noted there was no measurable tension on the side of the wall remote from the line of application of the eccentric load. Mean strains at mid-height of the wall were between 15 to 20 per cent higher than those given by the four overall shortening dials at the corners of the wall.

The wall was again reset to give a final axial loading to failure and the load-shortening and load-strain relationships are shown in Fig. 10 and Fig. 11. Both relationships were linear up to a load of 140 tons and at this load the average strains from the corner dial gauges and from the strain gauges at mid-height gauges were 435×10^{-6} and 490×10^{-6} respectively.

The changes in the thickness of the wall as the load increased are given in Fig. 12. There were no cracks visible to the naked eyes at the vertical through joint at 140 tons, but small changes in wall thickness were just measurable at this load. There were marked changes of the rate of shortening and the increase in strain on one side of the wall at loads above 140 tons (see Figs. 10 and 11). These would suggest that, in fact, some breakdown in bond at the vertical through joint had started at about 140 tons. The thickness gauge No. 6, near the top at one end of the wall, showed a heavy increase at 220 tons (see Fig. 12), and at this load a vertical crack at the mortar joint interface was visible to the naked eye at this end of the wall. The crack extended from the top of the wall downwards for about five courses of brickwork.

The maximum load of 303 tons attained by wall No. 2 was accompanied by vertical cracks on the front and back faces of the wall and local crushing of the bricks in the upper quarter of the wall. The wall after test is shown in Fig. 3(a).

The permissible design load for axial loading according to C.P. 111 (1948) would be 50.5 tons or 11.2 tons per ft. run of wall, obtained by dividing the maximum load carried by the test wall by a load-factor of six (see Clause 803 C.P. 111).

Wall No. 3

The overall shortening and the strains measured at mid-height in wall No. 3 during an initial eccentric loading up to the agreed proof load of 130 tons are given in Figs. 13 and 14. The modulus of elasticity at 80 tons calculated from these results was 1.50×10^6 lb/in², and was therefore little different from that obtained in a similar way for the duplicate wall No. 2, i.e. $1.53-1.57 \times 10^6$ lb/in². There was again only a small tension at 130 tons measured on the side of the wall the more remote from the axis of the load.

The overall shortening and the local strains measured during the final axial loading to failure are given in Figs. 15 and 16, and the change in thickness with increasing load is shown in Fig. 17.

An examination of the results from the axial loading test would suggest that there was probably a partial break-down in bond at the vertical through joint at about 140

tons, but cracking at the ends of the wall was not visible to the naked eye until a load of 180 tons was attained. These cracks at mid-thickness of the wall were observed at each end and extended down from the top of the wall for four or five courses of brickwork. Consideration of Figs. 15 and 16 and the corresponding figures for wall No. 2 indicates that after the initial breakdown in bond, the presence of the ties in these walls did not prevent one of the two stretcher bond leaves deforming more rapidly than the other.

Faint vertical cracks were visible on the front and back surfaces of the wall at about 300 tons and the greatest width of the end cracks was in excess of 1/100 in. As the maximum load of 327 tons was attained there was considerable crushing of the brickwork in the upper half of the wall.

The maximum load of 72.7 tons per ft. run obtained in this test, when used in conjunction with Clause 803 of C.P. 111 (1948) would give a permissible design load of 12.1 tons per ft. run, for a wall of this construction intended for axial loading conditions.

Wall No. 4

In a preliminary test carried out with the wall axially loaded up to 100 tons, the average shortening under load gave a value for the modulus of elasticity of the brickwork of 1.60×10^6 lb/in².

The wall was then subjected to an eccentric loading with an eccentricity of $1\frac{1}{2}$ in. at the top and bottom of the wall. The behaviour of the wall up to the specified proof load of 140 tons is shown in Figs. 18 and 19. The load-shortening and load-strain relationships were almost linear up to 100 tons and at this load the modulus of elasticity calculated from the average stress and the average strain, assuming a linear distribution of strain, ranged from 1.57×10^6 to 1.65×10^6 lb/in². This general agreement between the above values for the modulus of elasticity would suggest that for loads up to 140 tons, the stretcher bond wall under the eccentric test load had behaved as a 'solid' wall. There was no indication of cracking at this load, which was then reduced to 20 tons and the sets measured are indicated in Figs. 18 and 19.

The load was completely removed and the wall was repositioned to be central in the testing machine. The results from the final (axial) loading to failure are summarised in Figs. 20, 21 and 22. It will be seen there were slight differences in the values of overall shortening and local strains obtained from gauges on opposite sides of the wall. There was, however, no marked change in the rate of shortening or strain up to about three-quarters of the maximum load attained.

The first cracks to be seen by the naked eye were noted at a load of 180 tons. These were observed at the top of and at each end of the wall, and extended downwards over five courses of brickwork. It will be seen from Fig. 22 that, in fact, partial failure of the bond at the interface of the mortar of the vertical through joint and the bricks could have occurred at a somewhat earlier load, possibly about 140 tons. There was a marked widening of the vertical crack passing through gauge position No. 3 at 260 tons, and it is interesting to note that throughout this test the smallest width of end cracks was at mid-height of the wall.

As the maximum load of 394 tons was approached, vertical cracks which formed on the front and back faces of the wall were accompanied by local crushing of the brickwork. The appearance of the wall after test is shown in Fig. 3(b).

The failing load of 394 tons considered with regard to Clause 803 of C.P. 111 (1948) would give a permissible design load for axial loading of 65.7 tons or 14.6 tons per ft. run of wall.

Wall No. 5

The wall was subjected to an initial loading comprising an eccentricity of load top and bottom of $1\frac{1}{2}$ in. The results of the measurements of overall movement and local strains for loads up to the specified proof load of 140 tons are

(a) Mortar mix = 1 : 1 : 6   (b) Mortar mix = 1 : $\frac{1}{4}$: 3.

Fig. 3: Typical damage to walls.
Views after axial loading tests to failure.

given in Figs. 23 and 24 respectively. Their relationships with applied load was linear up to 100 tons and the modulus of elasticity was calculated at this load on the basis of average stress and average strain, assuming a linear distribution of stress. Values obtained in this way ranged from $1\cdot59 \times 10^6$ lb/in² from overall strains and $1\cdot47 \times 10^6$ lb/in² from the local strains at mid-height of the wall.

The wall was unloaded, and repositioned in the testing machine to give axial loading in a final run to failure. The results are given in Figs. 25, 26 and 27.

Cracks at mid-thickness of the wall became visible to the naked eye at a load of 260 tons. They occurred at each end of the wall at the interface of brick and the mortar of the vertical through joint, and extended from the bottom of the wall upwards through five courses of brick-

work. The effect of these cracks is reflected in Fig. 27 showing changes in the thickness of the wall as measured by gauges 1 and 4. It will be seen also that cracks of the order of 0·001 in. in width were possibly present at loads of the order of 180 tons. At 340 tons the width of the widest end crack was no more than about 0·005 in., and it would seem that the metal ties were fully operative as transverse reinforcement. Vertical cracks formed and were visible on the front and back faces of the wall at 385 tons. and crushing of the brickwork occurred near the top of the wall when the maximum load of 409 tons was approached.

Using the recommendations of Clause 803 of C.P. 111 (1948), the permissible design load for axial loading from the result of this test would be 68·2 tons or 15·2 tons per ft. run of wall.

TABLE 3

Comparative data from axial loading tests on storey-height 9 in. thick walls

Type of Wall	Type of Brick	Crushing Strength of Mortar lb/in²	Crushing Strength of Brick lb/in²	Strength of wall		Strength Ratio
				Stress at Maximum load lb/in²	Maximum load Tons/ft.	Wall / Brick
Walls of normal bond. Fairface on one side only Type (a) Type (b)	Single-frog clay bricks.	3000 2100	3000 12200	1250 3330	60·0 155·0	0·42 0·27
Walls without headers. Fairface on two sides	Double-frog clay bricks Double-frog clay bricks	2010 2160 600 530	8290 8290 8290 8290	1810 1880 1395 1510	87·4 91.0 67·4 72·7	0·22 0.23 0·17 0·18

Comparative data

A summary of the failing loads is given in Table 2 and some comparative data from axial loading tests on storey-height normally bonded walls of nominal 9 in. thickness are included in Table 3. It would seem from this limited information that for the stronger mortar, i.e. 1:¼:3 mix, the 'strength ratio', i.e. wall strength/brick strength for brick with a crushing strength of 8290 lb/in² in normal bond would be somewhat in excess of 0·27, and certainly higher than the strength ratio actually obtained from the two tests on the special bond walls of the 1:¼:3 mortar mix described in this report.

As regards walls built with the weaker mortar and the bricks with a crushing strength of 8290 lb/in², no direct test data from normal bond walls of solid bricks are available. Consideration of other test results however would suggest that the 'strength ratio' would be within the range of 0·20–0·25 for walls of normal bond, and again somewhat higher than that obtained in the case of the stretcher bond test walls.

Conclusions

(1) There were indications that initial damage was in the form of cracks along the vertical mortar through joint near the top of the wall at loads ranging from about 140 to 180 tons. These cracks became visible to the naked eye at loads ranging from ½ to ¾ of the failing loads of the four axially loaded walls.

(2) The metal ties positioned across the vertical joint, particularly those near the top and bottom of the wall, resisted the widening of the cracks at this mortar joint as the load increased, and the widest crack was generally less than 1/100 in. at loads quite close to the maximum load attained.

(3) In the case of the four walls eventually loaded axially to failure, an initial load cycle with eccentric loading caused no visible damage at the specified proof loads; but in each case sets were recorded after removing the proof load.

(4) The mean failing load under axial loading for the two walls built with a 1:1:6 mortar mix was 315·0 tons or 70·1 tons per ft. run.

(5) The failing load of the single wall loaded eccentrically was 245 tons or 54·5 tons per ft. run and the ratio of this load to the mean failing load for two similar walls (1:1:6 mortar mix) loaded axially was 0·78.

(6) The mean failing load under axial loading for the two walls built with a 1:¼:3 mortar mix was 401·5 tons or 89·2 tons per ft. run.

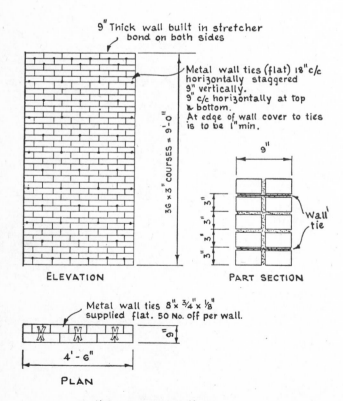

Fig. 4: Details of test walls.

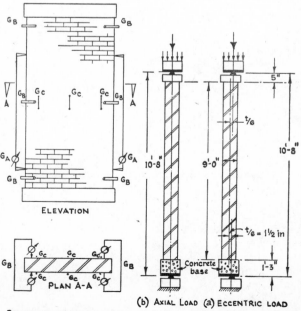

Fig. 5: Test details.

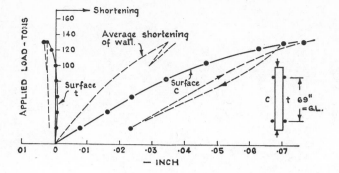

Fig. 6:
Shortening and extension measured on surface of wall No. 1 (Eccentric loading).

Fig. 7:
Strains on surface measured at mid-height of wall No. 1 (Eccentric load).

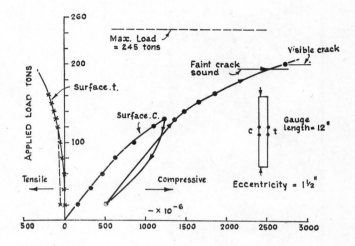

Fig. 8:
Shortening measured on surfaces of wall no. 2 subjected to eccentric loading.

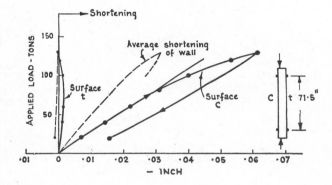

Fig. 9:
Strains at surface measured for eccentric loading at mid-height of wall No. 2.

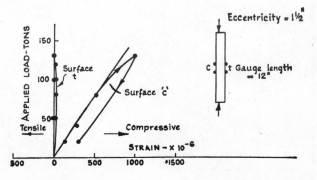

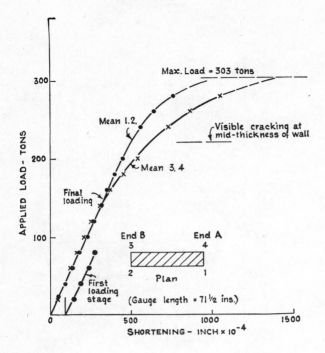

Fig. 10:
Load-shortening relationships for wall No. 2 under axial load.

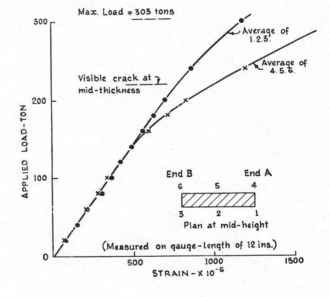

Fig. 11:
Strain measured on surface at mid-height of wall No. 2 under axial load.

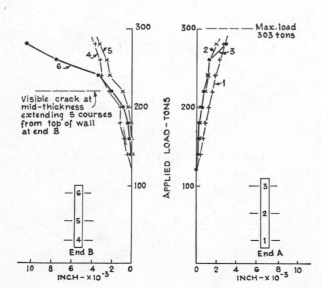

Fig. 12:
Increase in thickness measured at ends of wall No. 2 subjected to axial loading.

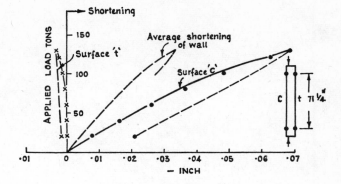

Fig. 13:
Shortening and extension measured on surface of wall No. 3 subjected to eccentric loading.

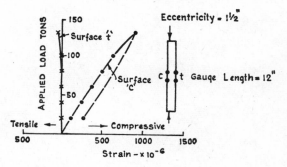

Fig. 14:
Strains at surface measured for eccentric loading at mid-height of wall No. 3.

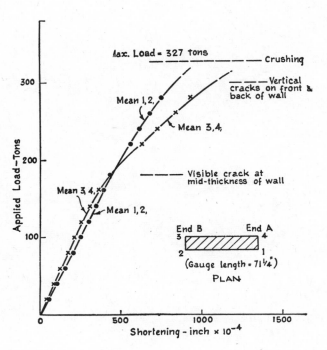

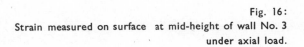

Fig 15:
Load-shortening relationships for wall No. 3 under axial load.

Fig. 16:
Strain measured on surface at mid-height of wall No. 3 under axial load.

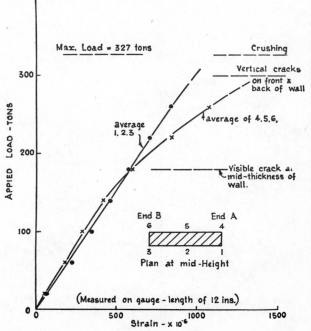

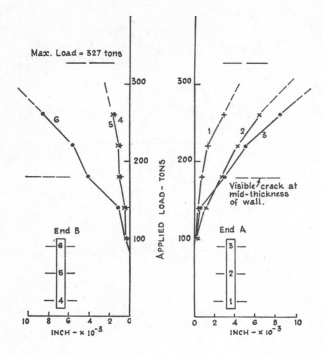

Max. Load = 327 tons

End B

End A

Visible crack at mid-thickness of wall.

Fig. 17:
Increase in thickness measured at ends of wall No. 3 subjected to axial loading.

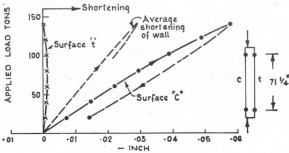

Fig. 18:
Shortening measured on surface of wall No. 4 subjected to eccentric loading.

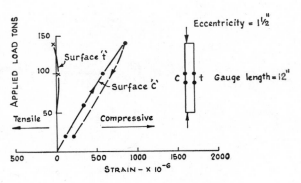

Fig. 19:
Strains at surface measured for eccentric loading at mid-height of wall No. 4.

Eccentricity = 1½"

Gauge length = 12"

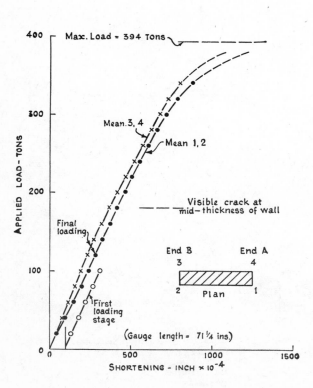

Max. Load = 394 Tons

Mean 3, 4

Mean 1, 2

Visible crack at mid-thickness of wall

Final loading

First loading stage

End B   End A

Plan

(Gauge length = 71¼ ins.)

Fig. 20:
Load-shortening relationships for wall No. 4 under axial load.

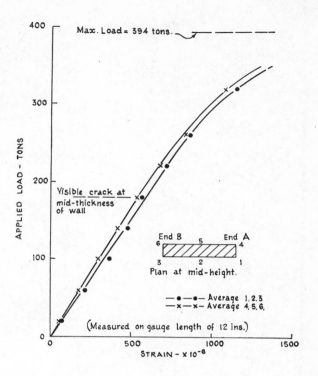

Fig. 21:
Strain measured on surface at mid-height of wall No. 4 under axial load.

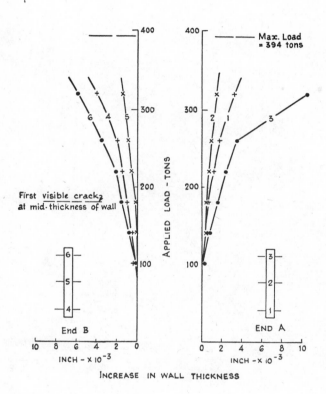

Fig. 22:
Increase in thickness measured at ends of wall No. 4 subjected to axial loading.

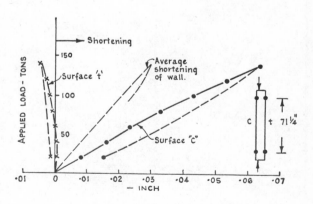

Fig. 23:
Shortening and extension measured on surface of wall No. 5 subjected to eccentric loading.

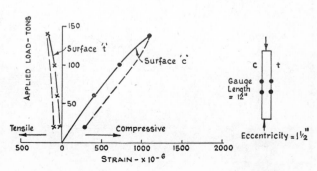

Fig. 24:
Strains at surface measured for eccentric loading at mid-height of wall No. 5.

105

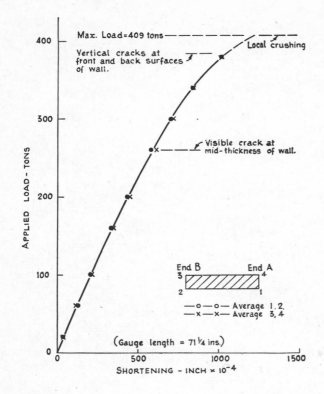

Fig. 25:
Load-shortening relationships for wall No. 5 under axial load.

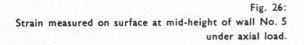

Fig. 26:
Strain measured on surface at mid-height of wall No. 5 under axial load.

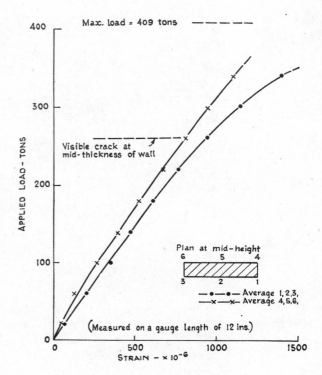

Fig. 27:
Increase in thickness measured at end of wall No. 5 subjected to axial loading.

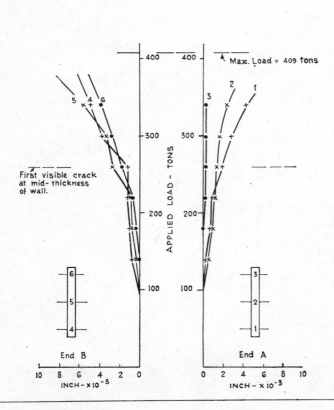

The shear strength of some storey-height brickwork and blockwork walls*

By L. G. SIMMS, B.Sc., A.M.I.C.E.

Building Research Station, Department of Scientific and Industrial Research

Shear strength of walls has hitherto been regarded as a relatively minor property of the structural material of the wall, and there is little published information available particularly as regards walls built with bricks currently available in this country. When designing for the effects of wind loading, therefore, little use has been made of the potential shear strength of buttress walls in non-framed buildings and of masonry infill in framed construction. This paper presents some relevant data obtained from tests on storey-height brickwork and blockwork walls.

Introduction

THERE has been increasing interest during the last decade in load-bearing brickwork and block-work walls for non-framed multi-storey construction, and for infill walls in steel or reinforced concrete frames to reduce side sway in tall structures. In the former, the mass of the walls and the weight of the floors they carry is not always sufficient to ensure stability against horizontal wind loading normal to the wall surface. Hence, when the main load-bearing walls of a constructional layout are disposed primarily in one direction, current practice is to provide secondary walls at right-angles to the main walls, so that resistance to wind loading in any direction can be made dependent on the shearing strength of some walls rather than on the bending strength. As regards the structural use of infill walls, some exploratory tests[1] have shown that when a frame with brick infill is subjected to loading in the direction of the plane of the frame, the shear strength of the brickwork makes an important contribution to the stiffness and strength of the composite assembly.

Shear strength has hitherto been regarded as a relatively minor property of walls and, as a result, there is little information available on the subject, particularly as regards walls built with bricks and blocks currently available in this country.

The primary object of the work described in this paper was to obtain test data to indicate the likely order of shear strength of some storey-height brickwork and blockwork walls.

Shear Strength

In normal brickwork laid to a bond with mortar in all joints, forces in the plane of the brickwork lead to tensile and shear stresses which in part are resisted by the adhesion between the brick and the mortar at the interfaces of the two materials.

Simple tests, such as the pulling apart of two bricks laid flat and joined together with mortar, are used to

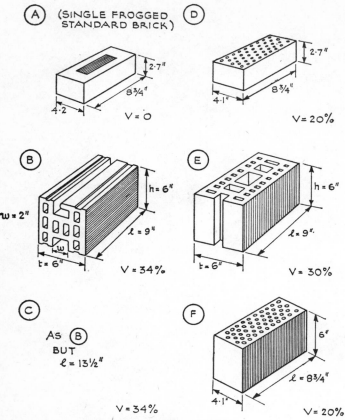

(A) (SINGLE FROGGED STANDARD BRICK)   (D)

V = O   8¾"   4·2   2·7"

V = 20%   8¾"   4·1"

(B)   h = 6"   w = 2"   ℓ = 9"   t = 6"   V = 34%

(E)   h = 6"   ℓ = 9"   t = 6"   V = 30%

(C) AS (B) BUT ℓ = 13½"   V = 34%

(F)   6"   ℓ = 8¾"   4·1"   V = 20%

Fig. 1. Types of brick and block.

acting on the horizontal and vertical joints in the brickwork are neither normal tensile nor pure shear stresses, and it is generally accepted that an approach to the behaviour of structural masonry under a wind loading is best obtained from some form of 'racking' test on a complete wall. A test of this kind in its simplest form is one in which a horizontal load is applied in the plane of the wall to an upper corner of a storey-height wall whilst the wall is prevented from sliding or rotating. This racking-strength test is generally considered to provide a measure of the shearing strength of masonry walls.[6, 7]

Types of building unit

The types of clay building unit used in the construction of the test walls are shown in Fig. 1. They comprised a solid single-frogged brick and a perforated brick, both of standard shape; a type of hollow clay block that has about 34% of horizontal cavity, and a perforated clay block that has a 30% vertical perforation. The main dimensions of these units and their crushing strengths are given in Table 1. These results indicate that the strength of the clay itself was of the order of 3,000-4,000 lb/in² for units A, B, C and E and 9,000-10,000 lb/in² for units D and F. The strengths given in the final column represent the strength of the unit when loaded parallel to its length which corresponds to the direction of loading in a racking load test on walls, i.e., parallel to the bed joints in the wall.

Test walls

The test walls were 8 ft high and generally 8 ft long. With one exception, they were built with mortar of compressive and tensile strengths, at the age the walls were tested, of approximately 1,000 lb/in² and 150 lb/in² respectively. The mortar used for a single wall with the high-strength perforated block of Type F had a compressive strength of 3,000 lb/in² and a tensile strength of 260 lb/in² at the age of test. Apart from one 9-in wall, all walls were built in stretcher bond.

get some idea of the order of the tensile bond strength; similarly a three-brick assembly in which a load is applied to the central brick can indicate the order of shear bond-strength. For solid bricks of standard shape, simple tests[2, 3, 4, 5] have been used to compare the probable tensile and shear bond-strengths of brickwork built with different types of brick, different mortars, conditions of curing and degrees of workmanship. However, where wind effects introduce forces at the end of a masonry wall the resulting stresses

TABLE 1

Crushing strength of bricks and blocks used in the construction of the test walls

Type	Description of building unit used in test walls	Crushing strength (calculated on gross area)	
		Load applied normal to surface of unit forming the bed face in the wall (lb/in²)	Load applied normal to surface of unit forming the perpend. face in the wall (lb/in²)
A	Single frogged clay brick of standard shape 8¾ in × 4 3/16 in × 2⅝ in	3200	1700
B	Hollow clay blocks 9 in × 6 in × 6 in, with 10 horizontal cells giving 34% voids	850	2750
C	Hollow clay blocks 13½ in × 6 in × 6 in, with 10 horizontal cells giving 34% voids	1050	2730
D	Perforated clay brick of standard shape 8¾ in × 4⅛ in × 2⅝ in. Voids 20%	7750	2000
E	Perforated clay block 9 in × 6 in × 6 in. Voids 30%	2740	590
F	Perforated clay block 8¾ in × 6 in × 4⅛ in. Voids 20%	7160	1600

For the bricks with horizontal cavities, i.e. Types B and C, the mortar bed covered only two-thirds of the thickness of the one-brick thick wall. The same area of mortar bed was purposely used for the walls constructed with the perforated bricks of Type E, i.e. the brick-layer left a 2-in wide gap in the horizontal mortar bed when building the 6-in thick test walls with this brick. As regards the vertical joints of the walls, these were generally mortared in the usual way and bond of these walls was called 'normal bond'; a few walls were built using no mortar at all in the vertical joints, the bricks being merely butted together and the bond here was called 'semi-bond.' A purposely-made central gap in the horizontal mortar bed is a relatively new brick-laying technique intended to give adequate resistance to rain penetration in an unrendered wall. The omission of all mortar in the vertical joints of walls is, on the other hand, a special technique used in Switzerland for a particular type of perforated clay oversize brick; it is understood that this technique is not used for all walls since there is some doubt as to whether the partial omission of the mortar would reduce the resistance to wind loading of internal buttressing walls. These special walls were included in order to determine the effect of the semi-bond technique on shear strength.

One wall was constructed with Type E perforated blocks, without mortar in the vertical joints, and was subsequently plastered on both sides. The plastering was carried out under supervision and comprised a $1:1:6$ cement : lime : sand rendering coat followed by a floating coat of similar materials, the finishing coat being neat gypsum plaster. The thickness of this three-coat plaster was $\frac{3}{4}$ in each side, bringing the overall thickness of the plastered wall to $7\frac{1}{2}$ in.

Test arrangements

In the racking-test arrangement adopted, storey-height walls were built on a concrete beam seated on the lower boom of a stiff test frame capable of taking a horizontal load up to 200 ton. In one test arrangement a rigid stop was located at a lower corner of the wall to prevent sliding, and a horizontal thrust was applied to the diagonally opposite corner by means of a hydraulic jack, loading the end of the wall through a stiff steel plate 12 in long and 6 in wide. Two steel rollers were located at the loaded corner between the top of the wall and a brick abutment clamped to the test frame; in this way horizontal displacement of the wall could occur without undue friction or tilting of the wall. It was realised that complete prevention of tilt as the wall was loaded was not practicable, and to ensure that any slight initial tilt could occur without damaging the wall, a layer of building paper was interposed between the mortar on the underside of the first course of the wall and the concrete base on which the wall was built. In this way the wall could tilt very slightly as a whole where this was necessary to take up any slight initial movement between wall and the upper brick stop.

This arrangement, which is shown in Fig. 2 (a), did not perhaps represent very closely the effect of wind loads on buttressing or shear walls in non-

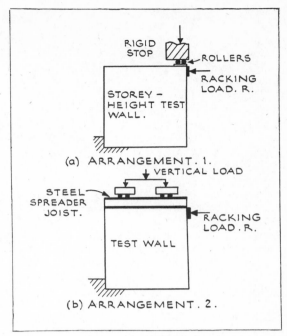

Fig. 2. Test arrangements.

framed construction. On the other hand, it was considered that there was some similarity between the forces introduced in the test wall and the forces in infill walls when a composite assembly, i.e., infill plus frame, is subjected to horizontal loading. In order to simulate better the conditions in the lower-storey shear walls of non-framed construction a few tests were carried out with the test arrangement 2 of Fig. 2 (b). In these tests vertical load was distributed along the top of the wall to represent the effects of dead load from the construction above and the test wall was then subjected to the horizontal racking load. The two test arrangements are also shown in the photographs of Fig. 3.

In some tests measurements were made of the lengthening of the tension diagonal of the test panel as the racking load was applied.

Results

Series I: strength of walls of solid bricks. The main results are given in Table 2, where the 'shear' stress is the racking load divided by the nominal area of the mortar in the horizontal bed joint of the test wall.

The five walls subjected to the racking load of test arrangement 1 all failed in the same way, and the wall shown in Fig. 3 (a) indicates the typical form of damage at failure. The pronounced diagonal crack formed suddenly and extended from a zone at or near the load plate at the upper corner to the masonry in the lower corner of the wall adjacent to the stop. The load causing this diagonal crack was the maximum racking load. The extension of the tension diagonal, measured over a gauge length of about five feet, indicated that immediately prior to failure the order of tensile strain for the solid brick walls was 60×10^{-6}. The 'shear' stress at failure for walls of length/height ratio not greater than unity ranged from 75-120lb/in², but for the 15-ft long storey-height wall the failing stress was considerably less. It should be noted that, with test arrangement 1, the wall shape will control the ratio of the average compressive stress to the 'shear' stress, and it is considered that the low shear strength of the long wall specimen is the effect of the lower value of this ratio as compared with that of the 8-ft square walls.

The local compressive stress under the load plate distributing the racking load is included in Table 2. This stress at failure was in all tests much lower than the crushing strength of the brick normal to the perpend joint (see Table 1).

For the four walls tested under arrangement 2 the compressive stress, due to the applied vertical load uniformly distributed along the top of the wall, was about 150 lb/in². This stress is of the order of the maximum permissible stress for a 9-in wall built of medium strength brickwork, as might be used for a shear wall in the lower storey of a tall non-framed building. Three of these walls failed by the formation of a diagonal-tension crack which was similar to that for the other arrangement of loading and was as shown in Fig. 3 (b); in two walls this crack was accompanied by local crushing of the brickwork over a small area of wall at the lower end of the diagonal crack. However, test wall No. 9, which was 15-ft long, failed by crushing in the lower restrained corner adjacent to the rigid stop, and no diagonal crack was observed. The approach to primary failure by crushing of the brickwork of the 15-ft long walls is indicated by comparing the calculated stress at failure under the racking load plate and the strength of the brick itself tested on end (see Table 1).

Series II: strength of walls of hollow and perforated blocks. The results of tests on eight walls built with various types of hollow or perforated units are summarised in Table 3.

With the exception of walls built with brick Types D and F, the shape of the unit required the mortar to cover only two-thirds of the full horizontal cross-section of the wall, and the 'shear' stress given for these walls is calculated using the area of the mortar, or two-thirds of the full cross-sectional area of the wall.

The maximum load-bearing capacity of five of the walls was obtained when the walls failed by the sudden formation of a diagonal crack. Immediately this crack became visible it extended over about the mid-half of the diagonal, from the loaded corner to the lower corner adjacent to the stop. The strain in the tension diagonal immediately before the crack formed was very variable and ranged from 20 to 60 x 10⁻⁶. The crack damage shown in Fig. 3 is again typical of the appearance of the walls at the end of the test.

The shear strength of walls which failed in this way represented the order of stress possible when failure is the result primarily of breakdown of the bond between the brick and the mortar, of intermediate strength (mix 1 : 1 : 6), used in construction of the test walls. It will be seen that the 'shear' stress (calculated on the area of the mortar of the bed joint) ranged from about 75 to 130 lb/in², and was of the same order as the strength of the 8-ft square walls of solid bricks.

As regards the two walls of brick Type E, the maximum load of one wall was attained with the formation of a diagonal crack and local crushing under the racking load plate, but failure of wall No. 15 was by crushing without the diagonal crack. It is significant that the compressive stress under the load plate at failure was in both walls about the same as the strength of the brick itself when loaded in the direction of its length (see Table 1 and Table 3).

TABLE 2

Shear strength of walls of solid bricks

Test arrangement	Type of mortar	Type of brick	Wall No.	Nominal dimension of wall H × L × T	Ultimate racking load (tons)	Stress at failure (lb/in²) 'Shear' stress*	Stress at failure (lb/in²) Local stress†	Mode of failure
1	Mix: 1 : 1 : 6 Strength: 1000 lb/in²	A	1	8 ft × 4 ft × 4 3/16 in	11.1	120	490	Diagonal tension
			2	8 ft × 8 ft × 4 3/16 in	13.0	75	580	do.
			3	8 ft × 8 ft × 8 3/4 in	28.0	75	600	do.
			4	8 ft × 8 ft × 2 5/8 in	11.9	100	830	do.
			5	8 ft × 15 ft × 4 3/16 in	15.0	45	670	do.
2	Mix: 1 : 1 : 6 Strength: 1000 lb/in²	A	6	8 ft × 8 ft × 4 3/16 in	18.0	100	800	Diagonal tension and crushing
			7	8 ft × 8 ft × 4 3/16 in	18.0	100	800	do.
			8	8 ft × 15 ft × 4 3/16 in	47.0	140	2100	Diagonal tension
			9	8 ft × 15 ft × 4 3/16 in	50.1	150	2200	Crushing

*Racking load divided by the area of mortar in the bed joint of the wall.
†Local compressive stress under racking load plate.

TABLE 3

Shear strength of walls 8 ft. high and 8 ft. long built with hollow or perforated bricks or blocks

Test arrange-ment	Type of mortar	Type of brick or block	Wall No.	Thickness of wall	Ultimate racking load (tons)	Stress at failure (lb/in²)		Mode of failure
						'Shear' stress*	Local stress**	
I	Mix: 1 : 1 : 6 Strength: 1000 lb/in²	B Hollow block horizontal cells	10	6 in†	14.1	80	440	Diagonal tension
		B do.	11	6 in†	13.0	75	440	do.
		C Hollow block horizontal cells	12	6 in†	14.0	80	440	do.
		C do.	13	6 in†	17.1	100	530	do.
		D Perforated brick	14	4⅛ in	23.0	130	1030	do.
		E Perforated brick	15	6 in†	17.0	100	530	Local crushing
		E do.	16	6 in†	22.1	130	690	Diagonal crack and crushing
	Mix: 1 : ¼ : 3 Strength: 3000 lb/in²	F Perforated brick	17	4⅛ in	47.9	270	2100	Crushing

†Mortar in two parallel strips covering two-thirds of wall thickness.
*Racking load divided by area of mortar in bed joint of the wall.
**Local compressive stress under racking load plate.

The combination of the high strength perforated brick Type F and the strong 1 : ¼ : 3 mortar gave a relatively high value of the 'shear' stress at failure of the one wall (No. 17) built with these materials. Failure was by local crushing of the wall under the load plate and the calculated local stress was well above the strength of the brick (1,600 lb/in²) given in Table 1.

Series III: strength of walls built in 'semi-bond'. The loads at first crack and the maximum loads attained by six walls built in brick types B, C and E in both normal bond and semi-bond are compared in Table 4. The initial damage resulting from the application of a racking load to the semi-bond walls was in every case in the form of a diagonal crack similar in appearance to the cracks in the walls (normal bond) of Fig. 3.

The diagonal crack in the walls of semi-bond formed at relatively low loads, ranging from 3.5 tons to 8 tons, but the walls continued to carry increased loads up to the maximum values which were two to three times the loads causing the diagonal crack. Apparently this build-up in strength after cracking was possible only by friction and wedging action with perhaps some arching in the area of undamaged masonry below the crack.

It will be seen from Table 4 that the load ratio for walls of semi-bond to those of full-bond at the formation of the diagonal crack ranged from ¼ to ½; the ratio, however, was much higher at the maximum load and varied from 0.8 to almost unity.

The single wall of brick Type E, built in semi-bond and plastered both sides to give an overall thickness of 7½ in, failed at a load of 27 ton when a diagonal crack was accompanied by local crushing under

the racking load plate. Compared with the walls built in normal bond with this brick, which had a strength of about 20 ton, there was little difference in strength after adjusting for the greater overall thickness of the plastered wall (see Table 4).

Discussion of results

Primary failure by crushing of the brickwork occurred in only three of the seventeen walls built in normal bond, in spite of the fact that the loading arrangement gave a heavier concentration of racking load than would be expected in practice. It would appear, therefore, that generally shear strength will be controlled by the bond between the brick and the mortar except in those walls built with bricks of very low crushing strength in the direction of the racking load in the plane of the wall. This conclusion that bond strength is a dominant factor affecting the racking strength of walls has also been reached elsewhere[4, 5, 8].

It has been stated that bond-tension and bond-shear tests on small two-brick or three-brick assemblies are primarily of value in comparing the bond strengths likely to be obtained with different bricks or different mortars, rather than providing an absolute measure of the bond strength of brickwork under combined stress conditions. However, a few subsidiary tests were carried out with two-brick and three-brick assemblies, using a Type A standard brick dipped in water prior to use and a mortar approximately similar to that used in the walls. The bond strength in tension ranged from 40-100 lb/in², and the bond strength in shear from 80-140 lb/in². These ranges overlap the actual

Fig. 3a.

Fig. 3. Appearance of $4\frac{1}{2}$-in. thick brick walls at failure under racking load: (a) Test arrangement 1 (with rigid stop); (b) Test arrangement 2 (with vertical load uniformly distributed on top of wall).

Fig. 3b.

TABLE 4

The racking strength of some storey-height 8-ft long walls and the effect of omitting the mortar in vertical joints
(Mortar strength 1000 lb/in²)

Type of building unit	Thickness of wall (in)	Walls of normal bond		Walls of semi-bond		Strength ratio: Semi-bond / normal bond	
		Load at formation of diagonal crack (ton)	Maximum load (ton)	Load at formation of diagonal crack (ton)	Maximum load (ton)	At diagonal crack	At maximum load
B Hollow clay brick, with 34% horizontal perforation (9 in × 6 in × 6 in)	6	7.0 12.0 Av. 9.5	14.1 13.0 Av. 13.5	3.5 4.0 Av. 3.8	13.0 12.0 Av. 12.5	0.40	0.93
C do. (13½ in × 6 in × 6 in)	6	14.0 17.1 Av. 15.5	14.0 17.1 Av. 15.5	8.0 8.0 Av. 8.0	15.0 11.0 Av. 13.0	0.52	0.84
E Perforated brick, with 30% vertical perforation (9 in × 6 in × 6 in)	6	No diagonal crack 22.1	17.0 22.1 Av. 19.5	5.0 5.3 Av. 5.2	18.4 19.4 Av. 18.9	0.24	0.97
E do. with ¾ in. thick plastercoat each side	7½†	—	—	27.0	27.0	0.98*	1.10*

†Overall thickness of wall.
*Adjusted for difference in thickness of wall.

'shear' stress from the racking test on walls built with the same type of solid brick.

Results from other sources already referred to have given values less than the lower end of the above range for bond tension, and greater than the upper end of the range obtained from the bond tests in shear. Care must, of course, be taken when comparing the results of such tests obtained in different laboratories, since the results can be very considerably affected by the condition of the bricks when used, by the subsequent storage of the specimens and, in the case of bond-shear, by stress normal to the surface.

However, results from different sources[4,5,8] are at least consistent in one respect. They all show that if the strength of the mortar is doubled or trebled the bond strength is increased by only one-third or one-half. This trend may be useful in assessing the result from the single wall built with the high-strength perforated brick and the strong mortar. With this type of brick the crushing strength in the direction of the racking load is likely to be approaching the maximum for all perforated bricks of non-standard shape. At the same time, the shear stress for this wall of 270 lb/in², which is so much greater than the value of 130 lb/in² obtained from the wall with another high-strength perforated brick and an intermediate strength mortar, is likely to be near the optimum for all perforated bricks. Hence, it seems that a shear stress of, say, 300 lb/in² is unlikely to be exceeded with walls of perforated bricks.

The range of 'shear stress' at failure of the walls built in normal bond of medium strength units and mortar, i.e. 75-150 lb/in², covers the order of shear strength of masonry walls obtained from tests in other countries. Differences between laboratory-built walls subjected to test loadings and site-built walls resisting the effect of wind loading are un-avoidable, but the results of these tests give some indication of the lower end of a range of shear strength of walls in buildings when subjected to horizontal loading in the plane of the wall.

It has already been emphasised that the shear strength of masonry walls is a function of the bond between the brick and the mortar, and it is therefore likely to be affected more by variation in quality of the brickwork than compressive strength. It is suggested that where 'shear' walls are used in the design of load-bearing brickwork they should be constructed under continuous supervision to ensure a good class of brickwork. Generally a normal layout will provide opportunities for an ample reserve of walls resisting shear. However, it is considered that shear walls should be clearly identified in the structure as walls not to be included in any structural modifications during the life of the structure unless such changes are checked by calculations.

With the above provisos in mind, and also the margin of safety usually associated with load-bearing brick walls in this country, a permissible shear stress of 15 lb/in² for a 1 : 1 : 6 mortar and perhaps 20 lb/in² for stronger mortar is suggested as reasonable. Where the compressive stress due to the dead load above a shear wall is considerable the tests described in this paper and elsewhere have suggested that the permissible shear stress could be increased. Thus the permissible stress in the lower storeys of multi-storey construction could be perhaps one-third of the vertical stress produced by the dead load but should not exceed 30 lb/in². It might be noted here that should the horizontal load cause such a wall to crack there would still be a useful margin of safety against sliding.

As regards the behaviour under racking load of the walls with semi-bond, the inference from a single test on a wall 8 ft high and 8 ft long, built with a type of perforated brick, is that the application of a strong high-quality three-coat plaster to both sides of the wall could prevent the cracking observed at low loads in the unplastered wall. However, in practice, severe cracking at $\frac{1}{4}$ to $\frac{1}{2}$ the ultimate racking load of walls of normal bond may occur in plastered walls in semi-bond if the plaster is merely a very weak two-coat plaster. The importance of the increase in strength of the semi-bond walls after the formation of the diagonal crack (see Table 4) is difficult to assess. The extent of the increase could be influenced by the shape of the brick and the shape of the wall itself; it may also be closely connected with the fact that in these tests the racking load and the reaction to this load were concentrated over small lengths of the masonry. It would seem, therefore, that the basis of comparison between the walls of normal bond and of semi-bond should be at present in terms of the load causing the initial cracking.

Conclusions

The following tentative conclusions were drawn from the results of the tests:

(1) Failure of a storey-height wall under racking load is usually due to a breakdown of bond, leading to the formation of a diagonal crack in the wall.

(2) Primary failure may sometimes be due to local crushing under the load. In the tests, this load was more concentrated than is likely in practice, but, nevertheless, it seems undesirable to build shear walls with units having a low strength in the direction of the racking load.

(3) With the margin of safety usually associated with brickwork, the 'shear' stress (horizontal load $\div$ area of mortar in bed joint) may range from 15 lb/in^2 for a medium-strength brick and mortar, to an upper limit of 30 lb/in^2 applicable for special conditions.

(4) Although the ultimate racking strength of a wall may not be reduced as a result of omitting mortar in the vertical joints, the load at which diagonal cracking occurs is very much less and hence it is desirable that these joints should be properly filled in all shear walls.

Acknowledgments

This paper describes work forming part of the programme of the Building Research Board of the Department of Scientific and Industrial Research, and is published by permission of the Director of Building Research. The author wishes to acknowledge the assistance of Mr. W. J. Reed in the experimental work.

References

[1] R. H. WOOD. 'The stability of tall buildings.' *Proc. Inst. Civ. Eng.* Vol. 11, September, 1958.

[2] PARSONS, STAND AND McBURNEY. 'Shear tests on R.B. Masonry beams.' *Nat. Bureau of Standards Research Paper, No.* 504, 1932.

[3] PLUMMER AND BLUME. '*Reinforced brick masonry and lateral force design.*' Published by Structural Clay Products Institute, Washington, D.C., November, 1953.

[4] J. R. BENJAMIN AND H. A. WILLIAMS. 'The behaviour of one-storey brick shear walls.' *Proc. Amer. Soc. C. Eng. Journal of Structural Divn.*, *Vol.* 84, No. S.T.4, July, 1958, Part I.

[5] S. V. POLYAKOV. *Masonry infilling in framed buildings.* Moscow, 1956.

[6] WHITTEMORE, STANG AND PARSONS. 'Building materials and structures.' *Nat. Bureau of Standards Report B.M.S.5*, 1938.

[7] C. B. MONK. *S.C.R. wall tests.* Structural Clay Products Research Foundation, Research Paper No. 1, 1953.

[8] C. C. FISHBURN. *Effect of mortar properties on the strength of masonry.* Nat. Bureau of Standards Monograph 36, 1961.

Wind Forces on Non-Loadbearing Brickwork Panels

by
R. E. BRADSHAW, A.M.I.C.E., A.M.I.struct.E.,
and
F. D. ENTWISLE, F.R.I.C.S., F.I.A.S., A.M.I.struct.E., A.M.I.mun.E.

Introduction

OVER THE LAST few years, very severe damage has been caused to buildings by gales—particularly in Yorkshire in February 1962—and there is now a growing interest in the stability of external infill panels subjected to wind loading.

This note discusses the problems associated with lateral loading on wall panels of this type built of brickwork, and puts forward an approximate method of design for safe panel sizes and thicknesses to resist given wind loading. It is not concerned with the lower storeys of brickwork which form the supporting structure as well as the wall panel, since these walls are subject to vertical compression and hence are less likely to develop critical tensile stresses. However, the uppermost two storeys of such buildings should be investigated.

The limited research carried out to date on lateral loading of wall panels shows that, provided the panel is adequately supported at the edges, failure will usually be by bond at the brick-mortar interface, although tension failure of a well bonded panel may take place in the brick itself or in the body of the mortar when weak materials are used.

There are three main types of bond failure at the brick-mortar interface:

(i) When bending is in the vertical direction, the horizontal joint will open (Fig. 1), i.e. tensile bond failure.

(ii) When bending is in the horizontal direction the bricks may slide across the mortar joint (Fig. 2), i.e. shear bond failure.

(iii) When bending is in the horizontal direction a well bonded panel may fail as shown in Fig. 3, the tension crack passing through the brick and perpend joint.

Among the factors (Ref. 1, 2, 3) influencing bond strength are:

(a) absorption or suction rate of the bricks;

(b) the initial water content and water retentivity of the mortar;

(c) type of mortar (cement/sand, cement/lime/sand, cement/sand with plasticizer, etc.) and cement content;

(d) type of brick (solid, perforated, frogged);

(e) thickness of mortar bed;

(f) workmanship.

The tensile bond strength is markedly reduced by the use of bricks of high suction rate or mortars weaker than 1 : 1 : 6 cement/lime/sand. Typical test results for tensile bond strength of mortar to solid bricks range between 10 lbf/in² and 80 lbf/in² for varying strength mortars, assuming good workmanship, correctly moistened bricks and mortar of reasonable consistency. Values as low as 4 lbf/in² and even 2 lbf/in² have been recorded when using a brick of high suction and a dry mortar. The shear bond strength may be four times as great as the tensile bond strength; during a series of tests on small panels of brickwork using cement/lime/sand mortars (Ref. 2), the modulus of rupture was a maximum of 220 lbf/in² with a failure as illustrated in Fig. 3.

It will usually be impracticable to carry out on site all the tests related to bond strength. Where a brick-to-mortar bond strength of 10 lbf/in² or more is required, it is recommended that panels of the kind illustrated in

Figs. 2, 3 and 4 be constructed and tested on site at 7 days, using the specified bricks and mortar to ascertain the approximate tensile bond strength before work commences on the building itself. Further tests may be carried out in the laboratory if considered necessary by the Engineer. By varying the water content of the mortar and adjusting the suction rate of the brick by dipping or spraying, a suitable combination of materials can usually be obtained. In cold weather it may not be advisable to wet the bricks, due to the danger of freezing. Additional water may be added to the mortar and in extreme conditions the water, sand and bricks should be heated (Ref. 10).

Design of Panels

The following notes and graphs are put forward as an approximate method for determining safe panel sizes and wall thicknesses to resist wind loading. *They are not in any way intended as an accurate stress analysis*, and it is proposed to improve and modify them as more data on bond strength and on the resistance of panels to lateral loading become available.

When large infill panels are used it will often be more economical to reinforce the brickwork to span horizontally and/or vertically rather than to increase the wall thickness. Reinforced brickwork cannot however be considered in this note, but may be referred to elsewhere (Ref. 13). There are broadly three panel support conditions:—

(a) When the panel is supported top and bottom, the sides being free, i.e. door or large window openings to either side. In this condition the wall will tend to span vertically under lateral loading and failure will normally be by tensile bond (Fig. 1). The Bending Moment will be equal to approximately $\frac{WL}{10}$ where 'L' is the vertical distance between restraints.

(b) When the panel is supported on all four sides. In this condition the wall will tend to span in two directions and failure may be by tensile bond (Fig. 1), shear bond (Fig. 2) or tension in the brick and perpend joint (Fig. 3), depending on the panel dimensions.

The maximum Bending Moment will depend upon the panel dimensions and the support conditions. Reference should be made to Tables 2 and 3.

(c) When the panel is supported at the sides—i.e. by return walls, brick piers or brick, steel or R.C. columns—and free at the top. In this condition the upper part of the wall panel will tend to span horizontally between return walls, piers or columns. The lower part will tend to cantilever from the base and failure may be by tensile bond (Fig. 1), shear bond (Fig. 2) or tension in the brick (Fig. 3) depending on the panel dimensions.

There is a growing conviction that it is wise to leave a movement joint at the top of a panel to accommodate the differential thermal and moisture movements of the frame and infill.

The designer may wish to ignore the two-way span effect due to support from three sides, since the maximum Bending Moment is usually high. In place of this he may provide a ring beam or horizontal reinforcement at the top of the wall to give top support, and consider as case (b).

Alternatively, the two-way span effect of the panel supported on three sides may be ignored and the panel considered as spanning horizontally between piers or return walls. In this case the panel may be designed for a *maximum shear bond stress of 20 lbf/in²* and where the panel spans horizontally between discontinuous supports the Bending Moment may be taken as WL/8. Where the panel spans horizontally between continuous supports the Bending Moment may be taken as WL/12.

'L' in both cases is the *horizontal* distance between supports. (Note that where panels are supported on four sides, 'L' is taken as the least panel dimension). A check should be made on the stability of supporting piers when these are used.

For the conditions (a) and (b) it is assumed that the maximum tensile bond stress does not exceed 10 lbf/in².

At mid-storey height the direct stress due to self-weight alone will be approximately equal to 5 lbf/in²* and the

---

* For brickwork weighing 110 lb/ft³, the stress due to self-weight is equal to 0·76 lbf/in² per foot of height.

For a storey-height of 9' 0", which is usual for multi-storey housing and offices, the stress at mid-storey height, due to self-weight alone, will be only 3·4 lbf/in². Where panels are supported on four sides this is compensated for by the additional tensile shear strength of 20 lbf/in².

When, however, the panel is supported top and bottom only and designed for a Bending Moment of $\frac{WL}{10}$ then the reduced resistance to bending should be taken into account. This can be carried out by designing for an increased wind load.

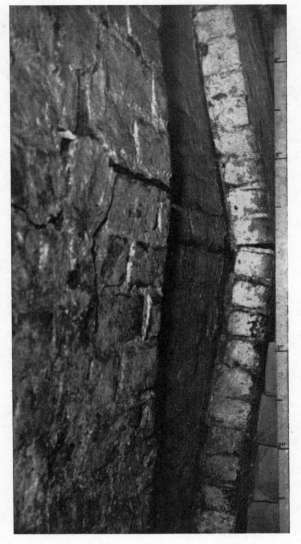

Fig. 1. *Tension failure of a brick retaining wall. High lateral earth pressures were caused by a bulldozer excavating and driving hard up to wall.*

Resistance Moments given in Table 1 have been calculated assuming a total tensile resistance to lateral loading of 15 lbf/in² (10 lbf/in² + 5 lbf/in²). This agrees with the revised C.P.111 (Ref. 5) as follows:

'Tensile stresses in brickwork or blockwork.

'In general, no reliance should be placed on the tensile strength of brickwork or blockwork in the calculations. The designer should assume that part of the section will be inactive and the remainder will carry compressive stress only.

'No tension should be relied upon at a damp-proof course or where water is present at the back of a wall.

'In some types of wall, tensile stresses in bending may be taken into account at the discretion of the designer. In such cases the walls should be built with bricks or blocks prepared before laying according to C.P.121.101 "Brickwork".

'For mortar not weaker than a 1 : 1 : 6 cement/lime/sand mix or its equivalent, the permissible tensile stress in bending should not exceed 10 lbf/in², when the direction of this stress is at right angles to the bed joints, and should not exceed 20 lbf/in² when the direction of tensile stress is at right angles to the perpend joints. The higher value should not be used where the crushing strength of the brick or block is less than 1,500 lbf/in².'

Reproduced with permission from C.P.111 (1964).

The New Zealand Standard (Ref. 4) permits tensile stresses of 5 lbf/in² for work constructed without continuous inspection and 10 lbf/in² for work constructed with continuous inspection.

When these stresses are due solely to wind and/or earthquake disturbances, they may be increased by one-third to 6.7 lbf/in² and 13.3 lbf/in² respectively.

These values are low when compared with some published figures and if site tests are carried out they may be modified accordingly, adopting a suitable load factor appropriate to the workmanship and site supervision.

The section moduli for varying wall thicknesses are also given in Table 1. In certain instances, the tension stress values and the resistance moments may be increased at the designer's discretion, and examples of special situations are given below:

(a) When loads from floors or roof are supported by the wall, so increasing the direct stress.

(b) When the panel height is greater than 15 feet, the direct stress due to self weight of brickwork will be more than the 5 lbf/in² allowed.

(c) When perforated bricks are used, the tensile and shear bond strengths, mortar to brick, may well be greater than for solid bricks, and a permissible stress greater than the 10 lbf/in² allowed may be justified. Tests are needed to ascertain this.

(d) When steel reinforcement is incorporated in the brickwork.

(e) When the brickwork is prestressed or poststressed, using horizontal or vertical rods or wires.

SHEAR: In determining the shear force along the perimeter of the panel, a wind force equivalent to 1.5 p should be taken (Ref. 7).

The strength of the various support conditions shown in Table 3 should be checked to ensure that they safely support the shear forces due to wind.

SLENDERNESS RATIO (S.R.): The maximum length for 'least panel dimension' given in Table 1 for S.R. of 18, 24 and 30 are based on:

(a) Effective height or length equal to the lesser of actual height or length of panel;

(b) Effective height equal to three-quarters of the actual height of panel. According to the new Code, C.P.111, (Ref. 5) the factor of 0.75 may be adopted when:

(i) the floor slabs restraining the wall are of reinforced concrete construction and bearing a minimum of 4″ on to the wall;

(ii) when the floors restraining the wall are of timber spanning *on to the wall* and where metal anchors are used (Ref. 5 and 6).

It is not intended to apply where the horizontal distance between restraints is used to ascertain the S.R., although for conditions such as F, G, H and L shown in Table 3 there is justification for doing so.

A limiting S.R. of 18 is normal for walls supported top and bottom only.

A limiting S.R. of 24 is reasonable for panels supported on four sides, and where the panel is approximately square (ratio a/b not more than 1.25) a higher limit of 30 could be adopted.

Recent research on slender walls at Edinburgh University suggests that S.R. may have less effect on the strength of walls than the reduction factors of C.P.111 imply (Ref. 5).

The tensile strength of 10 lbf/in² is a permissible stress and no reduction for S.R. is required.

Wind Loading

The basic wind pressures shown on graphs 1 and 2, ranging between 10 and 30 lbf/ft², are those values for p given in Table 3 of C.P.3, Chapter V (1952) Loading. (Ref. 7).

For plotting the graphs however, the calculations have taken into account the 0.7 reduction factor allowed for wall panels with normal openings (Ref. 7). The reduction is *not* applicable to buildings with both a ratio of height (to eaves level) to width of building less than one half and a pitch of roof less than 30°, nor for the design of individual panels. For such buildings or panels, the walls should be sufficiently strong to resist a total pressure outwards or inwards of 0.8 p (Ref. 7).

Calculations

The section moduli and resistance moments for the seven wall thicknesses considered are shown in Table 1. The five Bending Moments given in Table 2

$$\left(\frac{WL}{10}, \frac{WL}{12}, \frac{WL}{15}, \frac{WL}{18}, \frac{WL}{24}\right)$$ are approximate, and

similar to those given in Table 17 of C.P.114 (Ref. 8)

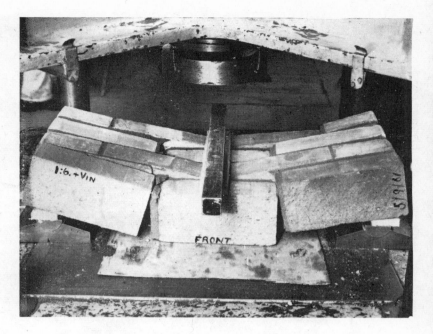

Fig. 2. *Failure of poorly-bonded panel, built with high suction bricks and mortar of moderate water retentivity.*

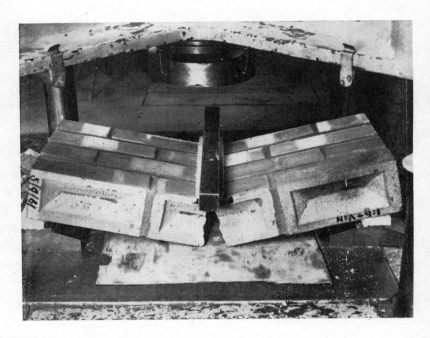

Fig. 3. *Failure of well-bonded panel, built with medium suction bricks and mortar of moderate water retentivity.*

for rectangular panels supported on four sides with provision for torsion at corners.

Two methods of panel edge support have been considered, viz. continuous and discontinuous. Some of the conditions providing such support are given in Table 3.

Where the maximum Bending Moment is at the continuous support, this value is given in Table 2.

Graph No. 1 shows the variation in Bending Moment with least panel dimension and the curves have been plotted for Bending Moments ranging between $\frac{WL}{10}$ and $\frac{WL}{24}$ and for wind pressures of 10, 20 and 30 lbf/ft².

Fig. 4. *Author's suggested site control test for tensile bond strength of mortar to brick.*

4a *shows the brick pier as constructed approximately 2′ 6″ high (ten bricks).*

4b *shows the brick pier supported over a clear span of 2′ 3″ after seven days curing, and under load.*

4c *shows the brick pier after failure.*

4a

4b

4c

Basis of rough calculation

(i) Self weight only

Brick pier turned on side after seven days curing and supported over a span of 2′ 3″. Assuming a deadweight of brickwork equal to 45 lb/ft², the tensile stress developed based on elastic theory will be approximately 10lbf/in².

$$\text{Bending Moment} = \frac{WL}{8} = \frac{9}{12} \times 45 \times 2.25^2 \times \frac{12}{8} = 255 \text{ lbf.in.}$$

$$\text{Section Modulus} = \frac{bd^2}{6} = \frac{9}{12} \times \frac{12 \times 4.125^2}{6} = 25.5 \text{ in}^3.$$

$$\text{Stress} = \frac{M}{Z} = \frac{255}{25.5} = 10\text{lbf/in}^2.$$

(ii) Pier under load

This site test is put forward primarily to compare the site variables such as moisture content of brick and mortar properties influencing tensile bond.

The greater the load supported by the pier then the greater the tensile bond strength mortar to brick.

The tensile bond stress due to the applied load may be calculated in a similar manner to that for self weight, using the appropriate bending moment. For the load applied uniformly over the length of the pier the Bending Moment will be $\frac{WL}{8}$. For the loading shown in Fig. 4b the Bending Moment will be approximately $\frac{WL}{6}$.

Typical Calculation for Bending Moment =

$$\frac{WL}{24} \text{ and } p = 10 \text{ lbf/ft}^2.$$

L = least panel dimension (height or length).
W = wind load = wind pressure p (10 to 30 lbf/ft²) multiplied by wind reduction factor 0.7 and multiplied by L.

When L = 7 feet and p = 10 lbf/ft² (i.e. W = 10 × 0.7 × 7′).

$$\text{Bending Moment} = \frac{WL}{24} = \frac{10 \times 0.7 \times 7^2 \times 12}{24} = 172 \text{ lbf.in./ft.}$$

Typical Calculation for Bending Moment =

$\frac{WL}{15}$ **and p = 30 lbf/ft².** When L = 12 ft. and p = 30 lbf/ft² (i.e. W = 30 × 0.7 × 12′) Bending Moment $= \frac{WL}{15} = \frac{30 \times 0.7 \times 12^2 \times 12}{15} = 2,420$ lbf.in./ft.

By selection of the appropriate Bending Moment from Table 2, Graph No. 2 may be used to determine:

(a) the wall thickness, given the panel dimensions and wind pressure;

(b) the least panel dimension* given the wall thickness and the wind pressure;

(c) the maximum wind pressure, given the wall thickness and the panel dimensions.

Example 1

Single-storey factory building

Height of panel to eaves, 15 feet, with concrete ring beam at eaves level. Distance between restraints is 20 feet (see Table 3, condition J—discontinuous support at sides).

Wind pressure p = 12 lbf/ft² (exposure D) $\frac{a}{b} = \frac{20}{15} = 1.33$ where a and b are the panel dimensions—b less than a.

From Table 2, with one support continuous and three

---

* When the least panel dimension is found, a check on the length/height ratio should be made to ensure that the correct Bending Moment is used.

Table 1

Wall thickness	Effective thickness (in.)	Maximum length for Least Panel Dimension						Section Modulus $Z = \dfrac{bd^2}{6}$ per 12" wall length	Resistance Moment R.M. $= fZ$ (f=15 lbf/in.²) per 12" wall length
		S.R. = 18		S.R. = 24		S.R. = 30			
		Effective length = actual length † $\overline{18}$	Effective length = $\frac{3}{4}$ actual length † $\overline{18}$	Effective length = actual length † $\overline{24}$	Effective length = $\frac{3}{4}$ actual length † $\overline{24}$	Effective length = actual length † $\overline{30}$	Effective length = $\frac{3}{4}$ actual length † $\overline{30}$		
4½"*	4½"	6' 9"	9' 0"	9' 0"	12' 0"	11' 3"	15' 0"	34 in.³	510 lbf.in.
10½"–11" cavity*	$\frac{2}{3}(4\frac{1}{2}" + 4\frac{1}{2}") = 6"$	9' 0"	12' 0"	12' 0"	16' 0"	15' 0"	20' 0"	68	1020
6¾"	6¾"	10' 0"	13' 6"	13' 6"	18' 0"	16' 9"	22' 6"	91	1365
4½" + 6¾" cavity*	$\frac{2}{3}(4\frac{1}{2}" + 6\frac{3}{4}") = 7\frac{1}{2}"$	11' 3"	15' 0"	15' 0"	20' 0"	18' 9"	25' 0"	125	1875
9"	9"	13' 6"	18' 0"	18' 0"	24' 0"	22' 6"	30' 0"	162	2430
4½" + 9" cavity*	$\frac{2}{3}(4\frac{1}{2}" + 9") = 9"$	13' 6"	18' 0"	18' 0"	24' 0"	22' 6"	30' 0"	196	2940
13½"	13½"	20' 3"	27' 0"	27' 0"	36' 0"	33' 9"	45' 0"	365	5468

*Wall thickness taken as actual thickness of 4⅛" for calculating Section Modulus. For calculating limiting panel dimensions for Slenderness Ratios of 18 and 24, the wall thickness is taken as the nominal 4½".

†Limiting dimensions marked 18, 18, 24, 24, 30 and 30 on graph No. 2.

Research at the Building Research Station (Ref. 11 and 12) has shown that for cavity walls built with ties to B.S.1243 ("Metal Wall Ties") any bending moments induced by eccentric or lateral loading will be shared by both leaves in proportion to their stiffnesses. This is because the ties, acting in tension or compression, ensure that both leaves take up the same are of bending.

Hence in Table 1 the Section Modulus for cavity walls has been obtained by adding the modulus for each leaf.

discontinuous, the appropriate Bending Moment is $\dfrac{WL}{12}$.

Referring to Graph No. 2 and assuming a Bending Moment of $\dfrac{WL}{12}$, for a wind pressure of 12 lbf/ft² a 10½"–11" cavity wall is not sufficiently strong. Any of the stronger wall thicknesses to the right would be suitable, and architectural considerations and other factors will determine which is the best.

Example 2

12-storey block of flats (10½"–11" cavity wall)

Storey height 8' 6". Distance between restraint is 10' 6" (see Table 3, condition G—continuous support at sides); $\dfrac{a}{b} = \dfrac{12.5}{8.5} = 1.48 =$ say, 1.5.

From Table 2 with four sides continuous the appropriate Bending Moment is approximately $\dfrac{WL}{15}$. Referring to Graph No. 2, the maximum wind pressure for a least panel dimension of 8' 6" is approximately 25 lbf/ft², equivalent to Exposure D. (Ref. 7).

Table 2 Bending Moments

PANELS SUPPORTED ON 4 SIDES

Type of Panel	Values of $\dfrac{a}{b}$ where a and b are the panel dimensions and b < a.			
	1·0	1·25	1·5	1·75 or more
3 or 4 sides continuous	$\dfrac{WL*}{24}$	$\dfrac{WL}{18}$	$\dfrac{WL}{15}$	$\dfrac{WL}{12}$
2 sides continuous	$\dfrac{WL}{18}$	$\dfrac{WL}{15}$	$\dfrac{WL}{12}$	$\dfrac{WL}{12}$
3 or 4 sides discontinuous	$\dfrac{WL}{18}$	$\dfrac{WL}{12}$	$\dfrac{WL}{12}$	$\dfrac{WL}{10}$

* These values are approximate and generally err on the side of safety. They are similar to those given in C.P.114, Table 17 (Ref. 8) for two-way span R.C. slabs with torsional resistance. Where the maximum moment is at the continuous edge this coefficient is given. When the wall panel is rigidly supported at the sides, there will be an increased resistance to lateral loading due to arching action within the wall thickness. This is beyond the scope of this note although research has been carried out on the arching of thin reinforced concrete slabs (Ref. 9).

──── REFERENCES ────

1: T. Ritchie and J. I. Davison – "*Factors affecting bond strength and resistance to moisture penetration of brick masonry*". National Research Council of Canada. Research Paper No. 192 of the Division of Building Research, July, 1963.

2: J. F. Ryder – "*The use of small brickwork panels for testing mortars*" Transactions of the British Ceramic Society, August, 1963.

3: V. A. Youl and E. R. Coats – "*Some studies in brick-mortar bond*". Australian Building Research Congress 1961, Paper 2, C.B.I.

4: New Zealand Standards Institute. Model Building By-law, Part X, Masonry Construction, July, 1959.

5: British Standard Code of Practice, C.P. 111 (1964). "*Structural recommendations for loadbearing walls*".

6: Department of Health for Scotland. Technical Memorandum No. 1, Revised 1960, "*Slender wall construction for houses*".

7: British Standard Code of Practice, C.P. 3 – Chapter V (1952) Loading.

8: British Standard Code of Practice, C.P. 114 (1957). "*The structural use of reinforced concrete in buildings*".

9: A. J. Ockleston – "*Arching action in reinforced concrete slabs*". Structural Engineer, June, 1958.

10: Winter Building, H.M.S.O., 1963.

11: N. Davey and F. G. Thomas – "*The structural uses of brickwork*". Proceedings, Inst. Civil Engineers, February, 1950.

12: F. G. Thomas – "*The strength of brickwork*". Structural Engineer, February, 1953.

13: R. E. Bradshaw – "*An example of reinforced brickwork design*". The British Ceramic Research Association, Special Publication No. 38, April, 1963.

Table 3

SUPPORT CONDITIONS FOR INFILL PANELS

Position	Support Condition	*Notes*
Roof Level	Discontinuous	(a) *In situ* or precast R.C. slabs bearing 4″ minimum on to the brickwork. (b) *In situ* or precast R.C. slabs not bearing on the wall should be tied by metal anchors at intervals of not more than 6′ as detail A and Ref. 5. (c) Timber anchored to wall using metal anchors of minimum cross-section $1\frac{1}{4}″ \times \frac{1}{4}″$ securely fastened to the joists and provided with split and upset ends or other approved means for building into the wall. The anchors should be provided at intervals of not more than 6′ in buildings of one or two storeys and not more than 4′ for all storeys in other buildings. For details see Ref. 5 and 6. 　　If a light roof construction is adopted, take precautions to prevent roof lifting due to wind suction.
Intermediate Floors	Continuous	(a) When wall continuous past edge of floor—as details A and B. For detail A provide anchors as noted for R.C., precast concrete and timber at roof level. (b) When *in situ*, R.C. cast on top of wall (see details C and D).
	Discontinuous	(c) When precast R.C. bearing on wall. (d) When brickwork constructed *after* R.C. framing and floors have been cast (see details C and D).
Ground Floor	Continuous	When brickwork below ground level retains fill on one side, stiffening piers may be required from Foundation to Ground Level.
Sides	Continuous	(a) When brickwork fully bonded to return and intersecting walls. (See details E and F). (b) When brickwork is continuous past R.C. columns or steel stanchions (see details G, H and I).
	Discontinuous	(c) When the brickwork not continuous past support (see detail J).

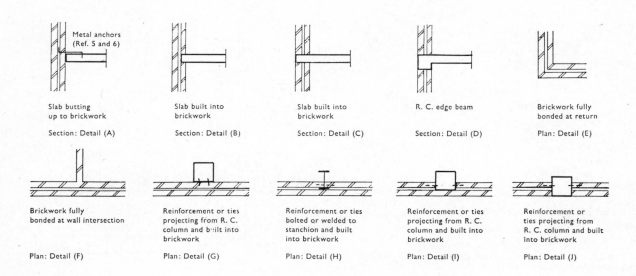

Metal anchors (Ref. 5 and 6)

Slab butting up to brickwork — Section: Detail (A)

Slab built into brickwork — Section: Detail (B)

Slab built into brickwork — Section: Detail (C)

R. C. edge beam — Section: Detail (D)

Brickwork fully bonded at return — Plan: Detail (E)

Brickwork fully bonded at wall intersection — Plan: Detail (F)

Reinforcement or ties projecting from R. C. column and built into brickwork — Plan: Detail (G)

Reinforcement or ties bolted or welded to stanchion and built into brickwork — Plan: Detail (H)

Reinforcement or ties projecting from R. C. column and built into brickwork — Plan: Detail (I)

Reinforcement or ties projecting from R. C. column and built into brickwork — Plan: Detail (J)

The details A to J inclusive are included to show some of the structural implications. They are not in any way intended to be comprehensive and other factors will require consideration.

For example, in detail A, if the floor slab were to deflect or the wall to expand due to thermal conditions the metal anchors would tend to lift and crack the floor screed.

In details I and J a vertical D.P.C. and protection for the ties—wall to column—may be required.

For notes on details A to J see Table 3.

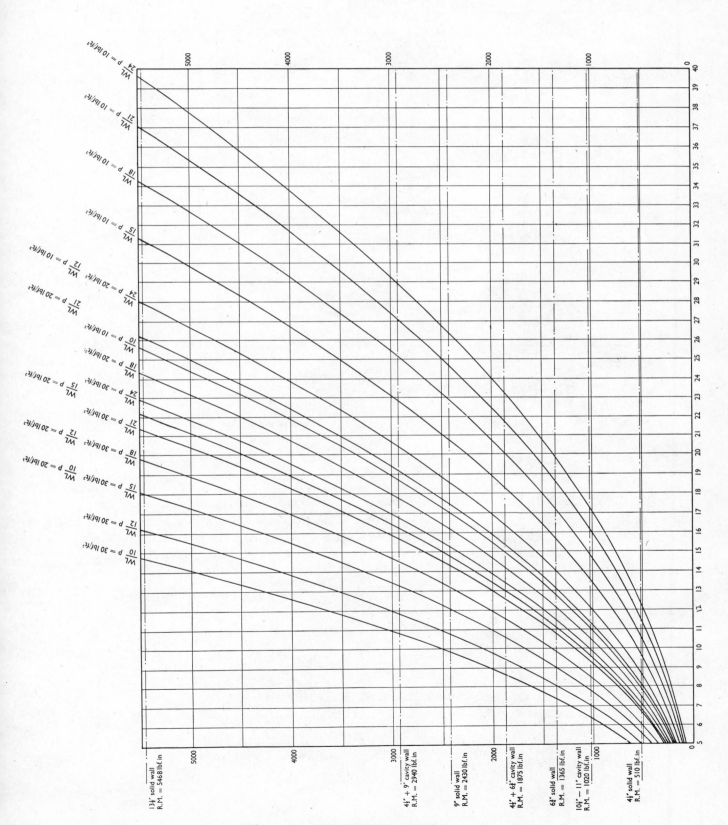

GRAPH No.1

Showing increase in Bending Moment with respect to panel dimension for various Bending Moments and wind pressure p.

PANEL DIMENSION IN FEET

For panels supported on four sides the least panel dimension is given. For panels supported at the top and bottom of the wall and designed for a Bending Moment of approximately $\frac{WL}{10}$ the panel height is given.

BENDING MOMENT in lbf.in.

GRAPH No.2

Set of Graphs showing least panel dimensions for varying Wind Loading and Bending Moment—based on a total tensile resistance to bending of 15 lbf/in² (10 lbf/in² tensile bond and 5 lbf/in² allowance for self-weight of brickwork)
For selection of appropriate Bending Moment refer to Table 2.

▲ 18
● $\overline{18}$ 24
❘ 30
■ $\overline{24}$

Limiting slenderness ratios. For explanation see Table 1.

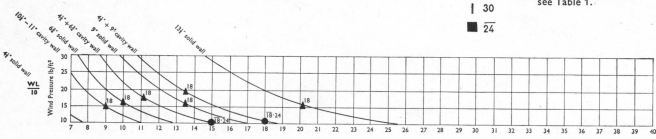

For panels supported on four sides the least panel dimension is given.
For panels supported at the top and bottom of the wall and designed for a Bending Moment of approximately $\frac{WL}{10}$ the panel height is given.

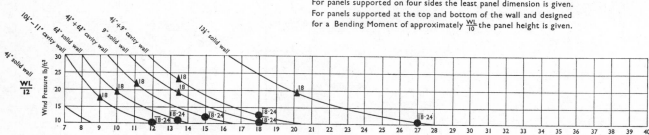

For panels supported on four sides the least panel dimension is given.

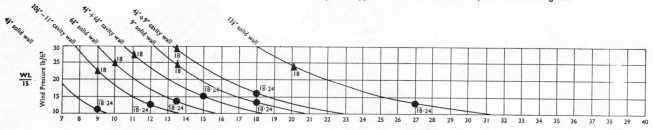

For panels supported on four sides the least panel dimension is given.

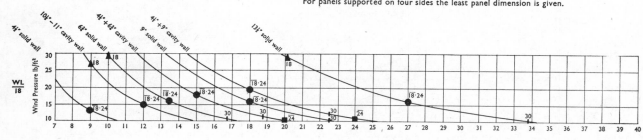

For panels supported on four sides the least panel dimension is given.

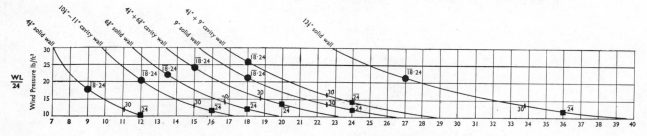

For panels supported on four sides the least panel dimension is given.

Movement Joints in Brickwork

by K. Thomas, C.Eng., A.M.I.Struct.E., A.R.T.C.

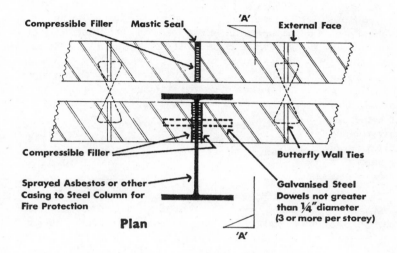

Plan

Compressible Filler — Mastic Seal — 'A' — External Face

Compressible Filler

Sprayed Asbestos or other Casing to Steel Column for Fire Protection

Butterfly Wall Ties

Galvanised Steel Dowels not greater than ¼" diameter (3 or more per storey)

Wall Anchorage to Steel Columns

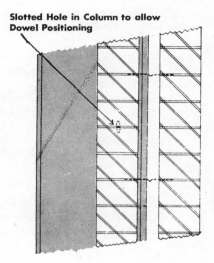

Slotted Hole in Column to allow Dowel Positioning

Section "A - A"

Note:
Galv'd Steel Dowel should be lubricated or have a Plastic Sheath to allow movement to take place in the Wall.

General

The skilful use of materials and correct detailing of a structure are the responsibility of the architect and engineer, and brickwork like all other components of a building must be considered from many aspects. This note deals with one very important one—the movement aspect.

Movement joints are too often omitted and whilst structural failure of brickwork is rare, unsightly cracks can be avoided if the appropriate measures are taken at the design stage. We are all very familiar with 50 to 100 year old boundary walls which run for miles with no apparent break. However, it must be remembered that such walls were constructed in lime mortar, capable of absorbing considerable amounts of dimensional change without the accompanying distress. The introduction of high strength mortars enable brickwork structures of greater loadbearing capacity to be built more rapidly, but less flexibility is present in the joints and consequently allowance for movement must be made.

Dimensional changes of brickwork may be due to variation in temperature or moisture content, chemical or frost action, movements in adjacent structural materials, or deflections and settlement. Movement may take place due to any one of the phenomena mentioned but is quite often due to some combination.

Thermal Movements

All building materials tend to move when subjected to temperature variation. Ideally such movements are reversible when considering individual materials in an unrestrained condition. However, brickwork is not merely a composite material made up of burnt clay, sand, cement, lime and/or plasticizer which is held, to some extent, or permitted to slide on a d.p.c. but also laid with varying degrees of skill which undoubtedly will affect the end product and hence its movement characteristics. Temperatures used for calculating expansion should be the average wall temperatures. For solid walls, these may be temperatures at the centre of the wall; in cavity wall construction, there may be differential thermal movement between the inner and outer leaves and in such situations, provision should be made for maximum thermal movement by considering the average temperature of the outer leaf.

Although the movement of internal walls is likely to be much less than that for external walls, one must consider inside temperatures as well as the heat transfer characteristics of the construction between the outside and the inside, particular attention being paid when designing internal rooms around refrigerated areas or boiler houses. Special care should also be taken when designing thin walls with a Southerly aspect as in some parts of this country, surface temperatures can reach 120°F. Such temperatures can give rise to a thermal gradient through the wall and in cavity construction, even between the two leaves, which may cause excessive bending in addition to longitudinal expansion. The coefficient of thermal expansion for unrestrained fired clay brickwork is to a large extent, dependent on the type of brick and mortar used; the value which is generally taken as representing the average thermal expansion of brickwork is 5·6 times 10^{-6} per °C (3 times 10^{-6} per °F) in a horizontal direction and vertically may be up to one and a half times this value.

Vertical thermal movements in walls are generally reversible but horizontal movements may only be reversible if the wall does not crack as a result of the expansion or contraction. This depends upon whether the wall is built on a soft d.p.c., as the degree of restraint imposed by the d.p.c., appears to be the critical factor.

(See B.R.S. Digest No. 12, First Series)

Moisture Movement

Fired clay products like many other building materials, exhibit reversible dimension changes dependent on their moisture content. In addition fired clay, whilst cooling in the kiln, begins to take up a permanent expansion which can go on but at a greatly reduced rate over several years. The magnitude of this permanent expansion of the unrestrained product, varies with the type of clay and the maximum firing temperature.

Much work has been carried out on moisture expansion, both in this country and abroad, with the result that many papers have been published on the subject in recent years, the net effect being to highlight a characteristic of burnt clay which has always existed and is easily allowed for in design and the not inconsiderable number of brick buildings that have been erected in the past, which exhibit no form of distress are an endorsement of this point.

It has been found that permanent expansion starts as the units commence to hydrate during cooling in the kiln. The work carried out to date shows that in all cases investigated at least 50% of the total expansion in two years has taken place within two days of the commencement of cooling. Consequently, providing that bricks are not built into the work whilst still hot, moisture expansion is unlikely to present a major problem; the normal delay of a day or two coming between the time the bricks are drawn from the kiln and their delivery on to the site and actually laying in the work, will allow the bulk of the moisture expansion to take place.

Cavity Walls

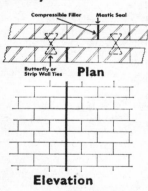

Staggered Joint

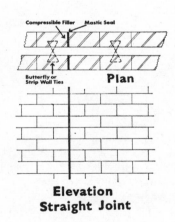

Straight Joint

13½" Solid Walls

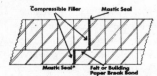

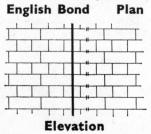

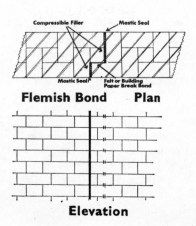

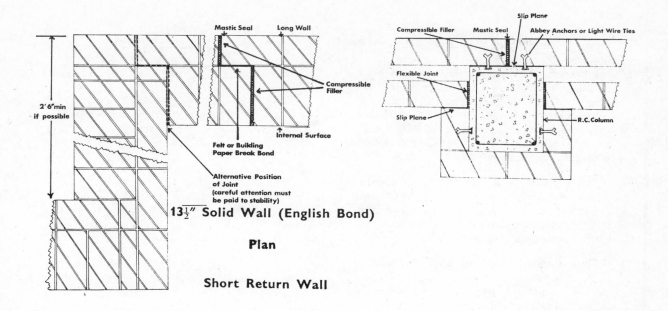

13½" Solid Wall (English Bond)

Plan

Short Return Wall

Plan
Wall Anchorage to R.C. Column
Similar Anchorage To R.C. Beams

Attempts have been made by many research workers, to determine figures for moisture expansion; however, each clay has its own inherent characteristic and many have different maximum firing temperatures, coupled with the fact that the degree of restraint of units in construction makes the problem of accurate determination extremely complex. At the present time, the work carried out by research workers indicates that any attempt to define a test method which would provide useful data for designers would be quite impracticable.

The designer on the other hand, quite naturally requires some guidance on moisture movement and the current method which has been satisfactorily adopted by both Canadian and American authorities, is to specify an expansion allowance of half the thermal expansion.

Sulphate Expansion

Sulphate attack and the accompanying mortar expansion rarely occurs and is only experienced in unprotected or badly designed parapets, retaining walls, and other structures normally liable to remain wet for long periods, and then only when bricks of high sulphate content are used in such situations. Sulphate failure can also occur due to brickwork being built in soils having a high sulphate content, and the precautions listed in B. R. S. Digest No. 123, First Series, should be observed where such soil conditions are known to exist, e.g., stacking of bricks on ground which has previously received chemical treatment or alternatively has been used for grazing. Airborne pollution is another possible contributory factor. Whatever the source of the sulphates, expansion of the mortar takes place, resulting in movement of the brickwork.

On the subject of bricks containing various sulphates, the foreword to B.S. 3921, "Bricks and Blocks of Fired Brickearth, Clay or Shale," states:

The committee has given serious consideration to the problem of framing a specification which is based on the knowledge that bricks containing undue amounts of calcium, magnesium, potassium and sodium sulphates are liable to produce complaints about walls built with them. The complaints may be of two kinds; sulphate expansion of Portland Cement mortar and efflorescence on brickwork.

Although cause and effect have been established broadly, considerable difficulty has arisen when trying to decide what are suitable maximum limits for the permissible contents of calcium, magnesium, potassium, sodium and sulphate individually or *in toto*. In some circumstances it would appear that bricks with a total soluble sulphate content of well under 1 per cent have given severe trouble in sulphate expansion; in others, bricks with soluble salt contents of as much as three times this amount have been used without arousing comment.

The same sort of evidence has been forthcoming on particular salts e.g., potassium sulphate. For instance there has been complete absence of complaints over extended periods when bricks containing 0.25 per cent soluble potassium have been used. Elsewhere trouble has arisen with bricks containing less than 0.25 per cent. In these circumstances it has been considered unreasonable to set a maximum of 0.25 per cent of soluble potassium for bricks in general.

The explanation of this conflicting evidence remains a matter of conjecture. It is well known, for example, that for sulphate expansion to occur it is necessary to have soluble sulphates, tri-calcium aluminate, and water in juxta-position.

Thus, sulphate expansion does not occur in brickwork where the bricks have negligible sulphate content, or the mortar has a low tri-calcium aluminate content, as in mortar made from sulphate-resisting cement, or when water is largely excluded by sound methods of building construction. Thus it is easy to visualize conditions in which bricks of moderate salt content could have given good service and other conditions in which bricks of less salt content could have performed badly. There are many other factors too, which obscure this issue.

In situations where sulphate expansion is likely to occur, it is recommended that sulphate resisting cement be used. (See B.R.S. Digest No. 123, First Series.)

Movement Due to Freezing

The movement of brickwork due to freezing is a secondary action, which can only take place after sulphate or frost attack of the mortar. After the primary failure, pores and crevices are opened up in the mortar, thus allowing water penetration and the subsequent expansion upon freezing. It is important therefore to protect brickwork during erection to prevent the intrusion of excessive moisture which is liable to cause frost failure in the mortar and expansion in the brickwork.

Differential Settlement

When differential settlement occurs between parts of a building, cracking is inevitable. Joints are of limited value in such situations unless closely spaced, as for example, in the C.L.A.S.P. (Consortium of Local Authorities Special Project) system where each individual unit is permitted to rotate, thus absorbing movement and preventing cracking.

The prevention of cracking due to differential settlement is largely a question of foundation engineering with loads distributed according to the bearing capacity of the soil and ground adequately compacted etc. Where such movement is anticipated, in some cases (e.g. when a new building abuts an existing one) it may be advantageous to make a complete break in the form of a temporary joint, until differential movement has ceased.

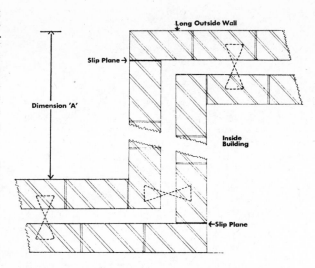

Plan

Short Return Walls (Cavity Wall)

Slip planes may be desirable in the positions shown if dimension "A" is 2'-6" or less.

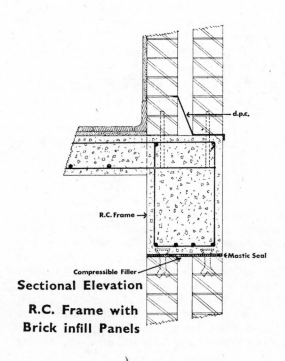

Sectional Elevation

R.C. Frame with Brick infill Panels

Large expanses of cavity brickwork are subject to differential movement due to various causes and Clause 308 (f) of C.P. 111 "Structural Recommendations for Loadbearing Walls" states:

External cavity walls. Where the outer leaf is half brick or its equivalent thickness in blocks, the uninterrupted height and length of this leaf of external cavity walls should be limited so as to avoid undue loosening of the ties due to differential movements between the two leaves. It is therefore recommended that the outer leaf be supported at intervals of not more than every third storey or every 30 ft., whichever is less. However, for buildings not exceeding four storeys or 40 ft. in height, whichever is less, the outer leaf may be uninterrupted for its full height. Consideration should also be given to the uninterrupted length of any leaf.

Differential Movement of Dissimilar Materials

When designing composite walls (i.e. walls involving the use of more than one basic type of unit), it is extremely important to consider the way in which the units will behave. Clay brickwork, for example, may expand whilst concrete and other siliceous products are likely to contract. It is of the utmost importance to realise that the rigid bonding of materials with diametrically opposed movement characteristics, can only result in trouble, unless the length of wall is extremely short, and whenever possible, walls faced with bricks dissimilar from the backing, should be of cavity construction using wire wall ties.

When cavity wall construction is used, it is also important at jambs of openings etc., where the cavity is closed with the brickwork, and at stop ends, to provide a vertical slip plane in the form of bituminous felt or polythene. This is desirable, regardless of whether a d.p.c. is necessary.

Panel Walls Enclosed in Frames

In-filling panels in frame buildings of steel or reinforced concrete, are liable to high accidental stresses, which may, cause cracking unless allowance is made for movement within the panel at the design stage.

Deflection or shortening of the frame (particularly in the case of reinforced concrete), can impose considerable forces on the brickwork panels and when coupled with eccentricity, a buckling action takes place in the brickwork. It is therefore recommended that in multi-storey buildings, the top of panel walls should not be built in rigidly to the frame, but be packed with resilient material capable of absorbing any movement. When this procedure is adopted the panels should be considered as either spanning horizontally, or cantilevering from the base. Where panels of brickwork are intended to add to the rigidity of the frame and act as a bracing medium

for wind etc., the above procedure should not be adopted. In such situations, the panels should be designed to accommodate not only the stresses from vertical loading but also those which may arise from the possible causes of movement within the panel and the framework itself.

Thermal movement of single storey buildings and the top storeys of multi-storey buildings can be rather critical due to wide variations of temperature daily. This coupled with the fact that the floors are subject to only minor variations in temperature can result in a racking action taking place. It is undesirable to rigidly tie walls vertically in such situations and dowel type fixing is preferred. This gives adequate support to resist lateral wind and other pressures and at the same time some degree of flexibility normal to the lateral forces.

Mortar

Although mortar forms only a small proportion of brickwork as a whole, its characteristics, nevertheless, do have a significant effect, particularly in relation to movements, both within the wall itself and with adjacent parts of the structure.

Mortars recommended for use in various situations are shown in Tables 1 & 2, the properties of such mortars exhibiting creep and plastic flow, which will undoubtedly tend to relieve high stresses and reduce the risk of cracking the brickwork. It is uneconomical and very unwise to specify mortars stronger than necessary. This is particularly so with low strength units, which have been known to show signs of distress due to such practices.

Selection of Materials for Movement Joints

It is essential, when designing a movement joint, to consider the type of movement which may occur. In the case of brickwork, movements may be due to expansion and contraction, and therefore a material which is resilient and easily compressible should be adopted. Certain types of fibre board and similar materials are not suitable for expansion joints. Tests have shown that in some instances a stress of 300 lbf/in^2 can be reached to achieve 50% compression of the material and on removal of the load the material does not return to its previous dimensions.

Strips of V-shaped copper have been used successfully for many years, but newer materials such as pre-moulded, extruded, closed cell rubber or plastics (polyurethane and polyethylene), alternatively rubber or plastic sections can be used. Under certain circumstances such as temporary joints etc., the use of lime/sand mortar may be considered sufficient for movement to take place (a 1 : 3 to 1 : 4½ mix by volume of non-hydraulic lime and sand).

The sealing of movement joints has always been a problem; the introduction of polysulphide based sealants has improved their performance considerably.

Spacing of Joints

It is impracticable to produce a comprehensive specification for the position and spacing of movement joints for all structures. Each building must be considered as an individual unit and provision made accordingly.

It is also vital that when joints are positioned the stability of the structure is in no way impaired. Joints are often concealed by rain water down pipes, etc., and care should be taken to ensure that fixing of lugs or brackets occurs only on one side of the joint. Rendering and plastering should not be carried over a joint, each coat being severed by a well defined cut before the work hardens. Alternatively, rendering can take place with a temporarily secured batten to the edge of the joint or a vertical d.p.c. When calcium silicate or other siliceous units are used for one leaf of a cavity wall in conjunction with fired clay products, provision may be necessary for drying shrinkage. This can often be taken care of by introducing permanent movement joints, or alternatively, in some instances by temporary joints using a low strength lime-sand mortar. After an initial drying out period the joint can be raked out and pointed. However, unless a straightforward vertical joint is used, this is rather impracticable unless pointing of the whole wall is to be carried out later as colour variation of the mortar will tend to make an undesirable feature of the joint.

Long Walls

As a general guide in long unrestrained walls of fired clay units, $\frac{3}{8}''$ movement joints should be provided at approximately 40 ft. centres. When cavity walls having one leaf of fired clay and the other of calcium silicate or other siliceous materials are used, movement joints should be provided at more frequent intervals of say 20–25 ft.

Offsets and Junctions

Joints at offsets and junctions are recommended as high concentrations of stress build up at these points due to movement. Short returns should also be avoided as it has been found that where the length of the return is not more than 2 ft. 3 in. cracking is likely to occur. It is suggested, therefore, that to avoid cracking of this kind the length of the return should be not less than 2 ft. 6 in. (See B.R.S. Digest No. 114, First Series.)

Corners

Joints are often sited at or near external corners and great care should be taken to ensure that such joints in no way impair the stability of the structure.

Door and Window Openings

High stresses tend to build up at door and window openings in long walls, especially in light framed factory buildings. It is therefore, recommended that movement joints should be considered in such situations.

Parapets

Joints in parapet walls should follow the lines in the main structure and should be carried through the parapet. If additional joints are considered necessary these should be positioned approximately midway between those running throughout the full height of the building. Stability must be carefully checked in such situations particularly, as no tension is permitted at the d.p.c. level.

Summary

Thermal Movement

Horizontally—Average coefficient of thermal expansion of brickwork is 5·6 times 10^{-6} per °C (3 times 10^{-6} per °F).

Vertically—1·5 times the above values.

Moisture Movement

Horizontally—Allow for an expansion of half the horizontal thermal value.
Vertically—Allow for an expansion of half the vertical thermal value.

The bulk of the permanent moisture expansion in the bricks will take place within the first day or two after starting to cool in the kiln; generally speaking, bricks should not be built into the works within two days of being drawn from the kiln.

Sulphate Expansion

This cannot occur unless moisture is present.

Differential Settlement

Generally a foundation problem.

Dissimilar Materials

Do not use one material to restrain another; when used adopt short lengths of wall.

Panel Walls Enclosed in Frames

A movement joint should usually be provided at the top of all panels; care should be exercised with peripheral fixing.

Mortar

Mortars recommended for different situations are given in Tables 1 and 2.

Selection of Materials for Movement Joints

Easily compressible materials are recommended.

Spacing of Movement Joints

Each building must be considered on its merits; for long walls, clay brickwork should have movement joints at approximately 40 ft. centres; calcium silicate materials at approximately 20–25 ft.

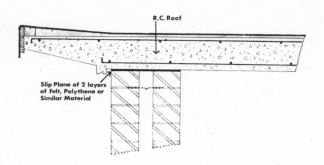

Sectional Elevation

Roof Expansion Joint

When lightweight roof construction is adopted vertical uplift must not be ignored.

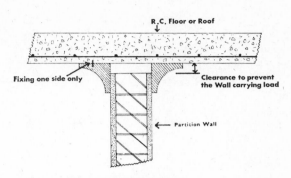

Sectional Elevation

Partition Wall

Fixing must be given due consideration as the Partition wall dimensions depend upon the degree of fixity.

Polysulphide Base Sealants

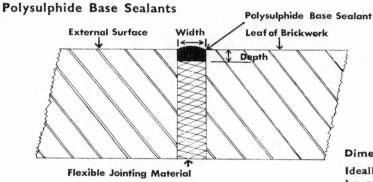

Movement Joint

Plan

Dimensional Specification for Joint Sealant

Ideally a bonding surface (depth) of $\frac{1}{2}''$ to $\frac{3}{4}''$ should be provided. When as usually is the case, the width is less than $\frac{1}{2}''$: The depth of sealant should be **NOT LESS** than one-half the width and **NOT GREATER** than the width.

Typical Plans indicating positions where movement joints may be appropriate

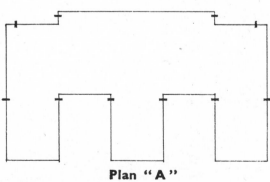

Plan "A"

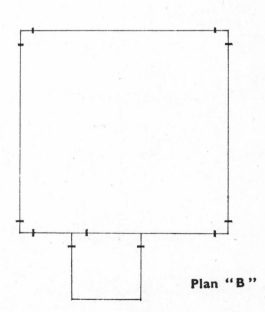

Plan "B"

Notes:
Position of joints may depend on the location of doors and windows in long walls.

Plan "B"
The positions of movement joints suggested on this plan at the corners are alternatives and it is not intended that two joints should be provided.

Table One
Selection of Mortar Mixes

Type of construction	Exposure conditions	Early Frost Hazard	Type of Bricks or Blocks	Mortar Designation
EXTERNAL WALLS Retaining wall	Any	Yes or No	Clay, concrete or calcium silicate	1
Parapet, free standing wall or below damp proof course	Any	Yes or No	Clay	1, 2 or 3
			Concrete or calcium silicate	3
Between eaves and damp proof course	Severe	Yes or No	Clay, concrete or calcium silicate	3
	Sheltered or Moderate	Yes	Clay, concrete or calcium silicate	3
		No	Clay	3 or 4
			Concrete or calcium silicate	4
INTERNAL WALLS AND PARTITIONS	—	Yes	Clay	3
			Concrete or calcium silicate	4
		No	Clay, concrete or calcium silicate	4 or 5

Table Two
Equivalent Mortar Mixes (Proportions by Volume)

	Mortar Designation (from Table 1)	Hydraulic-lime : sand	Cement lime : sand	Masonry-cement : sand	Cement : sand with plasticizer
Increasing strength but decreasing ability to accommodate movements caused by settlement, shrinkage, etc.	1	—	$1:0-\frac{1}{4}:3$	—	—
	2	—	$1:\frac{1}{2}:4-4\frac{1}{2}$	—	—
	3	—	$1:1:5-6$	$1:4\frac{1}{2}$	$1:5-6$
	4	$1:2-3$	$1:2:8-9$	$1:6$	$1:7-8$
	5	$1:3$	$1:3:10-12$	$1:7$	$1:8$

Direction of change in properties (within any one mortar designation)	⟶ Increasing resistance to damage by freezing. ⟶
	⟵ Improving bond and consequent resistance to rain penetration ⟵

Reproduced by permission of the Controller of H.M. Stationery Office from Building Research Digest, 2nd Series, No. 58.

References

1. British Standard Code of Practice. C.P. 111 (1964). "Structural Recommendations for Loadbearing Walls."
2. British Standard Code of Practice. C.P. 121.101 (1951). "Brickwork."
3. British Standard Code of Practice. C.P. 122 (1952). "Walls and Partitions of Blocks and of Slabs."
4. British Standard 3921:1965. "Specification for Bricks and Blocks of Fired Brickearth, Clay or Shale."
5. The British Ceramic Research Association, Special Publication Number 38, April 1963. "The Use of Ceramic Products in Building."
6. Proceedings of the British Ceramic Society, No. 4, July 1965. "Load-bearing Brickwork."
7. Cutler & Mikluchin. "Clay Masonry Manual." Brick and Tile Institute of Ontario. April 1965.
8. H. C. Plummer. "Brick and Tile Engineering." Structural Clay Products Institute. Washington, D.C.
9. Technical Note on Brick and Tile Construction No. 18. "Differential Movement—Cause and Effect." Part I of III. Structural Clay Products Institute, Washington, D.C. April 1963.
10. Technical Note on Brick and Tile Construction No. 18A. "Differential Movement—Expansion Joints." Part II of III. Structural Clay Products Institute, Washington, D.C. May 1963.
11. Technical Note on Brick and Tile Construction No. 18B. "Differential Movement—Flexible Anchorage." Part III of III. Structural Clay Products Institute, Washington, D.C. June 1963.
12. "Principles of Modern Building." Volume I, 3rd Edition. M.O.T. Building Research Station.
13. Digest No. 6. First Series. "The avoidance of cracking in Masonry Construction of Concrete or Sand-lime Bricks." Jan. 1957. M.O.T. Building Research Station.
14. Digest No. 123. First Series. "Sulphate Attack on Brickwork." June 1959. M.O.T. Building Research Station.
15. Digest No. 114. First Series. "Questions and Answers." Sept. 1958. M.O.T. Building Research Station.
16. Digest No. 4. Second Series. "Repairing Brickwork." Sept. 1960. M.O.T. Building Research Station.
17. Digest No. 58. Second Series. "Mortars for Jointing." May 1965. M.O.T. Building Research Station.
18. Digest No. 65. Second Series. "The Selection of Clay Building Bricks: I." Dec. 1965. M.O.T. Building Research Station.
19. Digest No. 66. Second Series. "The Selection of Clay Building Bricks: II." Jan. 1966. M.O.T. Building Research Station.
20. D. G. R. Bonnell & M. R. Pippard. National Building Studies Bulletin No. 9. "Some Common Defects in Brickwork." 1950. M.O.T. Building Research Station.
21. "Notes on the Construction of Movement Joints in Brickwork." Miscellaneous Paper No. 84. The Chalk Lime and Allied Industries Research Association.
22. "Notes on Movements in Buildings Affecting Brickwork." Miscellaneous Paper No. 86. The Chalk Lime and Allied Industries Research Association.
23. I. C. Freeman. Building Research Station. "The Moisture Expansion of Some British Building Bricks." British Ceramic Society. Autumn 1965.
24. R. Beard, A. Dinney and R. Richards. London Brick Company. Movement of Brickwork, Part 1. "Experiments on Fletton Brick Walls." British Ceramic Society. Spring 1966.
25. R. Beard, A. Dinney and R. Richards. London Brick Company. Movement of Brickwork, Part 2. "Some Case Histories." British Ceramic Society. Spring 1966.
26. R. Beard. London Brick Company. Movement of Brickwork, Part 3. "Preliminary Work to Investigate the Effect of Different Damp Proof Courses." British Ceramic Society. Spring 1966.
27. R. T. Laird and A. Wickins. Redland Brick Company. "Measurement of the Moisture Expansion of Works and Laboratory Fired Bricks and Difficulties in the Interpretation of the Results." British Ceramic Society. Spring 1966.
28. H. W. H. West, K. Pate, W. Noble, H. A. Mouat and V. J. Owen. British Ceramic Research Association. "Moisture Movement of Bricks and Brickwork." British Ceramic Society. Spring 1966.

Thermal Transmittance of Wall Constructions
A review of relevant information

by
R. BEARD and A. DINNIE

Fig. 2 **External View of Thermal Transmittance Laboratory.**

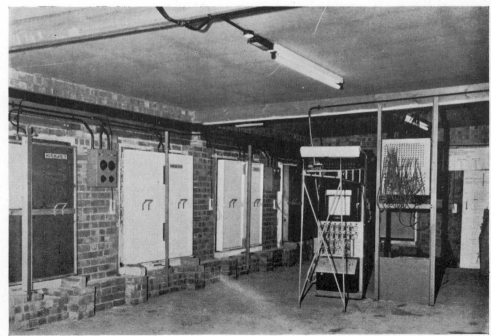

Fig. 4 **Internal View of Thermal Transmittance Laboratory showing Guarded Hotplates in position on test walls and instrumentation.**

Introduction

In a building which has reached stable temperature conditions, and with constant temperature outside, the heat input required to maintain the stable interior temperature is that which is equal to the total heat losses. Thus any reduction in the heat losses is reflected in a reduction of the heat required to maintain the interior temperature, or alternatively, higher interior temperatures can be maintained for the same total heat input. With the rise in cost of fuels and the demand for higher standards of comfort in our homes, it is clear why there has been an upsurge of interest in the thermal insulation of buildings.

Unfortunately, the idealised case outlined above is not representative of what actually occurs in practice.

The outside temperature is never constant for any length of time, and hence varying temperature gradients across the wall will induce a fluctuating demand of heat input to maintain a constant interior temperature. Heat is lost from a building in several ways; e.g. by conduction through the walls, windows, roofs, and floors, and by ventilation. It is the job of the architect or heating engineer to examine the effect of all of these factors in establishing the overall thermal performance of a building. If the estimate of the overall thermal performance of the building is to be realistic, then the architect or heating engineer must have at his disposal information on the properties of the various materials which is realistic and representative of the performance of the building

Nomenclature and Units

Term & symbol	Definition	Imperial System Units	Metric Equivalent Units
British Thermal Unit.	The quantity of heat required to raise the temperature of 1 lb of water through 1 deg Fahrenheit	Btu	J (joule)
Thermal transmission or rate of heat flow (Q)	The quantity of heat flowing in unit time.	Btu/h	W (watt)
Thermal conductivity (k)	The thermal transmission through unit area of a slab of a uniform material of unit thickness when unit difference of temperature is established between its faces.	Btu in/ft^2 h degF	W/m degC
Thermal resistivity $\left(\dfrac{1}{k}\right)$	Reciprocal of thermal conductivity.	ft^2 h degF/Btu in	m degC/W
Thermal conductance (C)	The thermal transmission through unit area of a material or combination of materials of any given thickness per unit of temperature difference between the hot and cold faces.	Btu/ft^2 h degF	W/m^2 degC
Thermal resistance (R)	The reciprocal of thermal conductance.	ft^2 h degF/Btu	m^2 degC/W
Thermal transmittance (U)	The thermal transmission through unit area of a given structure divided by the temperature difference between the air on either side of the structure.	Btu/ft^2 h degF	W/m^2 degC

The thermal transmittance, U, of a building element can be obtained by combining the thermal resistance of its component parts and the surface air layers and taking the reciprocal. For example in the case of the cavity wall:

$$U = \frac{1}{R_{is} + R_{il} + R_c + R_{ol} + R_{os}}$$

where R_{is} = resistance of inside surface
R_{il} = resistance of inner leaf
R_c = resistance of cavity
R_{ol} = resistance of outer leaf
R_{os} = resistance of outside surface.

Inserting resistance values for the component parts for say an 11-in Fletton brick cavity (unventilated) construction into the above formula could give the following:

$$U = \frac{1}{R_{is} + R_{il} + R_c + R_{ol} + R_{os}}$$

$$= \frac{1}{0.7 + 0.9 + 1.0 + 0.6 + 0.3}$$

$$= \frac{1}{3.5}$$

$$= 0.286$$

say   0.29

materials under the actual conditions in which they will be used.

Most building materials are porous to some extent, and will attain some equilibrium moisture content depending upon their environmental conditions. The moisture taken up will then replace the air in the pores of the material, and since water has a conductivity about twenty-five times that of still air at ambient temperatures, it can be readily appreciated that a small quantity of moisture can make a large difference to the thermal properties of porous materials. It is important therefore when considering porous materials that their thermal values should be known for the conditions which the materials attain in use.

This formula does not apply to non uniform walls incorporating heat bridges or sections of material of appreciably different thermal transmission.

Some people find the various terms confusing and therefore to show how the main terms are related to each other they are presented diagramatically in Fig. 1.

It is important to realise that the U-value is not a unique value for a particular structure, but is the heat transmitted from air on one side to air on the other side of unit area of the structure in unit time, per unit of temperature difference between the air on each side. Any variation, therefore, in the properties of the materials forming the structure which causes a different quantity of heat to flow will be reflected in a variation in the U-value. In addition the thermal capacity of the construction will influence the fluctuations in U-value brought about by changes in outside or inside air temperatures.

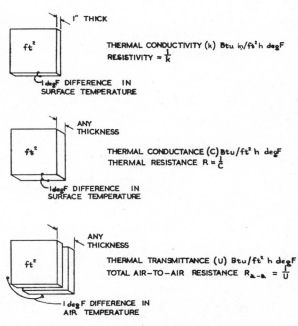

Fig. 1 **Terms used.**

The Inadequacies of Present Published Information

The thermal conductivities of many building materials are given in publications such as the Institution of Heating and Ventilating Engineers Guide[1] and the Building Research Stations' Book, "The Thermal Insulation of Buildings"[2], and from such information and values for surfaces and air space resistance, U-values can be calculated for many forms of constructions.

Most of the information given on the thermal properties of materials are the results of laboratory determined conductivity values obtained on prepared specimens of the materials. The conductivity value quoted is sometimes for an oven dry sample, sometimes for a sample conditioned in an atmosphere of 64°F and 65% RH, sometimes for a fully saturated sample and sometimes the conditioning of the sample is not specified at all.

Clearly a conductivity value of an oven dry sample is not appropriate for porous building materials; the conductivity of a sample conditioned at 64°F and 65% RH may not be appropriate for a material protected from the weather, such as the inner leaf of a cavity wall, and is certainly not appropriate for a material directly exposed to the weather, such as the outer leaf of a cavity wall; the conductivity of a fully saturated sample is also not appropriate.

Because of this, considerable care must be taken in choosing the appropriate information for calculating U-values, if these U-values obtained are to be realistically representative of the construction in practice.

Experimental Work—Thermal Transmittance Laboratory

The London Brick Company have carried out investigations into the thermal properties of various wall constructions, using both clay and concrete products on panels exposed to the natural environment. Tests of this kind take into account all the effects of the external environment, including the effect of the moisture on the thermal properties of the materials. The experimental technique has been described in detail elsewhere[3] and it is only necessary for the purpose of this Note to make brief reference to it.

Walls 4-ft long and 4-ft high, representative of the construction to be studied, are built and inserted into openings in the walls of the Thermal Transmittance Laboratory illustrated in Fig. 2. Heat is supplied from a guarded hotplate attached to the inside surface of the wall, and by means of thermocouples, the quantity of the heat passing into the wall and the temperature gradients set up across the wall as a whole and also across the components of the wall i.e.

inner leaf, cavity, outer leaf, etc.—are measured at half hourly intervals over the six winter months heating season. The temperature differences measured in the case of a cavity wall are illustrated in Fig. 3.

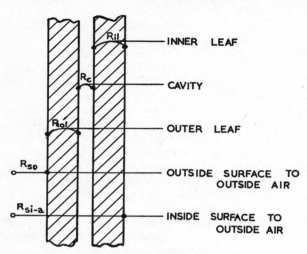

Fig. 3 **Temperature differences measured.**

It is not possible to measure the temperature difference between the inside air and the inside surface of the wall because of the proximity of the hot-plate. Thus it is not possible to calculate the thermal resistance of the inside surface and an accepted value of 0·7 is used in the calculation of the U-value.

Since thermocouples are used to measure the heat flow and temperature gradients, all the information is in the form of small electrical voltages which are recorded on a Potentiometric Recorder Integrator. The record is in chart form and digital form, the latter being photographed automatically at pre-selected intervals of time. An interior view of the Laboratory is shown in Fig. 4.

From the results obtained, daily resistance values of the components and U-values of the walls are calculated. The average values obtained for the six months winter heating season are taken to represent the thermal performance of the materials when used for similar positions in actual constructions.

Additional experiments have been carried out at different times to determine the moisture content of some building materials in various conditions of temperature and humidity and also the moisture content of brick and light-weight concrete in the inner and outer leaves of cavity constructions.

Results

(a) Seasonal mean Values

Measurements of the thermal performance of various constructions have been made since 1955. The mean seasonal results of the resistance of the components of each construction and the U-values are given in Table 1.

To illustrate the fluctuations in daily mean U-values, a typical record for three constructions is shown in Fig. 5.

Although Table 1 contains a large number of types of constructions, the information is more useful if presented separately for the various materials used in different positions so that the thermal properties of any type of construction could be examined by combining the relevant data from these results. Tables 2A and 2B show the average thermal resistance results for materials with respect to the position used in constructions. Also shown in these tables are the corresponding thermal conductance and, where appropriate, thermal conductivity. The corresponding metric units are given

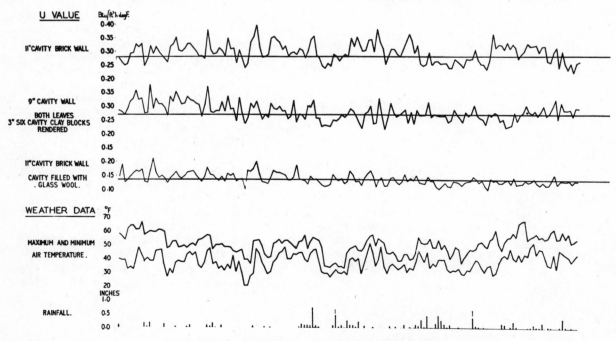

Fig. 5 **Chart showing fluctuations in Daily Mean U-values.**

for these average values. Results for north and south facing walls have been separated.

Conductivity values are only relevant to homogeneous materials and are therefore not given for hollow units.

(b) Moisture Content

All moisture content results are expressed on a basis of percentage by volume of dry material and are shown in graphical form in Figs. 6, 7 and 8. Figs. 6 and 7 give the results obtained on some fired clay and concrete products exposed to various conditions of temperature and humidity after first being oven dried (Fig. 6) and then saturated with water (Fig. 7). The results given in the graph in Fig. 8 were obtained from samples taken from the inner and outer leaves of two cavity constructions. Each curve is the plot of the average of the results for four samples.

Discussion of Measured Values

The results presented in detail in Table 1 are largely self-explanatory but there are a few points which merit consideration.

The thermal transmittance laboratory has been designed to enable test panels to be exposed as north facing, south facing, east facing, or west facing. Most of the experimental work has been done on north facing walls since walls in such an exposure receive no direct solar radiation. A few walls have been studied with south facing exposure.

(a) North Exposure

It will be noted (Table 1) that wall No. 4 has been tested throughout all the seasons and the variability of the thermal resistance of the inner leaf and the unventilated cavity is quite small. The outer leaf does show more variable results through the years, but then this no doubt reflects the variation in moisture content over different winters.

From Table 2A it can be seen that the thermal conductivity of common brickwork on the inner leaf position averages 4·89 Btu in/ft² h degF and that of facing bricks in the outer leaf position averages 7·09 Btu in/ft² h degF. These conductivities should be compared with the value of dry bricks measured at the National Physical Laboratory of 3·8–4·0 Btu in/ft² h degF.

It should be noted that it is possible to obtain some very low U-values. The values of 0·11 to 0·16 were obtained by using various forms of light weight insulating materials inserted in the cavity of normal 11-in brick cavity constructions (walls No. 2A, 19A, 21, 21C, 22, 22B, 24 and 31). Resin bonded glass wool and mineral wool enable U-values of 0·14 and 0·11 respectively to be obtained from normal brick cavity constructions. The resin bonded glass wool, wall No. 22, consisted of 1-in semi-rigid boards held against the inner leaf by a special metal spring clip. This arrangement enabled a 1-in air cavity to be maintained between the inside surface of the outer

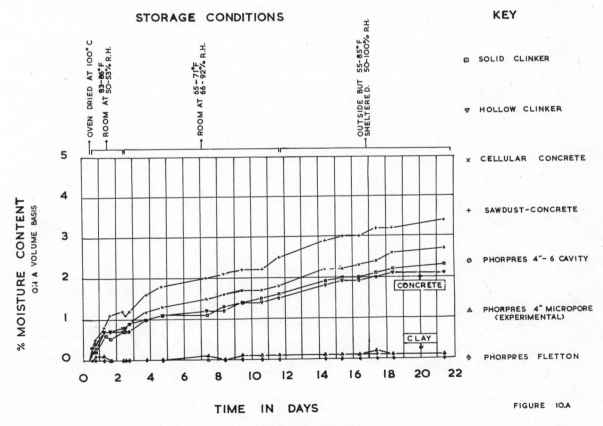

Fig. 6 **Regain of moisture of clay and concrete products.**

leaf and the insulating material, so avoiding the possibility of any rain penetration through the insulating material. A 1-in thick polystyrene board used in a similar position to the 1-in thick glass wool board gave a U-value of 0·16 (wall 21). Rockwool was blown into the cavity and filled its complete width in walls Nos. 24 and 31. No rain penetration was observed on these test walls. The plastic foams such as Ureaformaldehyde which also fill the cavity completely produce similar low U-values e.g. wall No. 22B.

Special mention must be made of plaster board dry lining on the internal surface of the walls. When aluminium-foil backed plasterboard was used and battened out on $\frac{7}{8}$-in wooden battens on the internal surface, U-values of the order of 0·14 were obtained on normal 11-in cavity construction (walls 19B, 19C and 21B).

(b) South Exposure

South facing wall results are of particular interest since it is thought that no previous measurements have been published for this exposure. Wall 2 (Table 1) has been tested in both the north and south facing exposures and it can be seen that in general the U-value for south exposure is less than for the north exposure. The average thermal resistances (Table 2B) of the inner leaf (0·90 south and 0·87 north) and outer leaf (0·58 south and 0·60 north) appears not to

be vastly different for the two exposures, but most of the increase in thermal resistance for the south facing wall is due to the thermal resistance of the external surface which has changed from 0·25 for the north exposure to the value varying from 0·48 to 0·94 for the south exposure (see Table 1).

It is interesting to reflect upon the cause of this increase in the external thermal resistance on the south facing wall. Unlike the north exposure, the south wall receives direct solar radiation which causes the temperature of the external surface of the outer leaf to increase. This creates a high temperature difference between the surface of the wall and the surrounding air and since the thermal resistance is calculated by dividing the temperature difference by the recorded heat input from the guarded hotplate attached to the inside surface of the wall, a high surface resistance is obtained.

A similar comparison between north and south facing exposure for the same wall was obtained for wall No. 14 where again a lower U-value and a higher external surface resistance was obtained on the south facing wall. These remarks also apply to wall 29.

Wall 31 with the high insulation obtained from the rockwool fill did not show any change in the U-value for the two directions of exposure but an increase in the external surface resistance was still noted for the south facing wall.

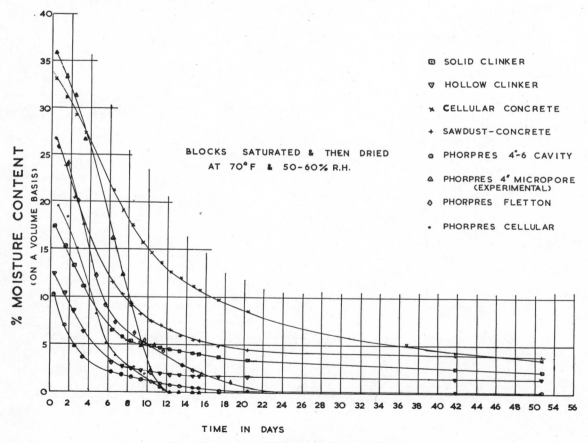

Fig. 7 **Drying rate of clay and concrete products.**

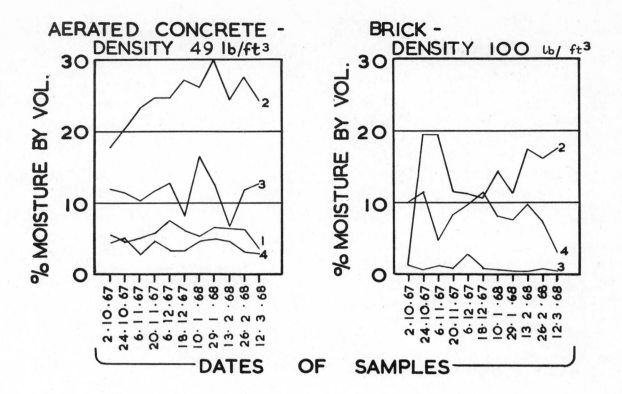

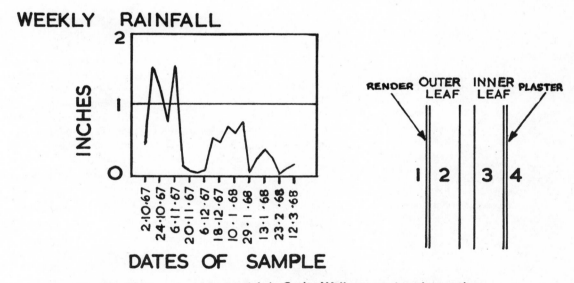

Fig. 8 **Moisture content of materials in Cavity Walls exposed to the weather**

It would seem therefore that as far as the materials used to construct the nner and outer leaf and light weight insulating maitrials used in the cavity are concerned, similar thermal resistance and conductivity values are obtained on the north and south facing walls, but the thermal resistance on the external surface of the south facing walls is greater than the corresponding value for north facing walls. This increase in the surface resistance tends to lower the U-value except that it becomes somewhat insignificant in the highly insulated walls.

(c) **Moisture Content**

All porous materials will absorb moisture from the atmosphere. In fact such materials will reach a stable equilibrium moisture content, depending upon the condition of the environmental atmosphere. Even in dry warm conditions some moisture vapour is present in the atmosphere and this causes porous building materials to reach some equilibrium moisture content. This point is illustrated in Fig. 6 which shows the regain of moisture for fired clay and concrete products exposed to various conditions of atmosphere after being oven dried for 72 hours at 100°C. At no time was liquid water admitted to the products. All cement based products attained an equilibrium moisture content above 2% (by volume) when stored in a sheltered outside position, whilst the fired clay products had an equilibrium moisture content of about 0·2% (by volume).

Similar results were obtained when the products were allowed to dry out in a warm dry atmosphere after being saturated with water, as shown in Fig. 7.

During the recent winter (1967/68) the moisture content of two cavity wall constructions was investigated by cutting 2-in diameter cores from the inner leaf and outer leaf and drying the cores to a constant weight in a ventilated oven at 100°C. The two walls investigated, situated in the south facing wall of the Thermal Transmittance Laboratory, were an 11-in cavity brick construction plastered internally with a 1:1:6 cement : lime : sand mix, and a similar cavity construction using aerated concrete blocks (49 lb/ft³) in both leaves, rendered and plastered, also with a 1:1:6 mix. Typical results are shown in Fig. 8 where each curve is the plot of the average of four samples.

It is interesting to note that the moisture content (percentage by volume) of the render (curve 1) and outer leaf (curve 2) of the aerated concrete block wall are very dissimilar, the outer leaf being about 25%. The inner leaf (curve 3) of the same wall being about 10% is drier than the outer leaf. The outer unrendered leaf (curve 2) and inner leaf (curve 3) of the brick wall were drier than the corresponding leaves of the aerated concrete block wall. Changes in the moisture content of the outer leaf reflected closely the rainfall whilst the brick inner leaf remained relatively dry at about 1% moisture content (by volume). Surprisingly however, the plaster (curve 4) on this leaf has a relatively high moisture content of between 5 and 10%.

Taken collectively the results shown in Figs 6, 7 and 8 demonstrate that for similar conditions of environment fired clay products attain a lower equilibrium moisture content than cement based products.

The effect of moisture content on the thermal properties of the products has been the subject of study and comment by other workers[4 5 6 7 8 9 10 & 11]

Measurement of the thermal conductivity of porous material containing some moisture is complicated by the migration of moisture to the cold surface. There is some merit, therefore, in observing the thermal properties of materials as used in practice without necessarily determining the moisture content.

Nevertheless, where moisture contents are known together with the appropriate conductivity value of the material in its moist condition, it is desirable to be able to convert this value to provide estimates of the conductivity at other moisture contents. Jacob[11] proposed a relationship, which is summarised in Table 3, for the effect of moisture content on the average thermal conductivity of inorganic building materials. Reference to Jacob's proposed relationship is made by Pratt[4] and Loudon and Stacy[5]. The latter reference states that work shows that errors greater than ±15% may apply for particular materials but nevertheless the relationship is useful for rough estimates of thermal conductivity of masonry materials.

Although the I.H.V.E. Guide[1] gives many conductivities and moisture contents for building and insulating materials there is an acute lack of information on the moisture contents attained by the materials in use. As indicated by the results in Fig 8 we probably have little idea of the actual moisture contents in practice and one is unable to make estimates of values which are likely to be realistic. This is clearly an area for further research.

General Comment

The study of thermal properties of building materials and thermal performance of structures as a whole can cover a very wide field, particularly when the effects of our varying climate are taken into account, as of course they should be. It is not possible to deal with all these matters in detail in this Technical Note hence the emphasis has been directed towards giving a general picture of the various aspects which are involved, indicating the importance of the effect of moisture on the thermal properties, and giving some measured values of the appropriate thermal conductivity, conductance and resistance values relating to building and insulating materials as they are used in wall constructions.

It must be emphasised however that the thermal performance of the building is of course only one of its functional requirements and the materials must be selected so that their properties meet all the functional requirements. It may not always be possible to realise the optimum thermal performance when considering all the other aspects. In this respect it is considered that fired clay brickwork and blockwork together with other highly insulating materials can provide a very satisfactory structure with desirable thermal properties and at the same time satisfy the other functional requirements of buildings particularly in respect of strength, durability, appearance and low maintenance.

TABLE 1

Mean Seasonal Results for North Facing and South Facing Walls

Wall No.	Test Season	Plaster	Inner Leaf	R_{il}	Cavity	R_c	Outer Leaf	R_{ol}	Render	$R_{os\text{-}oa}$	Sum of component resistances	$R_{is\text{-}oa}$	U Value
1 N	55/56	Yes	Fletton cellular brick	1·03	Air	0·92	Fletton cellular brick	0·76	—	—	—	2·81	0·29
1 N	58/59	Yes	Fletton cellular brick	1·05	Air	0·92	Fletton cellular brick	0·64	—	0·19	2·80	2·71	0·29
1A N	57/58	Yes	Fletton cellular brick	1·16	Ceramic fill 2⅝-in	2·14	Fletton cellular brick	0·60	—	—	—	4·09	0·21
2 N	55/56	Yes	Fletton common brick	0·91	Air	1·00	Fletton facing brick	0·71	—	0·25	2·70	2·90	0·28
2 N	58/59	Yes	Fletton common brick	0·88	Air	0·90	Fletton facing brick	0·67	—	0·89	3·05	2·87	0·28
2 S	59/60	Yes	Fletton common brick	0·82	Air	0·83	Fletton facing brick	0·51	—	0·88	3·59	3·15	0·27
2 S	61/62	Yes	Fletton common brick	1·00	Air	1·04	Fletton facing brick	0·67	—	0·94	3·65	3·63	0·23
2 S	62/63	Yes	Fletton common brick	1·02	Air	1·04	Fletton facing brick	0·65	—	—	—	—	0·23
2 S	66/67	Yes	Fletton common brick	0·92	Air	1·01	Fletton facing brick	0·77	—	0·48	3·18	3·17	0·26
2A N	56/57	Yes	Fletton common brick	0·81	Resin bonded glass wool 2¾-in 3·5 lb/ft³	5·21	Fletton facing brick	0·48	—	0·14	6·64	7·41	0·14
3 N	55/56	Yes	Fletton cellular brick	0·92	Air	0·92	Fletton cellular brick	0·83	—	—	—	2·77	0·29
3A N	60/61	Yes	Fletton cellular brick	0·91	Vermiculite fill 2·4-in 7·3 lb/ft³	4·06	Fletton cellular brick	0·59	—	0·15	5·71	5·28	0·17
4 N	55/56	Yes	Fletton common brick	0·97	Air	0·94	Fletton facing brick	0·80	—	0·24	2·95	3·00	0·27
4 N	56/57	Yes	Fletton common brick	0·98	Air	0·86	Fletton facing brick	0·52	—	0·22	2·58	2·82	0·28
4 N	57/58	Yes	Fletton common brick	0·97	Air	0·97	Fletton facing brick	0·62	—	—	—	2·84	0·28
4 N	58/59	Yes	Fletton common brick	0·91	Air	0·92	Fletton facing brick	0·54	—	—	—	2·71	0·29
4 N	59/60	Yes	Fletton common brick	0·89	Air	0·90	Fletton facing brick	0·57	—	—	—	2·65	0·30
4 N	60/61	Yes	Fletton common brick	0·91	Air	0·95	Fletton facing brick	0·55	—	—	—	2·76	0·29
4 N	61/62	Yes	Fletton common brick	0·92	Air	0·98	Fletton facing brick	0·54	—	—	—	2·75	0·29
4 N	62/63	Yes	Fletton common brick	0·93	Air	0·96	Fletton facing brick	0·55	—	—	—	2·72	0·29
4 N	63/64	Yes	Fletton common brick	0·96	Air	0·91	Fletton facing brick	0·65	—	0·27	2·76	2·75	0·29
4 N	65/66	Yes	Fletton common brick	0·93	Air	0·93	Fletton facing brick	0·63	—	0·23	3·05	3·04	0·29
4 N	66/67	Yes	Fletton common brick	0·98	Air	1·03	Fletton facing brick	0·81	—	0·25	2·75	2·75	0·27
4 N	67/68	Yes	Fletton common brick	0·92	Air	0·95	Fletton facing brick	0·63	—	—	—	—	0·29
5 N	55/56	Yes	"Phorpres" 4-in building block	1·68	Air	0·91	Fletton facing brick	0·72	—	0·22	3·53	3·91	0·22
5 N	62/63	Yes	"Phorpres" 4-in building block	1·73	Air	0·97	Fletton facing brick	0·73	—	0·33	3·76	3·75	0·23
5 N	63/64	Yes	"Phorpres" 4-in building block	1·67	Air	0·91	Fletton facing brick	0·62	—	0·30	3·50	3·48	0·24
6 N	55/56	Yes	"Phorpres" 3-in building block	1·37	Air	0·88	Fletton facing brick	0·59	—	0·29	3·13	3·17	0·26
7 N	56/57	Yes	"Phorpres" 3-in building block	1·02	Air	0·83	"Phorpres" 3-in building block	0·71	Yes	0·25	2·81	2·93	0·28
7 N	58/59	Yes	"Phorpres" 3-in building block	1·14	Air	0·89	"Phorpres" 3-in building block	0·89	Yes	0·23	3·15	3·53	0·24
8 N	56/57	Yes	"Phorpres" 6-in building block	2·10	—	—	"Phorpres" 3-in building block	—	Yes	0·30	2·40	2·48	0·31
9 N	56/57	Yes	Experimental 9-in hollow block	—	—	—	—	—	—	—	—	2·76	0·29
10 N	57/58	Yes	"Phorpres" experimental	1·77	Air	0·94	Fletton facing brick	0·53	—	0·21	3·45	3·63	0·23
11 N	57/58	Yes	10-in through the wall unit Continental design	3·45	—	—	—	—	Yes	0·21	3·66	3·90	0·22
12 N	57/58	Yes	Cellular concrete 4-in block 55 lb/ft³	2·17	Air	0·89	Fletton facing brick	0·74	—	0·26	4·06	3·81	0·22

continued overleaf

TABLE 1 (*continued*)

Wall No.		Test Season	Plaster	Inner Leaf	R_{il}	Cavity	R_c	Outer Leaf	R_{ol}	Render	$R_{os\text{-}oa}$	Sum of component resistances	$R_{is\text{-}oa}$	U Value
13	N	57/58	Yes	Hollow clinker concrete 4-in block, density 54 lb/ft³	1·12	Air	1·01	Fletton facing brick	0·57	—	0·23	2·93	3·15	0·26
14	N	58/59	Yes	"Phorpres" 4-in building block	1·31	Air	0·96	"Phorpres" 4-in building block	1·08	Yes	0·21	3·56	3·71	0·23
	S	61/62	Yes	"Phorpres" 4-in building block	1·52	Air	0·98	"Phorpres" 4-in building block	1·25	Yes	0·70	4·45	4·47	0·19
15	N	58/59	Yes	Experimental 9-in block, 2⅝-in height	2·37	—	—	—	—	Yes	0·13	2·50	2·55	0·31
16	N	59/60	Yes	Experimental 9-in block, 2⅝-in height	2·34	—	—	—	—	—	0·14	2·48	2·56	0·31
17	N	59/60	Yes	Experimental 9-in block, 8⅝-in height	2·97	—	—	—	—	—	—	—	3·11	0·26
18	N	59/60	Yes	Solid clinker concrete block, density 65 lb/ft³	·0·89	Air	0·84	Fletton facing brick	0·59	—	0·14	2·46	2·67	0·30
19	N	59/60	Yes	Fletton common brick	0·70	Air with ½-in polystyrene board against inner leaf	2·40	Fletton facing brick	0·53	—	0·14	3·97	3·92	0·22
19A	N	60/61	Yes	Fletton common brick	0·73	Polysytrene beads (0·5–1·0 lb/ft³)	5·31	Fletton facing brick	0·56	—	0·20	6·80	6·85	0·13
19B	N	61/62	—	Fletton common brick with foil backed plasterboard on battens	R_{il}=0·84 R_{pb}=4·01	Air	0·95	Fletton facing brick	0·60	—	0·22	6·62	6·63	0·16
19C	N	62/63	—	Fletton common brick with foil backed plasterboard on battens	R_{il}=0·88 R_{pb}=3·76	Air	0·92	Fletton facing brick	0·60	—	0·42	6·58	6·57	0·14
20	N	60/61	Yes	Fletton common and facing bricks (frog-up) in Flemish bond as 13½-in solid single leaf wall ($R_{13½}$=1·90)	—				—	—	0·30	2·34	2·36	0·33
21	N	60/61	Yes	Fletton common brick	—	Air with 1-in polystyrene board against inner leaf (R_{poly}=3·10)	4·21	Fletton facing brick	0·53	—	0·16	5·53	5·44	0·16
21A	N	61/62	—	Fletton common brick with plasterboard on battens	R_{il}=0·82 R_{pb}=1·45	Air	0·91	Fletton facing brick	0·60	—	0·18	3·96	3·51	0·24
21B	N	62/63	—	Fletton common brick with foil backed plasterboard	R_{il}=0·90 R_{pb}=3·61	Air	0·92	Fletton facing brick	0·63	—	0·39	6·42	6·42	0·14
	N	63/64	—	Fletton common brick with foil backed plasterboard	R_{il}=1·02 R 3·47	Air	1·03	Fletton facing brick	0·75	—	0·37	6·64	6·33	0·14
21C	N	67/68	Yes	Fletton common brick	6	Ureaformaldehyde foam	5·56	Fletton facing brick	0·44	—	0·06	6·72	6·86	0·13
22	N	60/61	Yes	Fletton common brick	—	Air with 1-in glass fibre board against inner leaf	4·67	Fletton facing brick	0·55	—	0·18	6·23	6·13	0·14
22A	N	61/62	Yes	Fletton common brick	0·91	Air	1·04	Fletton facing brick	0·62	—	0·28	2·85	2·89	0·28
22B	N	62/63	Yes	Fletton common brick	0·92	Ureaformaldehyde foam	4·86	Fletton facing brick	0·59	—	0·40	6·77	6·67	0·14
24	N	65/66	Yes	Fletton common brick	0·73	Rockwool fill 2½-in	7·45	Fletton facing brick	0·46	—	0·01	8·65	8·55	0·11
	N	66/67	Yes	Fletton common brick	0·81	Rockwool fill 2½-in	7·48	Fletton facing brick	0·46	—	0·06	8·69	8·69	0·11
	S	67/68	Yes	Fletton common brick	0·82	Rockwool fill 2½-in	6·52	Fletton facing brick	0·29	—	0·57	8·20	8·22	0·11
25	N	65/66	—	Foil back plasterboard	3·08	—	—	"Phorpres" 6-in building block	1·59	Yes	0·28	4·95	4·95	0·18

No.	Exp.	Period	Inner leaf	R_{il}	Cavity	R_c	Outer leaf	R_{ol}	Air	R_{os-oa}			
27	N	65/66	Foil backed plasterboard on 4-in × 2-in studs	2·49	—	—	"Phorpres" Claywall bricks (hollow)	0·47	—	0·21	2·74	3·15	0·26
	N	66/67	Foil backed plasterboard on 4-in × 2-in studs	2·63	—	—	"Phorpres" Claywall bricks (hollow)	0·86	—	0·19	3·68	3·21	0·26
28	N	66/67	4-in in situ concrete, 115 lb/ft³	1·10	1-in air 1-in polystyrene board against inner leaf	0·61 R_{poly}=2·57	Fletton facing brick	0·64	—	—	—	4·96	0·18
29	N	66/67	4-in cellular concrete block, 45 lb/ft³	1·94	Air	0·72	Cellular concrete block 45 lb/ft³	1·18	Yes	0·14	3·98	3·93	0·22
	S	67/68	4-in cellular concrete block, 45 lb/ft³	2·15	Air	0·63	Cellular concrete block, 45 lb/ft³	1·17	Yes	0·52	4·47	4·43	0·20
30	N	66/67	L.B.C. experimental block 56 lb/ft³	1·51	Air	0·82	Fletton facing brick	0·57	—	0·14	3·04	3·02	0·27
31	S	66/67	Fletton common brick	0·87	Rockwool fill 2 7/16-in	4·75	Fletton facing brick	0·55	—	0·59	6·76	6·69	0·14
	N	67/68	Fletton common brick	0·93	Rockwool fill 2 7/16-in	4·80	Fletton facing brick	0·55	—	0·20	6·48	6·69	0·14
32	N	67/68	Experimental polyurethane 9/16-in board on 3/8-in plasterboard on	5·30	—	—	Fletton Claywall bricks	0·56	—	0·02	5·84	6·05	0·15
33	N	67/68	Foil backed plasterboard on 4-in × 2-in studs	3·45	—	—	Fletton Claywall bricks	0·51	—	0·18	4·14	4·33	0·20
34	N	67/68	L.B.C. experimental block 56 lb/ft³	1·85	Air	0·95	Fletton facing brick	0·62	—	0·21	3·63	3·46	0·24

Note: Fletton common, facing cellular bricks laid frog down, except for wall 20

KEY TO TABLE 1

N   = North Exposure.
S   = South Exposure.
R_{is-oa} = thermal resistance of inside surface to outside air.
R_{il} = thermal resistance of inner leaf.
R_c = thermal resistance of cavity between leaves including any material in the cavity.

R_{ol} = thermal resistance of outer leaf.
R_{os-oa} = thermal resistance of outside surface to outside air.
R_{poly} = thermal resistance of polystyrene board.
R_{pb} = thermal resistance of plasterboard and the air space between plasterboard and inner leaf.

Air   = unventilated air cavity between leaves, 2-in–2½-in wide. For the purpose of this Note an unventilated air cavity is defined as a cavity between two leaves which is closed at the top of the wall and is not open to the atmosphere (no air bricks into cavity) other than by weep holes.

141

TABLES 2A AND 2B

Average Thermal Resistance, Conductance and Conductivity of Materials

TABLE 2A—North Facing Walls

Material	Position	Thickness	Test Season	No. of results	Resistance ft² h degF/Btu	Resistance m² degC/W (1 = 0·176)	Conductance Btu/ft² h degF	Conductance W/m² degC (1 = 5·678)	Conductivity Btu in/ft² h degF	Conductivity W/m degC (1 = 0·144)
Fletton facing bricks	Outer leaf	4 3/16-in	55/56–67/68	41	0·60	0·106	1·70	9·65	7·09	1·02
Fletton common bricks	Inner leaf	4 3/16-in	55/56–67/68	30	0·87	0·153	1·16	6·59	4·86	0·70
Fletton cellular bricks	Outer leaf	4 3/16-in	55/56–60/61	5	0·68	0·120	1·49	8·46	N.A.	N.A.
Fletton cellular bricks	Inner leaf	4 3/16-in	55/56–60/61	5	1·01	0·178	0·99	5·62	N.A.	N.A.
"Phorpres" 4-in building blocks	Outer leaf	4-in	58/59	1	1·08	0·190	0·93	5·28	N.A.	N.A.
	Inner leaf	4-in	55/56–63/64	4	1·60	0·282	0·64	3·63	N.A.	N.A.
"Phorpres" 3-in building blocks	Outer leaf	3-in	56/57–58/59	2	0·80	0·141	1·27	7·21	N.A.	N.A.
	Inner leaf	3-in	55/56–58/59	3	1·18	0·228	0·87	4·94	N.A.	N.A.
"Phorpres" 6-in building blocks	Outer leaf	6-in	65/66	1	1·59	0·280	0·63	3·58	N.A.	N.A.
	Inner leaf	6-in	56/57–63/64	2	1·73	0·302	0·58	3·28	N.A.	N.A.
4-in clinker concrete blocks, hollow 54 lb/ft³	Inner leaf	4-in	57/58	1	1·12	0·197	0·89	5·05	4·48	0·645
4-in clinker concrete blocks, solid 65 lb/ft³	Inner leaf	4-in	59/60	1	0·89	0·157	1·12	6·35		
L.B.C. 4-in experimental blocks	Inner leaf	4-in	57/58–67/68	2	1·81	0·319	0·55	3·12	N.A.	N.A.
L.B.C. 4-in experimental blocks	Inner leaf	4-in	66/67	1	1·51	0·266	0·66	3·75	N.A.	N.A.
4-in cellular concrete block, 45 lb/ft³	Outer leaf	4-in	66/67	1	1·18	0·208	0·85	4·83	3·40	0·49
	Inner leaf	4-in	66/67–67/68	1	1·94	0·341	0·52	2·95	2·08	0·30
4-in cellular concrete block, 55 lb/ft³	Inner leaf	4-in	57/58	1	2·17	0·382	0·46	2·61	1·84	0·26
4-in in situ concrete, 115 lb/ft³	Inner leaf	4-in	66/67	1	1·10	0·194	0·91	5·17	3·64	0·52
"Phorpres" Claywall bricks (hollow unit)	Outer leaf	3-in	65/66–66/67	2	0·67	0·118	1·70	9·65	N.A.	N.A.
Fletton Claywall bricks	Outer leaf	3-in	67/68	2	0·54	0·095	1·87	10·62	5·61	0·81
Experimental 8-in hollow block	Single leaf (rendered)	8-in	56/57	1	2·76	0·486	0·36	2·04	N.A.	N.A.
10-in through-the-wall unit, Continental type	Single leaf (rendered)	10-in	57/58	1	3·45	0·607	0·29	1·65	N.A.	N.A.
Experimental 9-in block, 2 5/8-in height	Single leaf	9-in	58/59–59/60	2	2·35	0·413	0·42	2·38	N.A.	N.A.
Experimental 9-in block, 8 5/8-in height	Single leaf	9-in	59/60	1	2·97	0·523	0·34	1·93	N.A.	N.A.
Fletton common and facing bricks laid frog-up as 13 1/2-in solid wall	Single leaf	13 1/2-in	60/61	1	2·04	0·359	0·49	2·78	6·44	0·93
Tile hanging on battens	Outer leaf		65/66	1	1·35	0·238	0·74	4·20	N.A.	N.A.

Walling

Cavity insulation / Dry lining

Insulation	Material	Element	Thickness	Date	No.	(1)	(2)	(3)	(4)	(5)	(6)
Cavity insulation	Ceramic fill (experimental)	Cavity	2⅝-in	57/58	1	2·14	0·377	0·47	2·67	1·23	0·177
Cavity insulation	Resin bonded glass wool, 3·5 lb/ft³	Cavity	2¾-in	56/57	1	5·21	0·917	0·19	1·08	0·52	0·075
Cavity insulation	Vermiculite fill, 7·3 lb/ft³	Cavity	2-4-in	60/61	1	4·06	0·715	0·25	1·42	0·60	0·086
Cavity insulation	Polystyrene board ½-in thick held against inner leaf + 1½-in cavity	Cavity		59/60	1	2·40	0·422	0·42	2·38	N.A.	N.A.
Cavity insulation	1-in cavity + 1-in polystyrene board against inner leaf	Cavity		60/61	1	4·21	0·741	0·24	1·36	N.A.	N.A.
Cavity insulation	1-in cavity	Cavity	1-in	66/67	1	0·61	0·107	1·64	9·31	N.A.	N.A.
Cavity insulation	1-in polystyrene board against inner leaf	Cavity		60/61 66/67	2	2·84	0·500	0·35	2·01	0·35	0·051
Cavity insulation	Polystyrene beads, 0·5-1·0 lb/ft³	Cavity	2-2½-in	60/61	1	5·31	0·935	0·19	1·08	N.A.	N.A.
Cavity insulation	1-in cavity + 1-in glass fibre board against inner leaf	Cavity		60/61		4·67	0·822	0·21	1·19	N.A.	N.A.
Cavity insulation	Rockwool fill, 6·75 lb/ft³ / Rockwool fill, 5·50 lb/ft³	Cavity / Cavity	2½-in / 2 1/16-in	65/66-66/67 / 67/68	2 / 1	7·47 / 4·80	1·320 / 0·840	0·13 / 0·21	0·91 / 1·25	0·32 / 0·43	0·052 / 0·062
Cavity insulation	Ureaformaldehyde foam	Cavity	2½-in	67/68	1	5·56	0·979	0·18	1·02	0·45	0·065
Dry lining	Experimental polyurethane 7/16-in board	Internal lining	1-in appr.	67/68	1	5·30	0·933	0·19	1·08	0·19	0·027
Dry lining	Plasterboard on 7/8-in wooden battens	Internal lining		61/62	1	1·45	0·255	0·69	3·92	N.A.	N.A.
Dry lining	Foil backed plasterboard on 7/8-in wooden battens	Internal lining		61/62-65/66	5	3·59	0·632	0·28	1·57	N.A.	N.A.
Dry lining	Foil backed plasterboard on 4-in × 2-in studding	Internal lining		65/66-67/68	3	2·82	0·490	0·35	1·96	N.A.	N.A.

TABLE 2B—South Facing Walls

	Material	Element	Thickness	Date	No.	(1)	(2)	(3)	(4)	(5)	(6)
Walling	Fletton facing bricks	Outer leaf	4 3/16 & 4⅛-in	59/60-67/68	7	0·58	0·102	1·88	10·67	7·80	1·12
Walling	Fletton common bricks	Inner leaf	4 3/16 & 4⅛-in	59/60-67/68	7	0·90	0·158	1·12	6·36	4·65	0·67
Walling	"Phorpres" 4-in building blocks	Outer leaf / Inner leaf	4-in / 4-in	61/62 / 61/62	1 / 1	1·25 / 1·52	0·220 / 0·268	0·80 / 0·66	4·54 / 3·75	N.A. / N.A.	N.A. / N.A.
Walling	Cellular concrete block, 45 lb/ft³	Outer leaf / Inner leaf	4-in / 4-in	67/68 / 67/68	1 / 1	1·17 / 2·15	0·206 / 0·378	0·85 / 0·47	4·83 / 2·67	3·40 / 1·86	0·49 / 0·27
Cavity insulation	Rockwool fill, 5·50 lb/ft³ / Rockwool fill, 6·75 lb/ft³	Cavity / Cavity	2 1/16-in / 2½-in	66/67 / 67/68	1 / 1	4·75 / 6·52	0·836 / 1·148	0·21 / 0·15	1·19 / 0·85	0·43 / 0·37	0·062 / 0·054

Notes: Fletton common, facing and cellular bricks laid frog down unless stated otherwise. N.A. = not applicable

TABLE 3
(Jacob's relationship for moist materials)

Moisture content (m) by volume %	0	1	2·5	5	10	15	20	25
Moisture factor F	1·0	1·3	1·5	1·75	2·10	2·35	2·55	2·75

Thermal conductivity (k_m) at a particular moisture content is modified to the thermal conductivity (k_0) at zero moisture content by dividing the moisture factor F, e.g.

$$k_0 = \frac{k_{2·5}}{1·5}$$

To translate to new moisture content, k_0 is multiplied by the appropriate value of F e.g. $k_{10} = k_0 \times 2.10$.

Acknowledgments

The work carried out in the Thermal Transmittance Laboratory forms part of the work of the Technical and Research Department of the London Brick Company Limited, and is published by permission of the Directors of the Company.

The Authors wish to acknowledge their thanks to Mr. T. G. W. Boxall, O.B.E., formerly Chief Technical Officer (now retired) who was responsible for initiating the work and who gave much valuable encouragement and advice, and to their colleagues at the Research Laboratories for their help and assistance.

References

1. The Institute of Heating and Ventilating Engineers Guide 1965.
2. Nash, G. D., Comrie, J., and Broughton, H. F. "The Thermal Insulation of Buildings." D.S.I.R. Building Research Station, 1955.
3. Dinnie, A., Beard, R., and Richards, R. "Measurements of the Thermal Transmittance of External Walls by a Guarded Hotplate Method." Trans. Brit. Ceram. Soc. Vol. 58, No. 2, Feb. 1959.
4. Pratt, A. W. "Heat Transfer in Porous Materials," Research Vol. 15, May 1962.
5. Loudon, A. G., Stacy, E. F. "The Thermal and Acoustic Properties of Lightweight Concretes." Journal of the Reinforced Concrete Association Vol. 3, No. 2, March/April 1966.
6. Jesperson, H. B. "Thermal Conductivity of Moist Materials and its Measurements." Journal I.H.V.E. Vol. 21, 1953–4.
7. Building Research Congress. Journal I.H.V.E. Vol. 20, April 1952.
8. Gurst, Evan, and Zuilen, D. Van. "The Influence of Moisture Content on the Thermal Conductivity of Building Materials."
9. Roux, A. J. A. "The effect of Weather Conditions on Heat Transfer through Building Elements."
10. Cammerer, J. S. "The Effect of Moisture on Heat Transmission through Building and Insulating Materials." Wärme and Kältetechnild No. 9, Sept. 1939.
11. Jacob, M. "Heat Transfer, Part One". London Chapman and Hall, 1949.
12. Billington, N. S. "Thermal Properties of Buildings." Cleaver-Hume Press Ltd., 1952.
13. Building Research Station Digest No. 35. (First Series), October 1951. "Heat Loss from Dwellings."
14. Building Research Station Digest No. 94. (First Series), November 1956. "Domestic Heating Estimation of Seasonal Heat Requirements and Fuel Consumption in Houses."
15. Building Research Station Digest No. 68. (Second Series), March 1966. "Window Design and Solar Heat Gain."
16. Building Research Station Digest No. 93. (Second Series)." Cellular Plastics for Buildings: 1" May 1968
17. Building Research Station Digest No. 23. (First Series), October, 1950. "Condensation Problems in Buildings."
18. Building Research Station Digest No. 132. (First Series), March 1960. "Condensation in Dwellings."
19. Building Research Station Digest No. 133. (First Series), April 1960. "Domestic Heating and Thermal Insulation."
20. Building Research Station Digest No. 9. (Second Series), April 1964."Dry-lined Interiors of Dwellings."
21. Building Research Station Digest No. 23. (Second Series), June 1962. "An Index of Exposure to Driving Rain."
22. Building Research Station Digest No. 91. (Second Series), March 1968. "Prevention of Condensation."
23. B.S. 874:1965. "Definitions of heat insulating terms and methods of determining thermal conductivity."
24. B.S.3533:1962 "Glossary of terms relating to thermal insulation."
25. B.S. 3837:1965. "Expanded Polystyrene Board for Thermal Insulation Purposes."
26. B.S. 2972:1961. "Methods of test for thermal insulating materials (and supplement No. 1 P.D. 4642)."
27. B.S. 3927:1965. "Phenolic foam materials for thermal insulation and building applications."
28. B.S. Code of Practice 3 Chapter VIII. "Heating and Thermal Insulation."
29. Building Regulations 1965—Part F. "Thermal Insulation." H.M.S.O.
30. Watson, A. "Measurement of moisture content in some structures and materials by microwave absorptions." Proceedings of the RILEM/CIB Symposium "Moisture problems in buildings." Helsinki, August 1965, Vol. 2, Paper 6–8.
31. Pratt, A. W. "Some observations on the variation of the thermal conductivity of porous inorganic solids with moisture content." D.S.I.R. Building Research Station, Current Papers, Research Series 30.

Loadbearing brickwork
– design for the fifth amendment*

B A Haseltine, BSc(Eng) ACGI DIC CEng MICE**
and
K Thomas, MSc CEng MIStructE FIOB ARTC

This document replaces the three earlier technical notes on the same subject.

* *The Building (Fifth Amendment) Regulations* 1970 *HMSO*
**Partner, Jenkins and Potter, Consulting Engineers*

1 INTRODUCTION

The tragic accident which destroyed a part of the Ronan Point block of concrete, system built flats in London on 16th May, 1968, when a gas explosion removed a loadbearing wall, has led to an agonising reappraisal of design methods by Structural Engineers and those concerned with building legislation. The report of the Inquiry into the collapse at Ronan Point[1] introduced a new phrase to engineering design – progressive collapse. This has been defined as the spread of local damage, however caused, to other parts of the structure remote from the point of mishap, probably affecting the overall stability. Following publication of the Report, engineers were bombarded with documents[2,3,4] detailing the precautions to be taken to aviod the possibility of progressive collapse when using large panel precast concrete construction, ie the Ronan Point type of building. RP/68/01 issued by the Institution of Structural Engineers also drew attention to the need to avoid progressive collapse in all types of large residential buildings where, for a variety of reasons, explosion damage could occur.

Loadbearing brickwork, which is usually used for the full range of domestic buildings up to 18 storeys in height, must clearly be designed and constructed on the same engineering principles as other structures. It must not be any more prone to progressive collapse than concrete construction. The reference to loadbearing brickwork in RP/68/01 did not help the designer on detailed points and a committee set up by the Institution of Structural Engineers to formulate guidance notes for the benefit of those designing loadbearing structures published its document 'Guidance on the Design of Domestic Accommodation in Loadbearing Brickwork and Blockwork to avoid Collapse Following an Internal Explosion'[5] in 1969. The committee recognised that there was very little information or experience of the susceptibility of brickwork structures to progressive collapse.

In an earlier Clay Products Technical Bureau Technical Note, Vol 2, No 6, engineers were shown how to comply with the Guidance Notes[5]. The Guidance Notes have now been largely superseded by The Building (Fifth Amendment) Regulations 1970[6], which have translated broad principles into a rigid regulation governing all types of structure. Two Brick Development Association Technical Notes were also published to illustrate the way in which the Fifth Amendment affects the design of loadbearing brickwork, and to show how the regulation can be met most economically for several plan types. This Note amalgamates and brings up to date the three earlier ones, which are now superseded.

It has been questioned whether loadbearing brickwork is still an economic proposition for buildings of more than 4 storeys. The answer is emphatically, yes. By sensible planning and attention to certain principles, loadbearing brickwork is still the cheapest form of construction and the method designers turn to when cost yard-sticks have to be met. The Fifth Amendment does not apply to the inner London GLC area (the old LCC area) but an amendment to the GLC By- laws is in force and is similar in concept to the Regulation applying throughout the rest of the country.

The new Regulations apply to *all* structures over 4 storeys in height. However, only loadbearing·brick structures are being considered in this Note. Up to 4 storeys, the degree of damage that can occur due to an explosion, for example, is limited, and is usually restricted to one occupancy. It is not uncommon to find that gas explosions in two storey houses completely destroy the structure, not because of any defect in the materials used, but because the walls are literally blown over as the roof is lifted up by the internal pressure.

With taller buildings, the degree of risk increases, and the effect of an incident in one occupancy can seriously affect others. Clearly, no design can allow for major disasters, such as exploding aircraft hitting a building, but the public at large have a right to expect that there is some inbuilt safeguard against accidents spreading damage to their homes from neighbours, above, below or sideways.

Unfortunately, too little work has been carried out on the effect of gas explosions in buildings, but the brick industry, with the support of the Brick Development Association, has organised a series of test explosions in a building specially constructed for the purpose. The experiments have been carried out by the British Ceramic Research Association, who have had the help of the Midlands Research Station of the Gas Council, and the Atomic Weapons Research Establishment, Foulness. A detailed report on this most valuable work is available from the British Ceramic Research Association,[7] but it is sufficient to say here that the explosion tests have established the effectiveness of venting through windows and lightweight cladding, shown that pressures in excess of about 23 kN/m² are very difficult to achieve in real situations, and established that, up to the peak pressure of 23 kN/m² obtained, brickwork may be damaged, but does not actually collapse.

Whilst the new regulations, unlike earlier guidance, do not recognise gas explosions as the reason for their existence, it is clear that the whole chain of events started with the gas explosion at Ronan Point, which removed an essential loadbearing wall, and it is protection against such incidents that is behind the design rules. Basically, the philosophy is that the removal of an essential part of a loadbearing structure must not lead to progressive collapse.

2 THE FIFTH AMENDMENT

The Regulation itself is a surprisingly short one, considering its wide implications, but by careful wording all types of structure have been embraced with the minimum of description. However, because of the brevity of the document, it is likely that a large number of different interpretations will arise and some initial difficulty may be experienced in arriving at a simple solution. The regulation is summarised below in non-legal language and the diagrams should help to make it more comprehensible. It is important to remember that there is now no reference to explosions and all thought that explosions lie behind the philosophy of the regulation must be suppressed, or implications that are not intended may be read into the clauses. A designer's whole approach to the problem must now be to consider an 'incident', and the incident only applies to one element of structure at any one time. Thus, the pressure specified also only applies to one element at a time.

The main points of the regulation are:

a Buildings under construction, or for which plans have been deposited with the Local Authority, are exempt.

b Buildings up to and including 4 storeys (including, unfortunately, basement storeys and piloti) are exempt.

c The loads to be used in designing a structural member which is essential to the structural stability of the building are:

(i) the dead loads

(ii) one third the live loads

(iii) one third the normal wind load

(iv) 34 kN/m², ie the well known 5 lb/sq in, in any direction, if certain conditions are not complied with. (see 2(d)(2))

Under these loading conditions, or when designing a structural member which remains after an incident, a factor of safety of 1.05 is all that is required. In practice, this means that, for brickwork 3.5 to 4 times the stresses given in CP.111* may be used in checking the strength of walls or piers for compliance with the amendment. The load factor of 1.05 is against structural failure. There is no requirement that the building should remain serviceable, and *damage can* occur outside the specified limits.

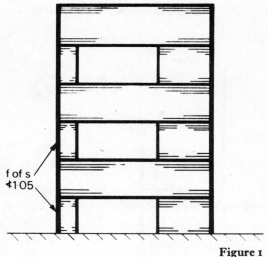

f of s ≮1·05

Figure 1

d Either

(1) Structural failure resulting from an incident must be limited to the storey of the incident, the storey above and the storey below and must not exceed a plan area of 70 m² or 15% of the floor area of the storey in question. For this alternative, the incident is defined as removing any one portion of a structural member at a time and the removal is considered to be carefully effected so that the surrounding structure is not damaged.

A portion of a structural member in this context is:

(i) a beam between supports

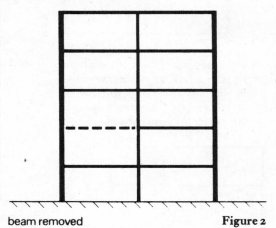

beam removed   **Figure 2**

(ii) a column between supports

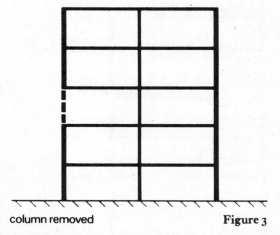

column removed   **Figure 3**

---

* *The load factors of 3.5 and 4 are based on test data provided by the Building Research Station. Over 200 wall and pier tests were carried out and the results used to formulate the basic permissible stresses in the current CP.111. Of the walls and piers tested, 95% had load factors of 4 or more, the remaining 5% having load factors above 3.5.*

147

(iii) a floor slab between supports or from an extremity to a support. A support, although not defined, is believed to include any wall having a mass of more than 151 kg/m².

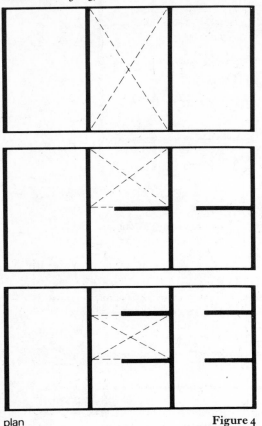

plan   **Figure 4**

(iv) a wall between supports, or from an extremity to a support, or from an extremity to another extremity. There is an overriding maximum for the length of wall to be considered removed of 2.25 times the storey height, eg 7.4 to 7.7 m for the normal domestic building.

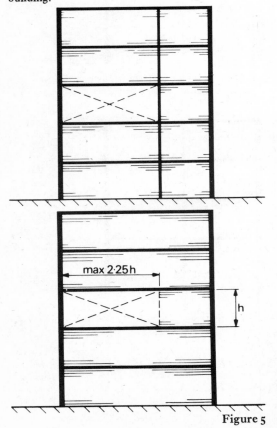

Figure 5

In all the above situations, the portion of a structural member need not be considered to be removed if

148

it is capable of withstanding a static force of 34 kN/m² in any direction.

OR

(2) If, for any reason, such as that the maximum damage permitted under (1) would be exceeded, a portion of a structural member cannot be allowed to be removed, then it must be designed to resist a pressure of 34 kN/m² in addition to loads (i), (ii) and (iii) in paragraph 2(c), but with a factor of safety of only 1.05.

There is no specific reference to tie steel, connections, or any of the engineering principles given in the previous guidance documents, but, of course, by restricting the damage allowed in the manner of d (1) and (2) above, the engineer is bound to provide for the sort of continuity aimed at by the previous documents. It cannot be emphasised too strongly that the regulation does not call for the whole building to be designed to resist a pressure of 34 kN/m². If it did, quite unrealistically strong structures would result.

e Technically, the top two storeys are exempt in all buildings. This comes about through structural failure being allowed in the storey of an incident, the one above and the one below. However, collapse of the top two storeys after an incident in the second from top storey would not be allowed, if the debris from the top two storeys overloaded the floors below so that more than one storey below the incident failed.

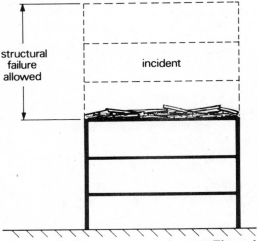

Figure 6

It has been suggested by various bodies, that the debris load of an insitu slab, or a properly tied precast floor, should be treated as once times its own weight (on the basis that about half the slab falls with a momentum of half its load onto the slab below) and that untied precast floors, which are in any event not recommended, should use a figure of three times their own weight.

3 THE GLC BY-LAWS

The GLC have amended their By-laws with the 'London Building (Constructional) Amending By-laws 1970'. The amendment is substituted for the old part 2.06 and, in principle, is similar to the Fifth Amendment. The definitions of structural members are more clearly defined, however, and, the word 'support' has been given a more specific meaning. When considering floors, in addition to the reference in paragraph 2(d) (1) (iii) above, a notional line of support may be assumed over substantial partitions, which are stated to be partitions having a mass of not less than 151 kg/m². For walls, the word support used in paragraph 2(d) (1) (iv) above is given the definition of lateral support in 1.04 in the By-laws, or, again, may be a substantial partition, weighing not less than 151 kg/m² and not less than 900 mm long. An important difference from the Fifth Amendment is that the maximum length of wall which needs to be considered removed does not apply to flank walls, where, if there is no substantial partition, the whole wall must be notionally removed. A further difference arises in the limit of area allowed to fail. If a larger section of structure is supported on *one* member, then the limit does not apply.

4 BRICKWORK AND THE REGULATION

Some plan forms are inherently stronger than others. A complex brick structure of the type shown in Fig. 7, having a good arrangement of walls in both directions, no thin walls without return ends, and an insitu floor slab, is clearly unlikely to be adversely affected by the removal of a length of wall. This type of structure is at the lowest end of an imaginary scale

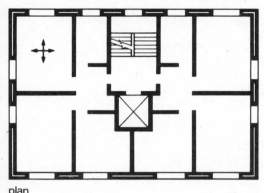

plan

Figure 7 *Well-stiffened brick cell structure which which would meet the requirements of the Fifth Amendment.*

of sensitivity to progressive collapse, and benefits from the ability of the structural elements to arch, cantilever or span over a damaged section. Whilst the principle of such plan forms is no longer described in the regulation, the need to adopt some stiffening over the straightforward crosswall structure with alternate timber floors, is clear. There is little chance, now, of using alternate timber floors in maisonettes over 4 storeys, but by careful planning, and a clear appreciation of the meaning of the Fifth Amendment, brickwork can easily, and economically, be used for housing and similar structures.

In the complex structure shown above, any 'portion of a structural member' in the form of an insitu floor slab (one at a time of course) could be removed, without leading to failure of more than 70 m² or 15% of the floor area. Similarly, if any one

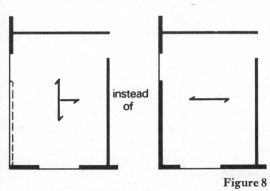

Figure 8

of the loadbearing walls was removed, using the definition in 2(d)(1)(iv), the floor slabs would, with a factor of safety of more than 1.05, span onto other walls, thus avoiding the critical amount of damage. In checking that the brickwork does not become

overloaded by a change of floor span, 3.5 times the stresses given in CP.111 can be used (see earlier).

Although it is not common to find a brick structure as stiff as that shown above, by substituting 102.5 mm walls for a partition, by designing with the enhanced stresses permissible when considering the structure under the Fifth Amendment, and, if necessary, by slightly increasing distribution steel, floor spans can be 'turned round' to take account of the removal of a wall. If two-way slabs were used, originally, there would probably be no need to increase the steel provided.

At the other end of the scale of sensitivity is a simple cross wall structure of the type shown in Fig 9. In this, there are no return ends on the walls, no

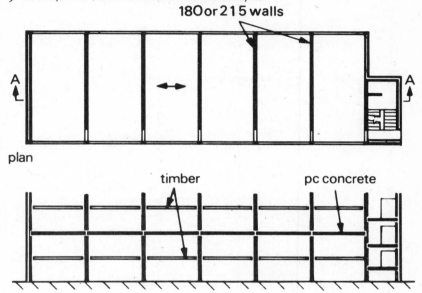

section A–A

Figure 9 *Simple crosswall structure, vulnerable to structural collapse. No end returns on walls, no spine walls, floors unconnected.*

spine wall (logitudinal stability being provided by a stair tower) and the floors are unconnected, simply supported precast units. The worst situation would be if alternate floors were of timber construction. Removal of a critical length of wall by any means in such a building is more likely than in the complex structure, because there are no substantial walls to limit the area affected.

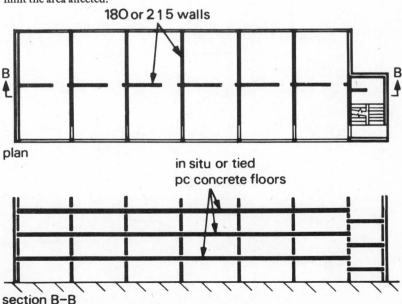

section B–B

Figure 10 *Improved resistance to structural collapse provided by the incorporation of spine walls and in situ floors.*

In Fig 9, if a critical length of wall is removed, the precast units or timber floor cannot contribute to arching action and progressive collapse might occur. Even here, however, the ability of deep brick walls to span or cantilever as beams contributes enormously to the structural stability. If the simple building in Fig 9 is stiffened, as shown in Fig 10, the ability of the building to withstand the removal of an essential structural member is enhanced. For example, a floor slab can clearly be removed, without exceeding the specified limits of failure. Indeed, it is usual, in analysing brick structures, to find that the floor slabs are never essential structural members.

If a section of crosswall (or end wall) is removed, there are five alternative approaches to the avoidance of collapse:

a Use brick walls for the cladding, and allow the slab to span from the spine wall to external wall in the ultimate case (ie with a factor of safety of 1.05).

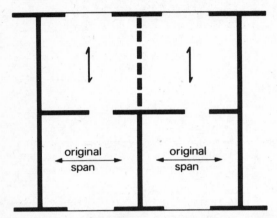

Figure 11 *On ultimate stresses, the brick infill cladding allows the floor slabs to span the other way in the event of a length of crosswall or end wall being removed.*

b Design the floor slab to span two bays on ultimate stresses (not usually economical, as the stresses do not go up as much as the bending moment). This approach cannot be used for an end wall. Alternatively, a design using the catenary concept, which is favoured by some Government departments, may be used. In this case, the floor slabs 'sag', perhaps 1 m or more, but do not fail because of the catenary formed by the reinforcement. If this approach is used, the reinforcement must be properly lapped over supports, and the necessary resistance to the

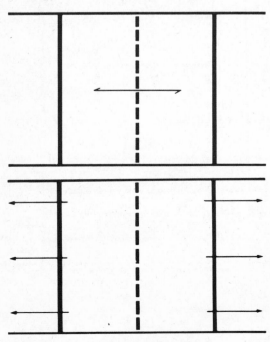

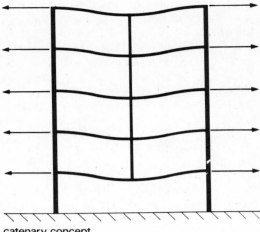

catenary concept

Figure 12 *The catenary approach. Allowing the floor to span between two points of support obviously entails serious damage to the storeys above.*

high tensions developed must be allowed for in an end span. The damage to the building above is considerable but, as it does not *collapse*, the regulation is met.

c Design the brickwork to cantilever over the opening, as shown in Appendix I. With insitu floors, this can still work, even if the spine wall is not exactly centrally placed. A tie rod at the end of the crosswall is essential in preventing the floor slab from flopping. It can be incorporated in the same way as reinforcement (see Fig 19).

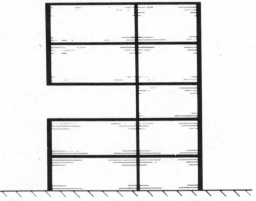

Figure 13 *Brickwork can be designed to cantilever over an opening. Working on ultimate stresses, the bond between brick walls and concrete floor slabs can be assumed to be sufficient to ensure composite action between them.*

d Provide a length of wall at the free end to resist a pressure of 34 kN/m² and design the brickwork above to span over the opening.

Calculations for this case are given in Appendix II.

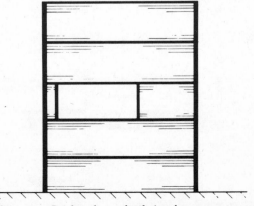

Figure 14 *Brickwork can be designed to span across an opening.*

e Use a wall of such strength that it can resist 34 kN/m² without failure (see later comments).

The various approaches used above can be applied to almost any type of brick structure. For any of them to work effectively, the floor must be capable of behaving, at a factor of safety of 1·05, in a way for which it would not normally be designed, eg to change its span, span further, act as the tension member of a beam etc. If an insitu floor is used, these additional functions can easily be allowed for with distribution steel, or extra bars in appropriate places. If precast floors are used for buildings over 4 storeys, the continuity, or ability to span in other directions, must be carefully considered. Members of the BPCF are aware of these problems and can advise users on the best solution. Fig 15 shows how precast floors can be linked together over supports, and Fig 16 is an example of a small upstand beam which would be necessary to stop a precast floor drooping over an opening which appeared in the wall below.

In many cases solution (a) above, or a suitable layout of walls in two directions, will be all that is needed, assuming that the floor can be arranged to 'change span' when any one loadbearing wall is notionally removed.

If solution (c) is appropriate, the calculation set out in Appendix I will help in the design. The horizontal tensile resistance needed is quite small and is easily provided in the floor slab. The questions of shear strength of brickwork, and the bond of the brickwork and floor slab have been raised. Tests, such as those carried out for composite action of brickwork and concrete in framed structures, suggest that in the near ultimate circumstances applying to this situation, the bond stress is not a problem. Similarly, if four times the shear values given in CP111 are used, the cantilevers can be justified near to the point of failure (see paper by Bradshaw and Thomas[8]). The brick industry are conscious of the need to provide back up information on cantilever design, and, although some work has been carried out on reinforced cantilevers, it is now proposed to test to failure ¼ full size and full size cantilevers, mainly to check on bond and shear strength.

If, as will arise in some structures, it is not possible to arrange the layout of the building to span or cantilever over notionally removed walls, recourse must be made to reinforced strong points. In these, under the action of their normal loading from above, including one third the live load, the factor of safety must be not less than 1·05 when the strong point is subjected to a horizontal pressure, in any one direction, of 34 kN/m². In Appendix II an example calculation for such a strong point is given. The calculation assumes that a joint is not provided between the strong point and the remainder of the wall. It is now accepted that any possible failure of a brick wall will be by yield-line fracture, and the area of extra pressure to be resisted by a strong point unseparated from a wall will be small. Of course, if a joint is provided between the strong point and the remaining wall, the only pressure to be allowed for will be on the strong point itself. The calculation is based on the likely method of failure of brickwork under combined bending and compression, a subject not adequately covered in CP111.

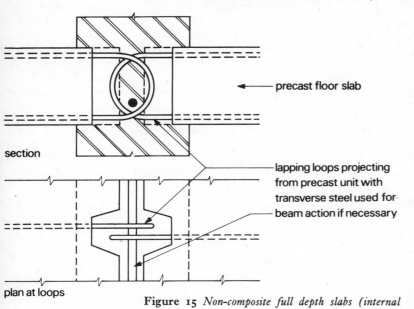

section

plan at loops

precast floor slab

lapping loops projecting from precast unit with transverse steel used for beam action if necessary

Figure 15 *Non-composite full depth slabs (internal wall 7 in and thicker).*

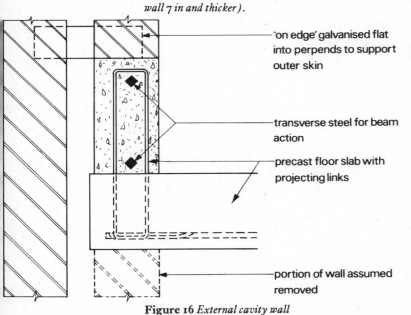

'on edge' galvanised flat into perpends to support outer skin

transverse steel for beam action

precast floor slab with projecting links

portion of wall assumed removed

Figure 16 *External cavity wall*

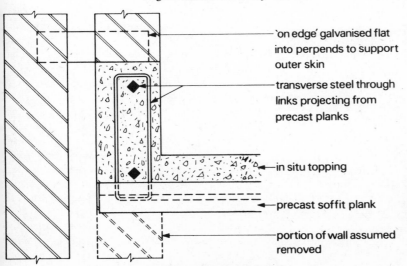

'on edge' galvanised flat into perpends to support outer skin

transverse steel through links projecting from precast planks

in situ topping

precast soffit plank

portion of wall assumed removed

Figure 17 *Composite plank floors.*

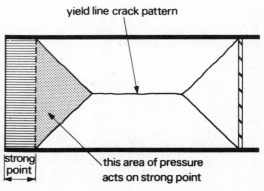

yield line crack pattern

strong point

this area of pressure acts on strong point

Figure 18

Should reinforcement be necessary, how it is actually put in a portion of a brick wall will depend on the thickness of wall. Some examples are given in the sketches. In external walls, protection of reinforcement from the effects of water must be provided. This may be in the form of cover, protection of the steel, or use of non-corrodable steel.

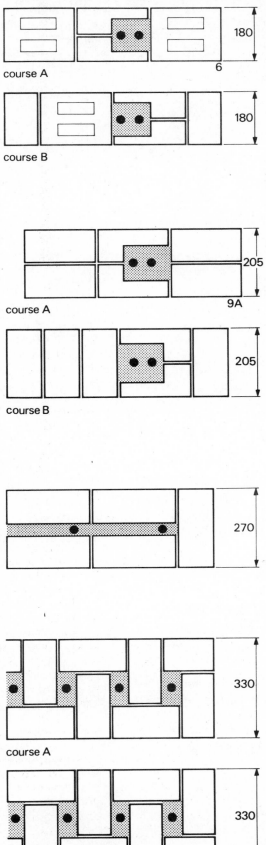

Figure 19 *Suitable methods of reinforcing brick walls, using standard specials.*

Clearly, justifying cantilevers, or providing strong points is an additional chore to the basic design of building, and the simplest approach of all would be to have all the walls capable of withstanding 3 kN/m², in which case, little else would need to be done. If all the walls are reinforced as strong points of course, the need to bridge over is eliminated, but this would be unrealistic. Fortunately, the test detailed in Section 5 have shown that brick walls have a high resistance to horizontal pressure under a certain amount of pre-load, simulating the load of a building above. As the top two storeys of a building are allowed to collapse, provided the debris load is allowed for, there are never less than two floor slabs and a roof, together with two storyes of wall load on the critical wall. The results given mean, that

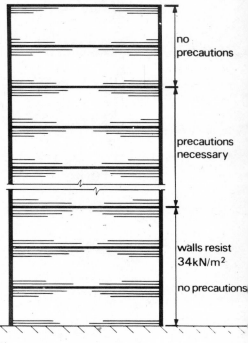

Figure 20

under some circumstances, the load two storeys down is sufficient to achieve a lateral resistance of 34 kN/m² on internal walls. Of course, at the level when the resistance of the wall is only 5% more than 34 kN/m², no further precautions are necessary assuming that the area of damage is not exceeded when the floor slabs are treated as the vulnerable 'portion of a structural member'. It is, therefore possible to limit the amount of building for which special precautions are required to a small number of storeys. Although, by designing brickwork to act compositely with concrete, the regulation can be complied with reasonably economically, it is a welcome relief to designers, to have major parts of building which are strong enough to meet the regulations without the need for detailed calculation.

However, the alternative path method made relatively simple by proper planning, is a sound engineering approach to a difficult problem and will be preferred by many eingineers to relying on the lateral resistance of a possibly vital wall. Furthermore when designing external walls on the upper floor to meet the Fifth Amendment, the alternative path method must be employed unless additional strength is built into the walls by means of strong points of reinforced brickwork.

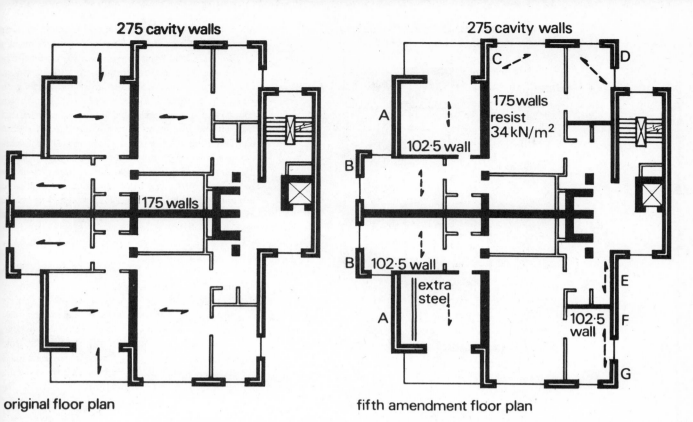

275 cavity walls

175 walls
resist
34 kN/m²

A

102·5 wall

B

B

102·5 wall

extra
steel

A

C

D

E

102·5
wall

F

G

275 cavity walls

175 walls

original floor plan

fifth amendment floor plan

Figure 21 *Only few low-cost modifications were required to adapt this building to the requirements of the Fifth Amendment.*

Fig 21 shows a building that was planned before the Fifth Amendment came into force. The two plans show the original design, and the amended one to bring it into line with the Regulation.

Firstly, the internal walls can be assumed to have a resistance of 34 kN/m² below the top two storeys since, with the spans involved, there is just sufficient vertical load. The floor slabs are thin precast planks providing formwork and tensile reinforcement, with a deep topping making up the required overall depth. Extra steel has been incorporated, as necessary, in the insitu concrete.

If wall A is considered removed, by substituting a 102·5 mm brick wall for the partition, the floor slab can span from the balcony wall to the 102·5 mm wall on distribution steel acting at near ultimate stresses. Similarly, wall B can be removed without causing structural collapse.

If either walls C or D are removed (only one at a time of course), by providing a small amount of reinforcement diagonally across the upper corner, the slab can span onto the short length of external wall, which would normally have been regarded as non-loadbearing.

Considering walls E, F and G, here again, a partition has been changed to 102·5 mm brickwork to form a 'support'. This has the effect of reducing the length of wall which has to be considered removed. Any one of the walls E, F or G must be capable of removal without causing structural collapse. Taking out wall E first, the brickwork above will span over the opening. If wall F is removed, the slabs – suitably reinforced – can carry the brickwork above over the opening. Similarly with wall G.

5 LATERAL LOAD TESTS

Following the work of gas explosions designed to test the resistance on brickwork to dynamic loading conditions, tests have been carried out in the wall frame at the Mellor-Green Laboratories of the British Ceramic Research Association, which more closely approached the static loading condition in order to obtain data on the lateral resistance of walls to be used in the design of brickwork to the Fifth Amendment. The walls, which are storey height (2·54 m) and 1·4 m long, are placed under a known precompression by means of the vertical loading system in the wall frame while the lateral load is applied by means of an airbag. A series of tests has been carried out to establish the relationship between lateral load and precompression and the results are reported below.

a Description of tests and results

Testing was carried out in a 8970 kN compression testing machine which consists basically of three portal frames 5 m high, attached to a common base grillage 2·4 m × 2·1 m. The cross-members and the grillage beams are 900 mm × 300 mm fabricated from 38 mm MS plate and the vertical members 300 mm × 300 mm × 283 kg/m universal columns. Minor modification of the machine enabled a heavy duty textile reinforced rubber bag to be placed alongside one face of each test wall, one side of the bag being completely in contact with the whole elevation of the wall, and the other side reacting on the inside of the machine frame columns. A piece of blind canvas was interposed between the bag and the wall to prevent the air-bag being cut by broken bricks as the wall failed. The wall arrangement is shown in Fig 22.

High-alumina cement mortar cappings were applied to the wall after the latter had been placed in the machine and the spreader beam was lowered to rest upon the capping mortar before it set, in order to achieve an intimate contact surface. Additionally, an underlay of 3 mm reinforced hard rubber was placed between the underside of the wall and the base platform of the test machine.

The wall testing machine has a nominal compressive capacity of 8970 kN derived from three hydraulic rams. The compressive load applied to a

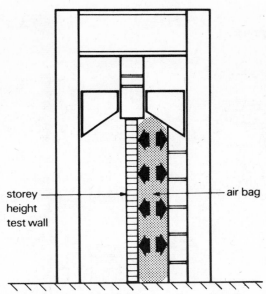

Figure 22 *Diagram showing the basic arrangement of the compressive strength testing machine when adapted to carry out lateral load tests.*

1·4 m wall during lateral testing – simulating the conditions at say 3 storeys from the top – is only 99·1 kN, which is minute compared to the capacity of the machine. A secondary hydraulic circuit was therefore installed to apply these low pressures so that the two outboard hydraulic rams are inoperative and precompression is read directly from the gauge for the single central ram.

During the tests, due to the deflection of the wall, the effective height tends to increase. This increases the pressure of oil in the hydraulic ram, and the compressive load applied to the test wall is maintained constant by bleeding oil from the ram leading to an upward movement of the spreader beam. In practice, this increase in the height of the wall would jack up the floor slab and thus effectively increase the compressive load beyond the design value, as elaborated below. In order to determine the upward movement of the spreader beam, measurements have been made using transducers and a digital voltmeter.

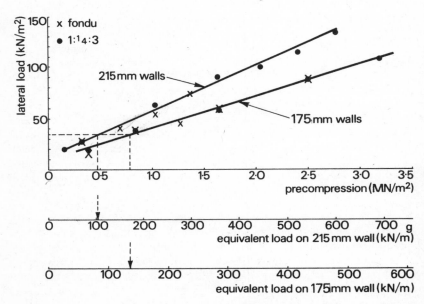

Figure 23 *Graph showing results of lateral load tests on 175 mm and 215 mm walls at various levels of precompression.*

To obtain the maximum number of results in the shortest time a series of walls were built in 1:3 high-alumina cement:sand and 1:¼:3 Portland cement

lime: sand mortars, the high-alumina mortars being tested at 2–3 days and the Portland cement mortars at 28 days. Good agreement was obtained between the results for the two mortars. Both the 175 and 215 mm walls have been built in pressed bricks with a nominal compressive strength of 34·5 MN/m² and 27·5 NM/m² respectively. The variation of lateral load with precompression is shown in Fig 23. In each test, deflections have been measured at the mid-depth of the wall by means of a simple moving pointer on a scale. The magnitude of these deflections is so large (up to 45 mm) that more sophisticated methods of measurement are not justified. The results for 215 mm walls are shown in Fig 23 and for 175 mm walls in Fig 23.

b Discussion

(i) Mode of failure: The tests described above have established clearly the mode of failure. It is now known that at low pre-loads (78 kN/m) failure is entirely geometrical and that the strengths of the mortar and the bricks play little or no part in the total resistance to lateral loading (Fig 24). It will be apparent from the diagram, that as the lateral load increases, a crack forms at mid-height of the wall thus causing a wedging action as the upper and lower sections of the

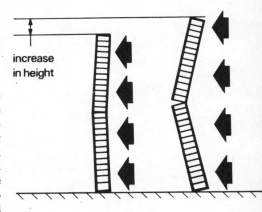

Figure 24 *At low compressive loads, the failure of brick walls subjected to lateral pressures is hinge-like. The height of the wall is increased as the upper and lower halves rotate.*

wall rotate. In the test machine only the upper beam is free to move and measurements have shown that at the low pre-loads indicated above, a vertical extension of about 8 mm occurs before failure. As the pre-load increases, the mode of failure changes introducing an element of crushing at mid-height of the wall on the laterally loaded face.

These patterns of failure seem to account for the shape of the curves in Fig 23. At low compressive loads the failure is wholly hinge-like, and the wall fails at the mid-depth into two intact halves; at somewhat higher precompressions there is some slight crushing failure of the bricks but up to about 1 MN/m² compression the relationship between lateral load and precompression is linear or almost so. At higher compressive loads local crushing failure of the bricks occurs and there is multiple catastrophic failure of the walls; over this range the graphs deviate from a straight line and appear to approach a maximum value. Eventually when the compressive load reaches the ultimate strength of the wall it will fail in compression without lateral load. The precise relationship at very high levels of precompression has not so far been investigated, but the range covered is adequate for practical purposes. At very low levels of precompression, the tensile bond of the joints probably influences the results, and the graph again deviates from a straight line.

(ii) Typical loadings – In simple crosswall structures the load per metre run is largely dependent on the span and type of floor used. Table 1 gives a typical range of loadings at mid-height at different numbers of storeys from the top (Fig 25) for internal walls, when insitu concrete floor slabs are used. These figures include an allowance for finishes plus ⅓ of the imposed load as allowed by the Fifth Amendment.

TABLE 1
Static load per metre on internal walls at mid height of storey.

Span (m)	Slab depth (mm)	Load kN/m		
		2 storeys down (at A)	3 storeys down (at B)	4 storeys down (at C)
3·05	125	80	107	135
3·66	125	97	131	165
4·57	163	118	159	200
5·49	175	143	192	241
7·32	225	209	281	352

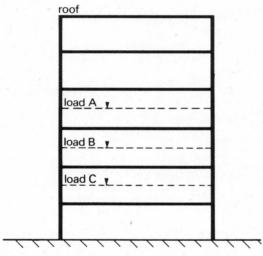

Figure 25 *Diagram and table showing typical static loads per metre run on internal loadbearing brick walls at mid-storey height when in situ concrete floor slabs are used.*

From Fig 23 it will be seen that a minimum pre-load of 0·56 MN/m² – 112 kN/m or 0·71 MN/m² – 151 kN/m is required for 215 mm and 175 mm walls respectively to achieve a lateral resistance of 34 kN/m² at failure. Without taking into account the factors discussed below, these pre-loads correspond to floor spans of about 4.57 m and 5.8 m respectively for 215 mm and 175 mm walls at a level immediately below the top two storeys. It is clear that at any lower level the requirements of the Fifth Amendment are readily achieved.

In addition to the loads in Table 1, the geometrical mode of failure demands that the portion of the building above the wall in question must be bodily lifted more than 8 mm before the wall is physically removed. For upward movement of this

where E = Young's Modulus of Elasticity, I = moment of inertia δ = the deflection, L = the span.

magnitude calculations using the formula $\dfrac{6EI\delta*}{L^2}$

prove that substantial additional pre-loads must be attracted to the wall due to the inherent stiffness of the reinforced concrete floor slabs above. For a 4 m span and 6 mm upward movement, this increase would be about 33%, ie at two storeys from the top the static pre-load would be increased from 97 to 128 kN/m.

The foregoing is based on the simple crosswall structure without any openings. Such openings will increase the loading on the wall in question and thus increase the lateral resistance. Similarly, when only part of a crosswall is under consideration, either

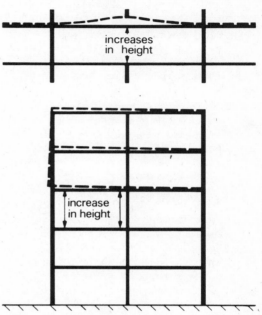

Figure 26 *If the height of the wall is increased (Fig 24), the portion of the building above it must be bodily lifted before the wall is physically removed.*

because the maximum length is exceeded, or a spine wall is included, considerable additional load will be attracted to the wall in question when it tries to fail (Figs. 26), and hence raises the building. Not only will the floors contribute, but the remaining crosswalls will attempt to hold the wall down.

c Effects of end restraints
It should be emphasised that these tests have been

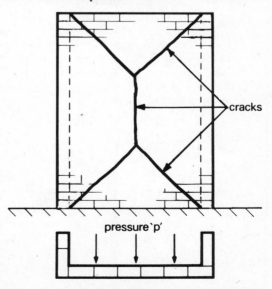

Figure 27 *Walls buttressed by returns typically fail in a yield line pattern of cracks, and have an increased resistance to lateral loads.*

carried out on walls restrained only at the top and bottom and the lateral load recorded must be expected to be a minimum. However, crosswalls should preferably be buttressed by spine walls and/or end returns. The present machine is limited in its capacity for testing walls with returns, but one has been built in the machine. This was a 215 mm wall, 2.10 m long overall with returns 588 mm long overall, with returns 588 mm long overall, built in 1:3 high-alumina cement:sand mortar and tested at 3 days. At a precompression of 78 kN/m run this wall failed in a yield line pattern of cracks at a lateral load of 62kk N/m² (Fig. 27). This is more than twice the equivalent 1.4 m wall without returns (28 kN/m²) although it must be observed that this is a very short wall in practical terms. Nevertheless, it seems likely that the effect of returns at one or both ends of a wall must be to stiffen it and thus increase in some degree the lateral load resistance and, also, of course, change the hinge failure to that of a yield line pattern.

In this sense therefore, the lateral loads shown in Fig. 23 must be regarded as conservative, although this is a single result and considerably more research is required before definitive recommendations can be given.

6 CONCLUSIONS

All building design is affected by the aftermath of Ronan Point with the issue of the Fifth Amendment. More thought must be given to planning and engineering to meet its requirements, but by careful design, loadbearing brickwork can meet the regulation with little or no increase in cost, and, accordingly, is able to retain its acknowledged position as the most economical structural material for housing, and residential types of buildings.

The 34 kN/m² resistance to lateral load required by the Building Regulations can be achieved by 175 and 215 mm walls under a preload of 151 and 117 kN/m respectively. These loads can be achieved at two storeys from the top of many loadbearing brick buildings, especially when due allowance is made for the load attracted to walls by the need to lift the building upwards in order to allow a wall to fail. Additionally, the effect of end restraint in the form of returns should increase the resistance to lateral loading.

For those who find difficulty in achieving the desired results on their own, the brick industry, the Brick Development Association or some companies, are available to give advice to those who contemplate using brickwork structures. By producing this document, the Brick Development Association hope that designers will be encouraged to use loadbearing brickwork in the knowledge that a standard level of interpretation has been given☐

APPENDIX 1
Calculation for a wall cantilevering over an opening

Assume a building 9 800 mm wide with 215 mm crosswalls at 4 900 mm centres having a storey height of 2 600 mm with a 150 mm in situ reinforced concrete slab.

The floor loading is:

Slab	3.6	
Finish	1.15	
Partitions	1.0	
Live Load		
($\frac{1}{3}$rd only)	0.67	
	6.42	kN/m²

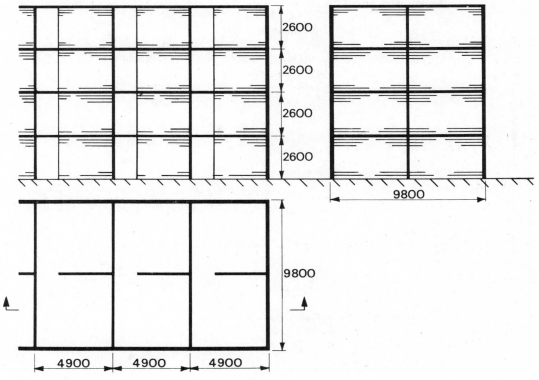

Figure 28

The front and back elevation cladding weighs 22.2 kN per bay. Weight of 215 mm brickwork 4.5 kN/m². An incident can occur on any floor and so any section of wall can be removed, though *only one* at any time.

Therefore, consider each floor cantilevering with the floor slab providing the tension and compression flanges. The brickwork provides the web, and the bond between brick and concrete is assumed to be sufficient (ref. Dr. Wood's [9] work).

The slab above the removed section of wall is assumed to span in catenary between crosswalls or hang in position by a tie-rod. If a precast floor was used, it would be held up either by a tie-rod between crosswalls or by a small upstand beam (see comments about precast floors on page 6). The compression force from the cantilever should, in either case, be capable of being provided by the brickwork.

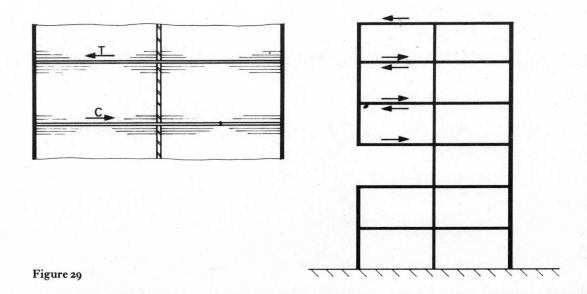

Figure 29

Load per m on one storey height cantilever
$$= 4.9 \times 6.42 + 2.45 \times 4.5 = 42.5 \text{ kN/m}$$

Cantilever moment $= 22.2 \times 4.9 + \dfrac{42.5 \times 4.9^2}{2} = 619 \text{ kNm}$

Lever arm of T and C $= 2 \ 600$ mm

Therefore $T = C = \dfrac{619 \times 1 \ 000}{2 \ 600} = 238 \text{ kN}$

Safe tension in high tensile steel $= 500 \text{ MN/m}^2$ (a compromise figure above the normal proof stress, but below the ultimate).
Therefore, area of steel required to resist tension

$$= \dfrac{238 \times 1 \ 000}{500} = 476 \text{ mm}^2 \text{ say 2 No. 20 mm high tensile bars}$$

In theory, this steel is only required at roof level, as all the other T's are cancelled by C's, but this amount of steel should be provided at every floor as a necessary tie.

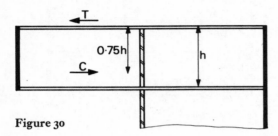

Figure 30

Compression in bottom boom of cantilever is resisted by the floor slab over, say, 1 250 mm, therefore compressive stress

$$= \dfrac{238 \times 1 \ 000}{1 \ 250 \times 150} = 1.27 \text{ MN/m}^2 \quad \text{and this is satisfactory.}$$

At the floor above the void, C has to be resisted by the brickwork.
The cantilever moment is the same as above.
Lever arm $= 0.75 \times 2 \ 600 = 1 \ 950$ mm

Therefore $C = T = \dfrac{619 \times 1 \ 000}{1 \ 950} = 317 \text{ kN}$

thus tension $= 317 - 238 = 79$ kN more than the upper case.

However, as tie steel for a force of 238 kN has been provided at every level, instead of just the roof, this additional tension can always be resisted. In practice, of course, the whole wall above a void will act as one deep cantilever.
Compression in brickwork in cantilever above damaged floor.

$$= \dfrac{317 \times 1 \ 000}{225 \times \dfrac{2 \ 450}{2}} = 1.15 \text{ MN/m}^2$$

A permissible stress is required in the brickwork (using a factor of 3.5) of $\dfrac{1.15}{3.5} = 0.33 \text{ MN/m}^2$.

which could be compared to the basic values in Table 3 of CP.111 (1964) and is well within the capacity of normal brickwork.

This calculation could also be used for a gable end, with revised loads and dimensions where necessary. The floor slab over a void would not, however, hang in catenary and support would have to be provided, eg a continuous tie rod.

APPENDIX 2
Calculation for a gable end strong point (column)

Use the same dimensions as Appendix I, but calculation is for the flank cavity wall 275 mm thick. Floor slab rests on both skins of wall.

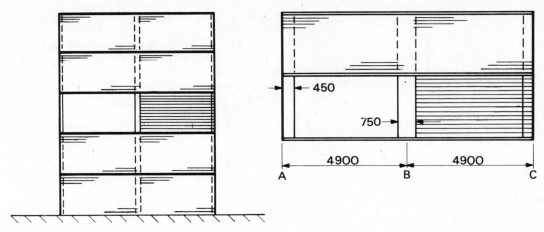

Figure 31

It is assumed that the load transfer from wall to column will occur floor by floor, as the level of the missing wall is not known. The slab over a removed wall must remain in position at least sufficiently to provide the tensile force for the storey height beam above. For this purpose, approximately a quarter span of the slab is assumed to act as a beam between columns. An upstand beam can be provided in the width of the wall where this amount of slab is not strong enough or where precast units are being used.

Load/m on slab beam $= 2.45 \times 6.42 = 15.7$ kN/m

Bending moment $= \dfrac{15.7 \times 4.675^2}{10 \text{ say}} = 34.3$ kNm (centre line spacing of columns $= 4.675$)

Moment of resistance at 2.25 overstress (this figure results from the increase in stresses allowed to maintain a load factor of 1.05) $= 2.25 \, Q.b.d^2 = 2.25 \times \dfrac{7}{4 \times 1\,000} \times \dfrac{4\,900}{4} \times \dfrac{125^2}{1\,000} = 75.3$ kNm

Therefore satisfactory.

Tension steel required $A_t = \dfrac{34.3 \times 1\,000 \times 1\,000}{500 \times 0.75 \times 125} = 731$ mm²,

ie 12 mm diameter high tensile bars at 200 mm centres top and bottom in the 1 250 strip.

Each storey height of wall, with its upper and lower floor slabs, forms a 'channel' beam loaded by one floor slab and its own weight.

Load/m $= 2.45 \times 6.42 + 2.45 \times 4.5 = 26.7$ kN/m

Bending moment $= \dfrac{26.7 \times 4.675^2}{10 \text{ say}} = 58.3$ kNm

Therefore, Tension $=$ Compression in 'floor flanges'

$= \dfrac{58.3 \times 1\,000}{2\,600} = 22.4$ kN

Although not now a requirement, as was the case in Circular 62/68, it is good practice to provide sufficient steel across the building to resist 21.3 kN/m

$= 21.3 \times 2.45 = 52.2$ kN

which is more than the tension due to beam action. Therefore provide steel for 52.2 kN force at a steel stress of 500 MN/m².

$A_t = \dfrac{52.2 \times 1\,000}{500} = 104$ mm² ie 1 No. 12 mm high tensile bar.

The load is transmitted to column by shear in the slab.

End reaction at A $= \dfrac{4.9}{2} \times 26.7 = 65.4$ kN

If column is 450 mm $\times$ 275 mm (see later) shear stress around two sides

$= \dfrac{65.4 \times 1\,000}{(450 + 275) \times 0.80 \text{ say} \times 125} = 0.90$ MN/m²

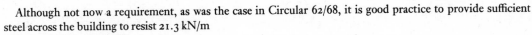

With 1:2:4 concrete, permissible shear stress 2.25 $\times$ 0.7 $= 1.58$ MN/m²

159

Columns Check external one (A) in this example. Calculation of others similar. Weight of 275 mm grouted cavity wall = 5.65 kN/m².

Load per floor = 65.4 + 22.2 (cladding weight) = 87.6 kN

Take column at 3rd level from top. Load on column after removal of wall between columns

$$= 2 \times 22.2 + 2 \times 65.4 + \frac{4.9}{2} \times \frac{4.9}{2} \times 6.42 \text{ (roof)} + \frac{450}{1000} \times 2.45 \times 5.65 \text{ (self weight)} = 219.9 \text{ kN}$$

When only loaded by the amount of building the column will take in the undamaged state, ie the whole wall uniformly loadbearing, the column must withstand 34 kN/m² horizontal pressure with a factor of safety of 1.05. In addition, a small area of 34 kN/m² pressure is transmitted from the adjacent wall.

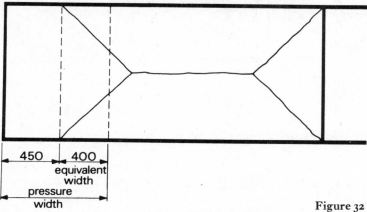

Figure 32

Moment due to pressure (column is continuous)

$$= \frac{(450 + 400)}{1\,000} \times \frac{34 \times 2.45^2}{12} = 14.4 \text{ kNm}$$

Load on column in undisturbed state (there is no reduction in load due to pressure as the pressure acts on *one* member at a time)

$$= 22.2 \times 2 + 3 \times 2.45 \times \frac{450}{1\,000} \times 5.65 \text{ (self weight)} + \frac{450}{1\,000} \times 2.45 \times 6.42 \times 3 \text{ (floor and roof)} = 84.3 \text{ kN}$$

Check section for tension using formula $\dfrac{W}{A} \pm \dfrac{M}{Z}$

$$\text{Stresses} = \frac{84.3 \times 1\,000}{450 \times 275} \pm \frac{14.4 \times 1\,000 \times 1\,000 \times 6}{450 \times 275^2}$$

= 3.22 or − 1.86 MN/m².

The tension is much too high, and reinforcement is required.

Try 2 − 16 mm diameter high tensile bars (this is 0.33% of the column area).

As the ultimate load is required, the load factor method is appropriate.

At failure the stress block shape is:

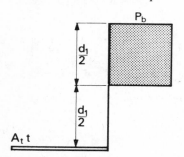

Figure 33

Lever arm = 0.75 d_1 A_t = area of steel   t = ultimate steel stress

Moment of resistance for steel = 0.75 $d_1 \times t \times A_t$

Moment of resistance for brick = $\dfrac{b.d_1}{2} p_b \, 0.75 \, d_1 = \tfrac{3}{8} P_b d_1^2 b$

Steel stress, say 500 MN/m² $d_1 = 138$ mm

∴ Moment of resistance for steel

$$= 0.75 \times \frac{138}{1\,000} \times \frac{500 \times 402}{1\,000} = 20.9 \text{ kNm}$$

For 28 MN/m² bricks set in 1 :½ :4½ mortar and assuming a factor of safety of 3.5 in the brickwork

$$M_{brick} = \frac{3}{8} \times \frac{138^2 \times 3.5 \times 1.75 \times 450}{1\,000 \times 1\,000} = 19.6 \text{ kNm}$$

Direct load on column = 84.3 kN

Slenderness ratio of column $= \dfrac{2\,600}{275} = 9.45$

Therefore reduction factor $= 0.86$

Area reduction factor $= 0.75 + \dfrac{0.124}{1.2} = 0.85$

Ultimate direct load capacity of column (P) assuming an ultimate compressive stress in the steel of 280 MN/m²

$$= \left(3.5 \times 450 \times 275 \times 0.86 \times \frac{\cdot 1.75}{1\,000} \times 0.85\right) \; + \; \left(\frac{402}{10^6} \times 280 \times 1\,000\right)$$

$$= 554 + 112$$

$$= 666 \text{ kN}$$

Combining the effect of bending and direct compression using the moment of resistance of the brick, as this is less than that for the steel.

$$\frac{W}{P} + \frac{M_{act}}{M_{brick}} = \frac{84.3}{666} + \frac{14.4}{19.6} = 0.86$$

If this ratio falls below 0.95 (allowing, therefore, the 1.05 factor of safety), the column design is adequate.

Therefore, the bricks and steel chosen provide a strong enough column.

APPENDIX 3
Wind pressures

The publication of CP.3, Chapter V, Part 2 (1970) has clarified the pressures to be used in wind design. However, there is very little guidance available on the design of walls to resist the wind forces. The practical problems which arise due to wind pressures in loadbearing brickwork structures are minimal as a result of the restraint of floors and return walls, etc, but this is not always easy to prove by calculation. With framed construction, the problem is more acute.

When walls are designed as infil panels as suggested by CPTB Technical Note Vol. 1, No. 6 'Wind Forces on Non-loadbearing Brickwork Panels' by Bradshaw and Entwisle[10], and the design pressures are calculated on the basis of CP.3, Chapter V, Part 2 (1970), unrealistic panel thicknesses will result unless higher tensile stresses are adopted than those suggested in CP.111. The limiting panel sizes suggested by Bradshaw and Entwisle are considered realistic and were determined empirically based on the satisfactory performance of walls in the Sheffield area during the 1962 gales (the 0.63 probability level). Recent research has shown that when panels are restrained on two or more edges, practical design can be achieved using the yield line theory and it is hoped to give details of this method of analysis as applicable to brickwork in a future Technical Note. Until this information is published, it is suggested that brickwork panels designed on the basis of Bradshaw and Entwisle's technical note should be calculated using a permissible tensile stress of 0.14 MN/m² when bending is assumed perpendicular to the bed joints, and 0.27 MN/m² when bending is assumed normal to the perpend joints. Even these figures appear to be conservative in the light of recent research.

REFERENCES

1 Report of the inquiry into the collapse of flats at Ronan Point, Canning Town, London. HMSO 1968

2 Appraisal and strengthening of existing high blocks: design of new blocks. Circular 62/68 MOHLG 1968.

3 Structural stability and the prevention of progressive collapse. RP/68/01, Institution of Structural Engineers. 1968.

4 Notes for guidance which may assist in the interpretation of Appendix 1 to Ministry of Housing & Local Government Circular 62/68. RP/68/02. Institution of Structural Engineers. 1968.

5 Guidance on the design of domestic accommodation in loadbearing brickwork and blockwork. RP/68/03. Institution of Structural Engineers. 1969.

6 The Building (Fifth Amendment) Regulations 1970. HMSO 1970.

7 Gas explosions in loadbearing brick structures. N. F. Astbury, H. W. H. West, H. R. Hodgkinson, P. A. Cubbage & R. Clare Special Publication No. 68, British Ceramic Research Association, 1970.

8 Modern Development in Structural Brickwork. Bradshaw R. E. and Thomas K. CPTB Technical Note Vol. 2, No. 3 January 1968.

9 National Building Studies – Research Paper No. 13. 'Studies in composite construction: Part 1, The composite action of brick panel walls supported on reinforced concrete beams'.

10 Wind Forces on Non-Loadbearing Brickwork Panels. Bradshaw R. E. and Entwisle F. D. CPTB Technical Note Vol. 1, No. 6, May 1965.

Some observations on the design of brickwork cladding to multi-storey rc framed structures

Donald Foster Dip Arch (Dist) Dunelm ARIBA

Consultant Architect, Structural Clay Products Limited.

1.0 INTRODUCTION

There have been, in recent years in the United Kingdom – certainly in the decade 1960–70 – a considerable number of examples of severe cracking of stone, concrete and brickwork cladding of multi-storey structures in which the supporting members are rc frames or walls.

The brickwork failures which have been observed (those, at least, which the writer has had to investigate) are characterised by horizontal cracking at storey height intervals, at or about floor level and always associated with squeezing out of the horizontal dpc cavity flashing at those levels. The worst damage and disfigurement has occurred when the edges of the floor slab or beam are covered by brick 'slips' – often called brick tiles – where the designer, as is most usual, has wished to avoid expressing the horizontal. In such instances the courses of brick slips have buckled or even been completely dislodged and they, and sometimes the bricks immediately above and below, have spalled badly although the spalling of the bricks above the slips has been noted to occur most readily when the cavity flashing has been bridged with mortar at the extreme external face. In all instances the cladding has been built in tightly between successive horizontal members of the frame.

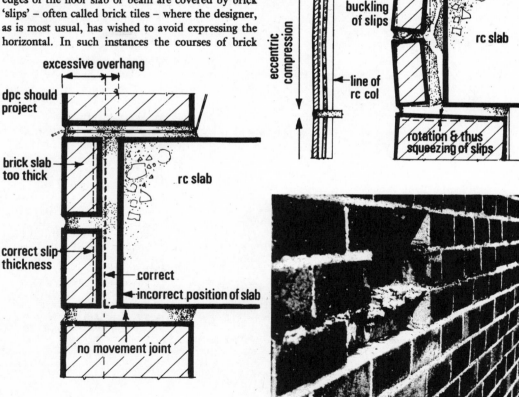

Figure 1 *Typical faulty construction*

Figure 2 *Typical failure*

It is now generally accepted that the principal cause of such failures is the squeezing of the cladding by the vertical shrinkage and creep of the concrete frame and that correct construction now demands the incorporation of movement joints – better described as pressure relieving joints – at storey height intervals. The purpose of this note is to discuss the implications ensuing from an adoption of this practice and to suggest appropriate details after first sketching the historical background.

2.0 BRIEF HISTORICAL BACKGROUND

With the advent of the steel frame, empirical rules governing sizes of panels (length and height in relation to thickness) of brickwork between surrounding beams and columns were eventually developed in the LCC area (now the GLC). These rules still persist because of lack of experimental data on the resistance of brickwork walls to lateral loads, * particularly where they are not subjected simultaneously to vertical loads.

It was necessary also to decide on a safe maximum overhang of brickwork where, as invariably happened then, designers wished to hide the frame. Here the choice was logical. Early LCC regulations allowed a wall to overhang its support a maximum of one third of its thickness and as external walls in brickwork were at that time (c. 1920) nearly always solid and not

less than a brick and a half thick or actually about $13\frac{1}{2}$ in ** the allowed actual overhang was $4\frac{1}{2}$ in. Thus it was possible to raise the brickwork vertically past the horizontal spandrel beam without cutting or use of bricks of reduced width. The modern equivalent of this construction – brought about by the continuing drive for reduction in cost, an increasing structural confidence which has led to thinner sections but inevitably also to greater deflections, the increasing use of the cavity wall and the still persistent desire to use brickwork as cladding – is, on a dimensional stability basis, poor by comparison. The same basic rule of an overhang of one third of the thickness of the supported wall still applies but now it is only 34 mm, ie one third of a very slender outer 102.5 mm leaf, compared to the $4\frac{1}{2}$ in (114.3 mm) or one third of the earlier 343 mm substantial wall cladding. Fig 3. Of equal importance was the supplanting of the steel frame by that of reinforced concrete, for economic reasons. Their widely differing properties are highly significant in this story. On the one hand is the steel frame, which is relatively accurately erected and which does not shrink and has negligible creep, whereas on the other is the rc frame which suffers extensive vertical shortening through creep and shrinkage and which is erected relatively inaccurately.

Finally the advent of the cavity wall introduced a continuous flexible horizontal damp proof flashing at the foot of each panel just above the courses of brick slips. Each such damp proof course is compressible and can therefore accommodate some, but unfortunately not enough, of the differential vertical movement. Events have shown that even two layers – about 3 mm or $\frac{1}{8}$ in total thickness – are insufficient.

It requires little imagination to appreciate the difference in dimensional stability between the earlier and present constructions. The latter seems positively fragile in comparison and there is little wonder that so many buildings have suffered disfigurement in the way described.

It is clear that any tendency to expansion of the brickwork cladding the steel framed structure would easily be restrained. The writer knows of no instance where horizontal cracking due to vertical differential movement has occurred in this type of structure yet the brickwork must have suffered the same thermal and moisture conditions as that in the present day rc structures. In the latter, however, the combination of large vertical strains from shrinkage and creep of the concrete columns, the thin sections, the inaccuracy of structure and the consequent highly eccentric loading of the outer leaf of brickwork and the brick slips has caused widespread trouble.

3.0 DESIGN CONSIDERATIONS
3.1 Lateral resistance
Horizontal pressure relieving joints must be located at the tops of the panel walls – ie between the upper surface of a wall and the soffit of the next slab above. The resistance of such walls to lateral loads must inevitably be weakened by the incorporation of soft joints because the support is then only from three edges – two sides and the base – instead of four, unless devices which can provide shear resistance and at the same time allow for differential movement are

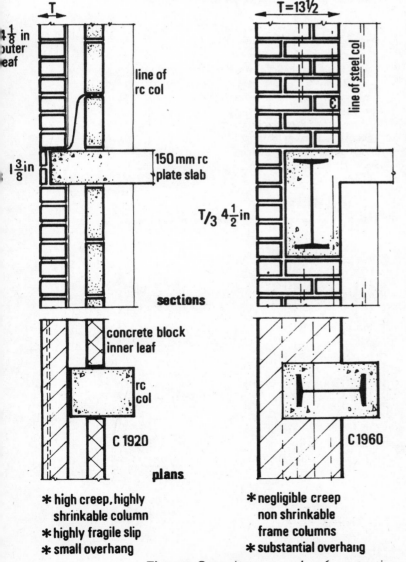

Figure 3 *Comparison c.1920 and c.1960 constructions*

* *Soon designers will be able to choose more rationally. Research on lateral resistance of non-precompressed walls is in hand at the BCRA.*

** *There was no standard length of brick then. A brick length plus joint was greater than 9 in – usually $9\frac{1}{4}$ in or more. Hence a brick and a half wall was often referred to nominally as a 14 in wall and was actually very nearly $13\frac{1}{2}$ in.*

163

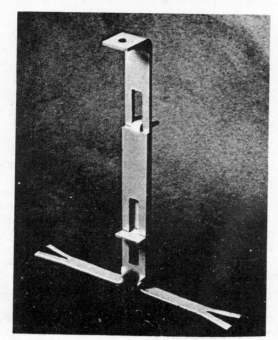

Figure 4 *A prototype stainless steel telescopic anchor (not yet in production) manufactured by George Clark (Sheffield) Ltd, and designed jointly by them and the Surveyor's Division, Department of Planning and Architecture, Sheffield Corporation.*

incorporated. Suggestions for telescopic anchors, Fig 4, have been put forward, but these are probably much too expensive for the normal run of building as well as being difficult to accommodate in the concrete floor or beam during construction. They or similar devices are unavoidable where an area of brickwork is isolated from columns by glazing and thus has no chance of lateral support. Fig 5. Discussion about the full range of possible combinations of brickwork and glazing, their relation to the frame and floors – ie visually suppressing or expressing the latter – and their corresponding resistance to lateral loading is beyond the scope of this note and in any event awaits the research noted earlier. It is clear, however, that in many designs there would be need for a soft joint at the top of the wall and that brick slips will then no

longer have direct vertical support. The designer must find alternative means of keeping them in place. Some possibilities are considered below:

3.2 Casting slips with slab

Although bond between slips and mortar is variable it is generally very much better than between mortar and slab. Grooving brick slips to improve adhesion is therefore only partially useful. However, brick slips will adhere firmly to concrete if cast with it and this has been tried many times. Adhesion has been shown to be more than adequate but it has proved very difficult to maintain correct vertical and horizontal alignment of the slips and to avoid the concrete staining their faces by leakage. It seems that the correct appearance of brickwork can only be maintained by laying the slips in the normal way.

3.3 Use of adhesives

One idea that immediately appeals is to use one of the many modern adhesives, such as epoxy resin, to replace the bedding mortar. Unfortunately these are so much more expensive than ordinary bedding mortar that they cannot easily be justified. Also adhesives like epoxy resin are meant to be used in very thin layers or beds but these are obviously unsuitable for brick slips where a thick bed is necessary to accommodate the inaccuracies of both bricks and slab alignment. Dilution of the epoxy with sand to give a thick bed is possible but usually the slips have to be held in place during setting. Other thick bed adhesives, developed from the very successful thin bed tile adhesives, have been tried with some success but to the best of the writer's knowledge, it is doubtful whether they could yet be given a long term guarantee by manufacturers, yet presumably at least a sixty year life is required without maintenance other than pointing. It seems, unfortunately, that despite the wide development of adhesive technology nothing has yet been developed which successfully reproduces the many favourable attributes of ordinary mortars yet gives the all important high adhesion.

3.4 Mechanical support

If this reasoning is accepted it follows that designers

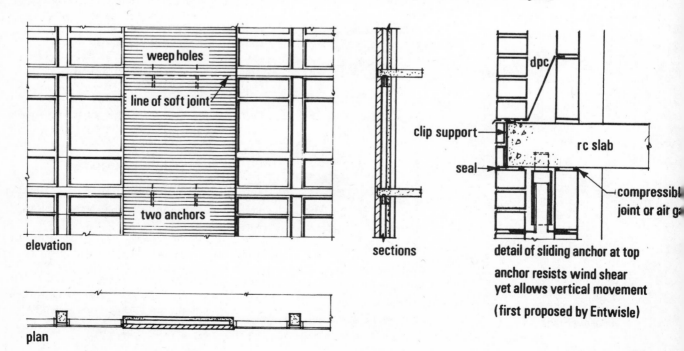

Figure 5 *Brickwork cladding isolated from columns*

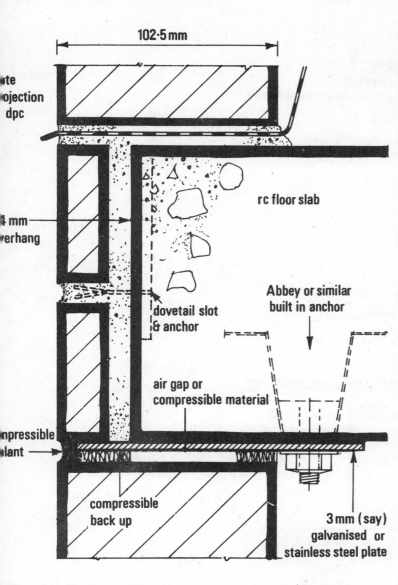

102·5 mm

te
ojection
dpc

4 mm
verhang

mpressible
lant

rc floor slab

dovetail slot
& anchor

Abbey or similar
built in anchor

air gap or
compressible material

compressible
back up

3 mm (say)
galvanised or
stainless steel plate

Figure 6 *Support by continuous plate method*

would be wise to consider some form of additional mechanical support. Two suggestions appear here. One consists of a projecting continuous steel plate, bolted to the soffit of the slab, upon which the weight of the units can be sustained. Fig 6. They are also bedded in mortar to the vertical face of the concrete slab but a dovetailed anchor or a high adhesive mortar can be used if the designer wishes to hold back the brick and mortar complex more positively. Using this 'belt and braces' approach the designer – and more important, the public – is safe even if such a mortar or adhesive were not to withstand the ravages of time and weather.

The other mechanical method suggested is intended to perform the double function of vertical support and holding back. Figs 7 and 8. One such anchor would be required per brick for the bottom course. Though only two courses of 'slips' are shown here it should be possible to devise similar anchors for three or four courses. It is probably not advisable to exceed this number.

3.5 Thickness of movement joint

It will be seen that with both suggestions the thickness of the supporting plate or clip effectively reduces the 10 mm thickness of the movement joint to about 6–7 mm and it is debatable whether this is adequate. The original thickness of 10 mm is governed, of course, by the need to match the joint with the normal bed joints. Any substantially greater thickness would undoubtedly destroy the monolithic appearance of a continuous sheet of brickwork though it would be acceptable in some circumstances.

It would clearly be helpful before deciding on thickness in any design to try and assess the maximum movement likely to occur. Edwards[1] has shown that drying shrinkage of concrete can be greater than 0.085%, or 850 microstrain (ie about one tenth of an inch (2.54 mm) in a ten foot (3.048 m) storey height) when it is made with some highly shrinkable aggregates and that it can be in a range of 0.066 – 0.085% (660 – 850 microstrain) when made with aggregates from a wide geological range. Creep at 10.3 N/mm^2 (1500 lbf/in^2) has been assessed by Hollington[2] as 300 microstrain giving an additional 0.036 in. (0.9 mm) per storey but evidently this latter figure could be doubled if a highly shrinkable aggregate is used – say 1.8 mm, giving a total creep and shrinkage of 4 to 5 mm. Elastic strain of the column (about 1 mm) plus thermal expansion of the brickwork (also about 1 mm for a temperature increase of 55 deg C) when added to this 4 to 5 mm gives a total of 6 to 7 mm, thus absorbing all the relieving space, even assuming the infilling to be capable of compression to zero, which it clearly is not. Nor has any account been taken of the possibility of long term moisture expansion of the brickwork which in the most adverse circumstances ie if built with bricks very fresh from the kiln, could amount to 0.05%, or about 1.5 mm per storey. Thus the total reduction could amount to about 8.5 mm per storey and this would require a total movement joint thickness of about 12–15 mm.

It must be appreciated that this total is the worst combination of laboratory determined coefficients derived from relatively small unrestrained specimens and that under site conditions – which in any event make accurate forecasting impossible – movement may well be less. For example, a true determination of total vertical shortening of the concrete columns due to shrinkage must take into account the environ-

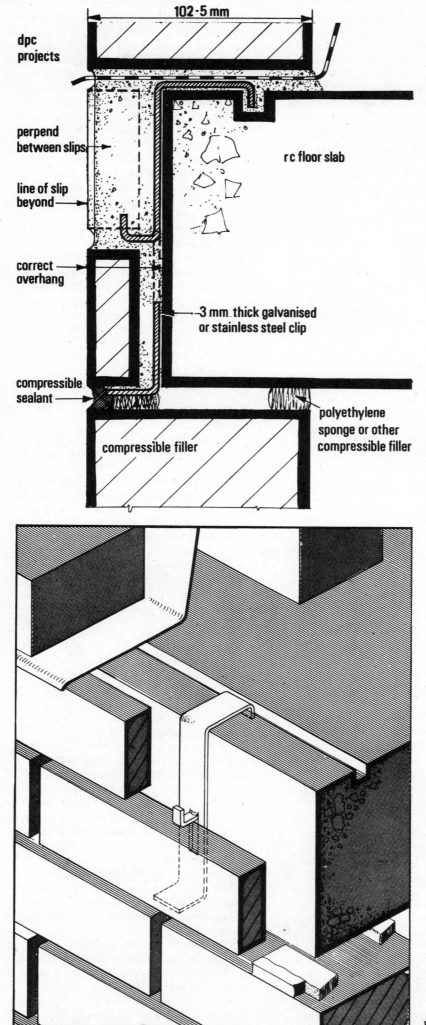

102·5 mm

dpc
projects

perpend
between slips

line of slip
beyond

r c floor slab

correct
overhang

3 mm thick galvanised
or stainless steel clip

compressible
sealant

polyethylene
sponge or other
compressible filler

compressible filler

Figures 7 & 8 *Support by clip method*

ment ie the conditions of humidity and the temperature of the surrounding air over a long period and the ratio of surface area to volume as well as the type and quantity of cement, the amount of water present and the shrinkage characteristics of the aggregate. Creep is affected by these factors, and by age at loading, and is also stress dependent. Any expansion of cladding due to thermal or moisture movement is easy to assess in comparison with creep and shrinkage of the columns but here again there can be no real accuracy outside a laboratory.

Clearly the practical designer can do no more than ensure that the materials and methods specified lead to minimum movement. He can insist, for example, that only aggregates of low shrinkage and bricks which are not kiln fresh are used and that movement joints are sufficient and frequent. He can specify that the cladding is not started until the frame is complete, thus avoiding the need for excessively thick movement joints at the top of the lower lifts of brickwork. Also, of course, he can vary the thickness of the movement joint for different storey heights. For multi storey domestic construction where the floor height is 2.6 m (about 8 ft 6 in) the 10mm joint shown is likely to be adequate even when the proposed metal supports are incorporated, provided the other safeguards mentioned are taken but if in doubt an increase to 12 mm is probably practical without marring the monolithic appearance of the brickwork.

Attempts to measure movements on actual structures have been comparatively rare. One American source[3] quotes total measured shortening of concrete columns in a complete actual structure as one inch per hundred feet of height – one tenth of an inch per 10 ft storey – which is as much as that derived above for reputedly the worst case of shrinkage alone, ie with the most highly shrinkable aggregate. Probably it is true that 'forecasts of movement are notorious for their inaccuracy. Generally, they are overestimated' ...

4.0 CHOICE OF MOVEMENT JOINT MATERIAL

There is a wide range of mastics and gaskets for movement joints matched by ample guidance [5] about their use and a growing number of British Standards[6]. It is generally advisable to consult manufacturers about any application. This note is therefore no place for a lengthy discussion of the various types of jointing material* but there are a few general observations which may help in the present context.

First, these pressure relieving joints will clearly not be subjected to substantial reversals of extension and compression because the bulk of the movement is compression arising from the certain shrinkage, creep and elastic contraction of the columns and the thermal and possible moisture expansion of the brickwork cladding. The only cyclic contraction will be due to thermal effects on the cladding and this has been shown to be a small percentage of the total. At about 10% it can be accommodated by a wide range of mastics in a butt joint. It can be reasonably conjectured, then, that the most expensive sealants – those such as polysulphide rubbers having the greater ability to tolerate cyclic movement over longer periods – will be unnecessary and that cheaper sealants such as oleo-resinous compounds can be used. These however, have a shorter expected life – say 2–10 years

*Note by K. Thomas: It is essential, when selecting a backing material for a movement joint of this nature, to ensure that it easily compressible. This problem was discussed in CPTB Technical Note Vol 1, No 10.

instead of 20 years – so initial cost must be balanced against maintenance.

A secondary consideration is that the proposed joints are in the external leaf of a cavity wall and they are therefore not here the sole barrier against rain penetration as they would be, say, in a single leaf form of construction. This is not to argue that resistance to penetration of the outer leaf is not desirable, which obviously it is, but that it clearly cannot be critical, otherwise the whole idea of a drained cavity system would be nonsense. The degree of exposure will clearly be some guide to choice in this context. In severely exposed situations wind driven rain has been known to spray across the cavity from fine fissures in the outer leaf. The best type of movement joint would then be needed but in normal exposure there would not be the same demand.

It can be seen that the performance of horizontal movement joints in brick cladding to the rc frame is probably not so demanding as with other cladding materials where the cyclic thermal variation is greater. This does not apply to vertical movement joints designed to take account of corresponding horizontal movement. These, at much greater intervals of about 12 – 13 m (say 40 – 43 feet) will have greater thermal diurnal expansion and contraction. It should be remembered that such joints are as necessary in brick slip courses as they are in the body of the brickwork.

5.0 CONCLUSIONS

Continuing advances in structural design methods, materials and techniques have resulted in buildings which are much less rigid and with members having much greater deflections and strains than hitherto. Differential movements or tendencies to movement between frames (particularly rc frames) and claddings are now much greater and have to be accommodated by properly designed movement joints. Given these, brickwork can offer in slender walls the most durable, economic and to many the most aesthetically pleasing of all claddings□

REFERENCES

1 Edwards, A. G. 'Shrinkage and other properties of concrete made with crushed rock aggregates from Scottish sources'. Journal of the British Granite and Whinstone Federation*, Vol 6 No 2, Autumn 1966. (Also Building Research Station Digest 35)

2 Hollington, M. R. Private communication

3 'Lessons from Concrete Structures'. American Concrete Institute, Monograph 1, ACI; Iowa State University Press, McGraw Hill.

4 Rodin, J. 'The implications of movement on structural design'. Symposium on Design for Movement in Buildings, The Concrete Society, London.

5 Building Research Station Digests 36 and 37

6 (i) British Standard 4254: 1967 Two part polysulphide based sealing compounds for the building industry.
   (ii) British Standard 3712, Parts 1 & 2: 1965 Methods of test for building mastics.
*Now the British Quarrying and Flag Federation.

The design of free-standing brick walls

K Thomas MSc CEng MIStructE FIOB ARTC
and
J O A Korff BSc (Eng) CEng FFB MICE*

*Assistant Divisional Engineer (Housing), Greater London Council,
Department of Architecture and Civic Design

1.0 EXPOSURE

Free-standing walls whether they be in a position of extreme exposure or in a sheltered area must be designed to withstand the effects of the elements. Unfortunately this type of construction usually receives the minimum of thought at the design stage, yet is open to more abuse generally than the more sophisticated elements of construction which may be adequately protected.

Exposure to wind, rain and frost should be assessed when building in an area with which the designer is unfamiliar, particular attention being paid to the combination of the driving rain index Fig. 1 and the severity of frosts.

It is vitally important that the correct details are used to produce an aesthetically acceptable construction and also to afford protection of the wall.

1.1 Selection of bricks

Brick is quite rightly accepted as an all-purpose building unit, but there are various types of brick each having a specific function, and these are defined in the appropriate British Standard[1],[2]

Selection of the correct brick for free-standing walls is perhaps more important than for many other situations as the wall is usually exposed to the weather on both sides. It is particularly important that satisfactory details are used at the top and bottom of the wall otherwise exposure in these areas may initiate deterioration of the wall.

Durability is extremely difficult to define and the standard (1) points out that crushing strenght is *not* necessarily an index of durability and may be misleading if used as such. It is true that bricks having a crushing strength in excess of 48.5 MN/m² are usually durable, but there are also bricks approaching this limit which decay rapidly if exposed to frost in wet conditions and others, very much weaker, which are durable. It is suggested that before bricks are used as copings, sills or in situations of relatively severe exposure that the manufacturers' advice be sought as

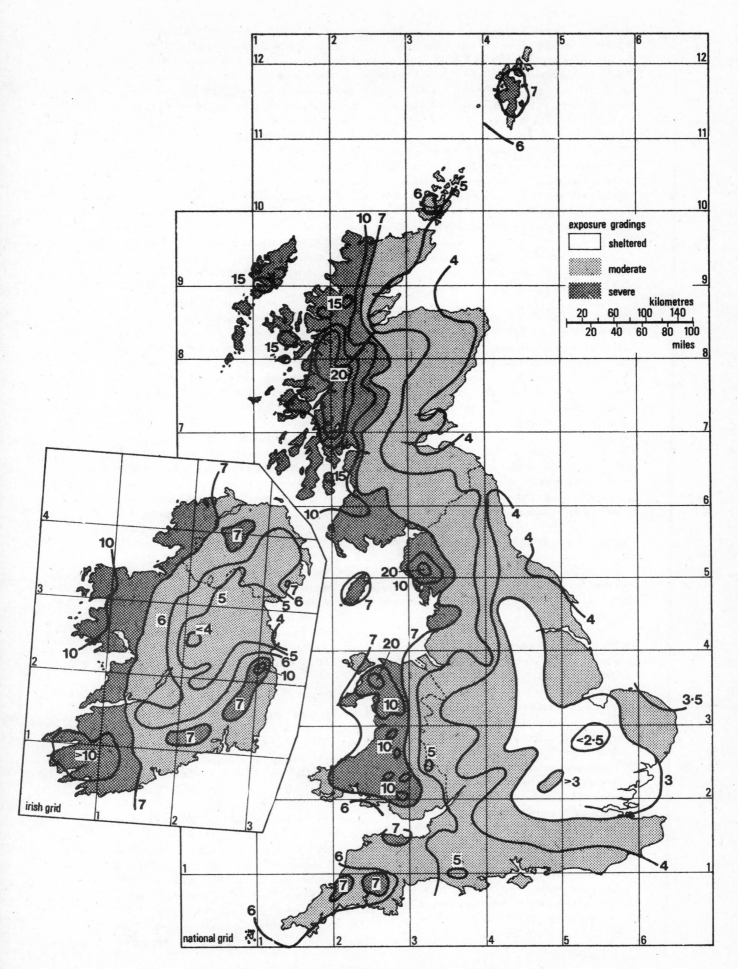

Figure 1 *Map showing index of exposure to driving rain in UK (from BRS digest 127[1]).*

to the most suitable brick for the purpose if they crush below 48.5 MN/m². When bricks are used in situations of extreme exposure it is recommended that bricks of special quality be used as defined in BS 3921 and that the use of sulphate resisting cement be considered for the mortar joints. Alternatively, a strong mix based on ordinary Portland cement is considered satisfactory by some designers (ie 1 : 0/¼ : 3 Portland cement : lime : sand).

Whenever possible it is suggested that at the base of free-standing walls two or more courses of dpc brick (defined in BS 3921) be used in lieu of the other types of dpc membrane which do not provide the same order of resistance to movement. The bricks should be laid in 1 : 3 cement : sand mortar as recommended in BS 743.[3]

Clay bricks should never under any circumstances be bonded with calcium silicate or concrete brick or blocks.

1.2 Construction of free-standing walls

It is vitally important that all external free-standing walls should have an adequate coping and that a damp proof course (preferably brick or slate) be provided approximately 150 mm above ground level. It is also essential that a continuous flexible dpc be included in the mortar joint directly under copings.

The problem in selecting an effective coping is two-fold:

a it must provide the wall with complete protection in a vertical direction and shed rainwater clear of the face as effectively as possible, and

b it must be aesthetically acceptable. Requirement (a) can readily be accomplished in a number of ways but (b) is an imponderable which not only depends upon individual taste (or lack of it) but also on the use and situation of the free-standing brickwork.

This section of the Technical Note is solely concerned with the performance aspects of copings and in this context the aesthetics of appearance are irrelevant.

1.3 Damp proof courses

As previously stated a brick dpc is preferable near the base of a free-standing cantilever wall for two reasons:

a it provides a certain tensile resistance to flexure and

b having a shear resistance of the same order as the wall itself, it provides good resistance to thermal movements and hence, fewer movement joints are required. At the top of the wall a continuous flexible dpc is recommended in the mortar joint directly under the coping; if this is omitted rainwater has free access via the mortar joints between the coping units.

Flexible damp proof courses wherever they occur should not be cut back from the face of the brickwork and pointed over to disguise their presence. It is essential that such dpc's should project, if possible 13 mm, to form a drip for rainwater thus protecting the brickwork below. Pointing over dpc's should never be permitted as this allows dampness to bypass the dpc and to some extent renders it ineffective.

1.4 Copings

The character and general appearance of a wall may be materially altered by the type of coping used and various examples are illustrated in Figs. 2–11.

The ideal coping projects either side of the brick-work, has 0 mm minimum overhang, including a throating of not less than 13 mm width and which is at least 13 mm from the face of the brickwork Fig. 2; alternatively an overhang as illustrated in Figs. 3 to 6 would be satisfactory. During construction care should be taken to ensure that the throating remains clean and is *not* filled or bridged by mortar. The top surface should also be sloped to shed rainwater to one side, preferably both, and most important of all it should be durable and preferably impervious. Some copings are pre-treated with a silicone solution to make them water repellent, such copings should be laid in the traditional way but a mastic joint will be necessary as the silicone treatment tends to prevent adequate bond with the conventional cement mortars.

If copings are of material other than fired clay, their thermal and moisture movement characteristics should be compared to those of the brickwork, and provision made for the resulting differential movements.[4]

Where brick on edge or similar copings are used it is recommended that suitable galvanised steel, stainless steel or non-ferrous metal end anchors be used to prevent dislodgement of end units. Fig. 12. If brick on edge copings are used they must be known to be durable in such a situation and ideally provide an overhang to shed rainwater clear of the wall.

1.5 Foundations

Boundary walls do not generally carry any vertical loads other than their own weight, and consequently the foundations do not always receive the necessary attention. Foundations for free-standing walls should be placed in undisturbed earth or well consolidated soil at a depth below the frost line. They may be mass or reinforced concrete depending upon site conditions.

1.6 Movement joints

Unless walls are constructed in a flexible mortar which does not contain cement (ie a lime : sand or similar mortar) movement joints are necessary if cracking of the brickwork is to be avoided. Details of the type and frequency of such joints are discussed in detail elsewhere.[4] Under no circumstances should copings bridge movement joints.

1.7 Mortar

In free-standing external walls the lateral strength of the brickwork is dependent on the tensile bond between the bricks and mortar unless a gravity design approach is adopted. It is therefore essential that a mortar providing the necessary tensile bond be used and selection of this mortar depends upon (i) the type of brick, (ii) the type of wall (ie boundary wall, parapet wall or retaining wall) and (iii) the degree of exposure and/or soil conditions.

For clay brickwork it is not usually necessary to specify a mortar mix richer than 1 : 1 : 6 Portland cement : lime : sand for normal boundary walls but, if the bricks have a high soluble sulphate content or the wall is in a position of extreme exposure, a sulphate-resisting cement is recommended. Some designers prefer stronger mortar mixes – see section 1.1.

Mortar plasticizers are often used in lieu of lime to aid workability but, as the entrained air bubbles can affect bond, their use is not recommended for free-standing walls unless a gravity design approach is adopted. Masonry cements also contain a plasticizing agent and are not recommended by the authors for the same reason. It should also be pointed out that these cements consist of approximately 75 % Portland cement and 25 % inert filler and plasticizing agent; they are often misused (ie weak gauging), specifiers assuming them to be 100 % Portland cement and not making the necessary adjustments to mix proportions. Stronger mortar mixes may be necessary for retaining walls and reinforced brickwork as recommended in CP 111.[5]

Calcium silicate bricks are generally laid in weaker mortars to confine cracking to the joints. Under normal circumstances a 1 : 2 : 9 cement : lime : sand mortar should be used but if the walls are constructed during cold weather a 1 : 1 : 6 may be more appropriate.

For more detailed information on mortars see Building Research Station Digest No. 58 (Second series), Mortars for Jointing.

2.0 STABILITY OF FREE-STANDING WALLS

All walls should be designed so that they have inherent stability against overturning. This may be achieved in different ways according to the type of wall and the function it has to fulfil. A straight free-standing wall of indefinite length may have adequate stability (subject to provision being made for normal movements) if its thickness is sufficient in relation to its height. Zig-zag and serpentine walls are more stable than straight walls due to their geometry. The tensile bond between bricks and mortar is of particular importance in free-standing walls, and for a given thickness and height straight free-standing walls constructed with bricks are likely to be slightly more stable than units manufactured with lightweight material. Walls can be divided into a series of panels which are stabilised by buttresses, columns, piers or intersecting walls.

Many free-standing boundary walls are not subject to defined lateral loads as, if designed on an engineering basis, the walls tend to be excessively thick. Experience has shown that straight (non-loadbearing) free-standing walls, whether external or internal, subject to wind pressures, should, unless stability calculations are carried out, have height to thickness ratios not exceeding the following:

Table 1 Height to thickness ratio related to wind speed

	Wind pressure N/m²	lbf/ft²	Height to thickness ratio
Up to	285	(6)*	Not exceeding 10
	575	(12)	7
	860	(18)	5
	1150	(24)	4

(intermediate values may be interpolated)

When damp proof courses incapable of developing adequate bond are used, the height (in this instance

NB *Table 1 was originally proposed by one of the authors for inclusion in a BSI document which has not yet been published.*

* *Pressures as low as 285 N/m² (6 lbf/ft²) can only be expected in the centre of London and nowhere else in the United Kingdom.*

the vertical distance from the position of effective lateral restraint below the dpc to the top of the wall) to thickness ratio should not exceed 75 % of the appropriate value in Table 1. The use of such dpc's are not generally recommended by the authors for free-standing walls.

Units weighing less than 961 kg/m³ are not normally used in external free-standing walls; however, should designers choose to use such units it will be necessary to reduce the height to thickness ratios quoted in Table 1.

American practice is a little more conservative and recommends that for straight garden-type walls, assuming ground level on each side of the wall to be the same, so that no stresses result from earth pressure, the wall must be designed to resist wind and impact loads. For 10 lbf/ft² wind pressure, it is recommended that the height above ground (h) should not exceed 0.75 of the wall thickness (t inches) squared ($h = 0.75 t^2$). This recommendation is based on the assumption that the overturning moment from wind does not exceed 0.75 of the righting moment from the wall dead load, and does not depend upon bond between the foundation and the wall.

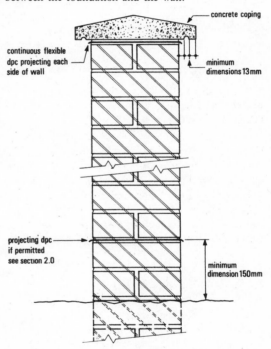

continuous flexible dpc projecting each side of wall

concrete coping

minimum dimensions 13mm

projecting dpc if permitted see section 2.0

minimum dimension 150mm

Figure 2 *Section through boundary wall.*

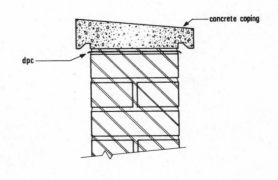

concrete coping

dpc

Figure 3

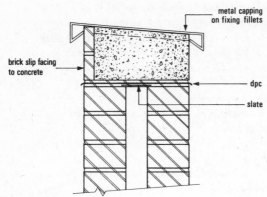

brick slip facing to concrete

metal capping on fixing fillets

dpc

slate

Figure 4

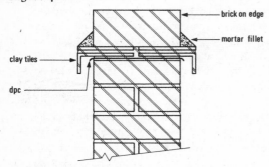

brick on edge

mortar fillet

clay tiles

dpc

Note: Brick on edge copings are only satisfactory with certain types of brick

Figure 5

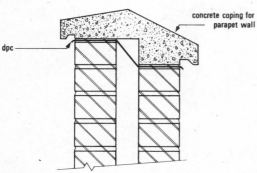

concrete coping for parapet wall

dpc

Figure 6

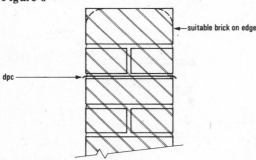

suitable brick on edge

dpc

Note: Brick on edge copings are only satisfactory with certain types of brick

brick on edge coping

Figure 7

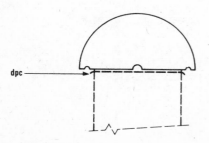

dpc

half round clay coping

Figure 8

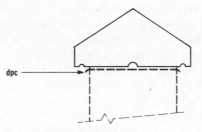

dpc

saddleback clay coping

Figure 9

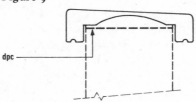

dpc

clayware coping

Figure 10

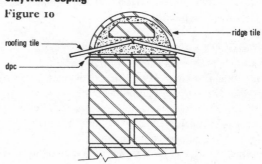

roofing tile

ridge tile

dpc

ridge tile coping

Figure 11

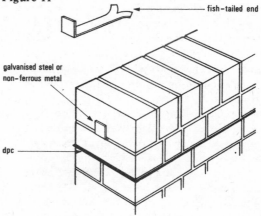

fish-tailed end

galvanised steel or non-ferrous metal

dpc

end cramp for brick on edge type coping

Figure 12

3.0 DESIGN METHOD FOR FREE-STANDING WALLS ADOPTED BY THE GREATER LONDON COUNCIL, DEPARTMENT OF ARCHITECTURE AND CIVIC DESIGN

3.1 Boundary walls
BASIC RULES

a Walls over 1 800 mm in height must be referred to the structural engineer for checking.

b Walls which may form part of future garages must be referred to the structural engineer for advice.

c Walls between 900 and 2 700 mm in height must be built in accordance with Figs. 13–19.*

---

* *Boundary walls in excess of 1 800 mm are required to have approval by the District Surveyor in the GLC area.*

d Walls with piers up to and including 900 mm height may be built without short returns (or staggers) in single leaf brickwork not less than 100 mm thickness.

e The dpc at the base of the wall shall be two courses of bricks having an absorption not greater than 4.5 % set in 1 : 3 sulphate-resisting Portland cement : sand mortar, alternatively, two layers of slate may be used.*

f It may be necessary to adjust the suction rate of some clay bricks.**

g Joints must be pointed as the work proceeds with a flush, weathered or ironed joint.

h Walls constructed of bricks of ordinary quality must be provided with an adequate coping. In areas where the incidence of frost is high, an overhanging coping must be used.

3.2 Mortar mixes

a Clay bricks of ordinary quality†: from foundation to 150 mm above ground level 1 : 3 sulphate-resisting Portland cement : sand mortar; from 150 mm above ground level to the underside of the coping 1 : 6 sulphate-resisting Portland cement : sand mortar with approved plasticizer (eg vinsol resin); coping 1 : 3 waterproofed cement (eg addition of aluminium stearate): sand mortar.

b Clay bricks with a low soluble sulphate content or clay bricks of special quality: from foundation to 150 mm above ground level 1 : 3 Portland cement : sand mortar; from 150 mm above ground level to the underside of the coping 1 : 1 : 6 Portland cement : lime : sand mortar or 1 : 6 Portland cement : sand plasticized mortar; coping 1 : 3 waterproofed cement : sand mortar.

3.3 Testing

Samples of all mortars are to be submitted regularly to the scientific adviser for testing.

Warning

a It is not recommended that staggered walls be used in conjunction with bituminous felt, pitch polymer or polythene dpc's as cracking is likely to occur due to rotation at the 'staggers' or short returns (see CPTB Technical Note Vol. 1, No. 10 'Movement Joints in Brickwork').

b With this type of construction it is particularly desirable to have movement joints at close centres (ie 6–12 m) carefully positioned, so as not to impair the stability of the wall.

---

* *Slate dpc's laid as recommended in BS 743 : 1951 – 'Materials for Damp-proof Courses'.*

** *If clay bricks have a suction rate in excess of 2 kg/m²/min, or the bricklayer considers they will be difficult to lay because of the suction rate, wetting of the bricks may be desirable. Alternatively, to overcome the problem of high brick suction a water-retentive mortar should be used, ie a mortar containing a proportion of lime will improve its water retentivity and resistance to suction.*

† *Clay bricks of ordinary quality may not be satisfactory in some situations outside the GLC area and if any doubts exist the brick manufacturer should be consulted.*

SCHEDULE OF RECOMMENDED GARDEN WALLS

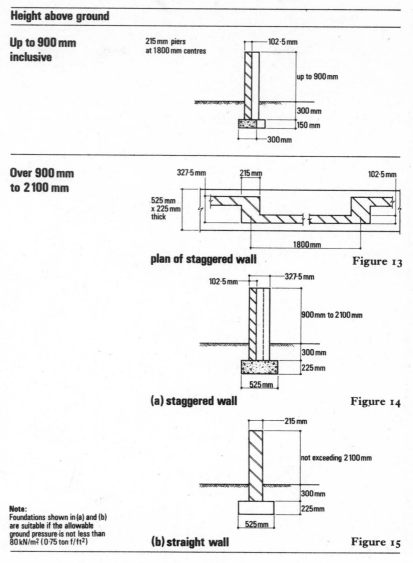

Height above ground

Up to 900 mm inclusive

Over 900 mm to 2100 mm

plan of staggered wall   **Figure 13**

(a) staggered wall   **Figure 14**

Note:
Foundations shown in (a) and (b) are suitable if the allowable ground pressure is not less than 80 kN/m² (0.75 ton f/ft²)

(b) straight wall   **Figure 15**

SETTING OUT OF STAGGERED WALLS

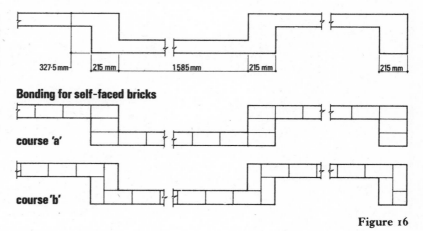

Bonding for self-faced bricks

course 'a'

course 'b'

Figure 16

Bonding for sand-faced bricks
(Where appearance of sand-faced header on the garden side is not acceptable. Alternatively, one sand-faced header can be sanded off on site.)

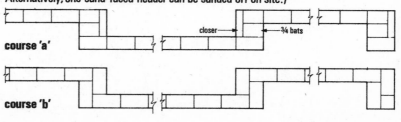

course 'a'

course 'b'

Figure 17

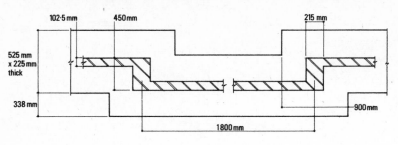

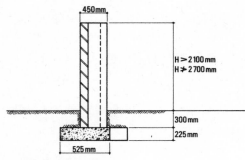

Figure 18   **(a) staggered wall**

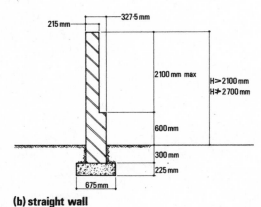

Figure 19   **(b) straight wall**

Masonry retaining walls

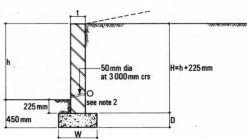

Figure 20

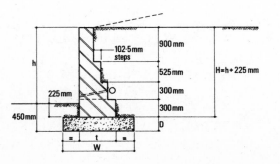

Figure 21

4.0 EARTH RETAINING WALLS

4.1 It is essential that retaining walls should be designed to withstand the likely pressures exerted by the retained material and/or the hydrostatic pressure which could occur.

Plain or reinforced brickwork retaining walls are economic except where the lateral pressures to be restrained are excessive or the height of retained material is beyond practical limits for this type of construction. In some instances the brickwork is used merely as a permanent shutter yet savings of 30 % or more are claimed for reinforced brickwork construction over reinforced concrete retaining walls.

4.2 General

a All free-standing retaining walls should be constructed of suitable bricks and have an adequate coping and a brick or slate dpc at the base as described in section 1.

b Ideally retaining walls should have an impervious lining (on the side adjacent to the retained material) to prevent moisture damaging the mortar and bricks and to prevent unsightly staining on the exposed face.

c In addition to (b) all earth-retaining walls should be provided with weep holes to allow for adequate drainage.

d An alternative to weep holes is drainage at the rear of the wall using land drains laid with open joints (French drain), surrounded by gravel or crushed stone.

e Sulphate-resisting Portland cement (or a strong ordinary Portland cement mix – see section 1) should be used for the mortar joints unless there is absolutely no risk of water penetration; this is particularly important in sulphate-bearing soil.

f Old bricks contaminated with gypsum plaster or bricks with a high soluble sulphate content should not be used for rubble at the back of the wall.

g In cavity wall construction stainless steel or non-ferrous metal ties are recommended.

h If the wall is likely to be permanently wet, bricks of special quality should be used.

i Thermal and moisture movement should be allowed for in all long lengths of wall.

4.3 Design method for free-standing walls adopted by the Greater London Council Department of Architecture and Civic Design

Table 2

h max mm	H mm	t mm	W mm 110 kN/m²	W mm 55 kN/m²	D mm
900	1 125	215	525	525	225
1 200	1 425	327.5	600	600	225
1 500	1 725	440	675	900	225
1 800	2 025	552.5	750	1 050	225

The above details are based on the following:
No surcharge and slope of retained earth not greater than 1 : 10.
Safe minimum bearing pressures:
Granular soil 110 kN/m² (1 tonf/ft²).

Cohesive soil 55 kN/m² (0.5 tonf/ft²).

No dpc (other than dpc bricks or slates).

Minimum crushing strength of bricks 20.5 MN/m².

Mortar mix – $1 : \frac{1}{4} : 3$ Portland cement : lime : sand □

Notes

1 In sulphate bearing ground or when the bricks have a high soluble sulphate content a sulphate-resisting cement should be used for the concrete and mortar; alternatively a richer mix based on ordinary Portland cement – see section 1.1.

2 In very wet conditions a French drain should be provided.

3 Brick veneers may be included in the wall thickness (t) if the normal wall tie spacings are halved.

4 Masonry cement and plasticizers should not be used.

5 The wall may be of uniform thickness throughout or be stepped as shown in Figs. 20 and 21.

6 Movement joints are to be provided at centres not exceeding 15 m.

REFERENCES

1 British Standards Institution. **'Specification for Bricks and Blocks of Fired Brickearth, Clay or Shale'**, BS 3921, Part 2: Metric Units: 1969.

2 British Standards Institution. **'Specification for Calcium Silicate (sandlime and flintlime) Bricks'**, BS 187, Part 2: Metric Units: 1970.

3 British Standards Institution. **'Specification for Materials for Damp Proof Courses'**, BS 743: 1970.

4 Thomas, K., **'Movement Joints in Brickwork'**, CPTB Technical Note, Vol. 1 No. 10, July 1966.

5 British Standards Institution. **'Structural Recommendations for Loadbearing Walls'**, Code of Practice 111: Part 2: 1970.

Workmanship factors in brickwork strength

A W Hendry BS(Eng) PhD DSc FIStructE FICE
Professor, Department of Civil Engineering and Building Science, Edinburgh University

1.0 INTRODUCTORY

In discussions of structural brickwork, the problem of 'workmanship' in relation to wall strengths is frequently raised and the possibility of large differences in strength between 'laboratory' and 'site' brickwork is pointed out. The intention of this note is to identify the main factors in workmanship and to gather together such information as exists which may help to appreciate their relative importance. We are not here concerned with gross errors or omissions on site, such as the use of the wrong bricks or mortar materials or with the variability of materials as such, but with the effect of various defects in the actual brick laying process on the strength of the resulting masonry.

The most obvious workmanship factors are as follows:

– Incorrect proportioning and mixing of mortar.

– Incorrect adjustment of suction rate of bricks.

– Incorrect jointing procedures.

– Disturbance of bricks after laying.

– Failure to build wall 'plumb and true to line and level'.

– Failure to protect work from the weather.

In practice, brickwork is liable to suffer, in greater or less degree, from all of these defects and it is clearly very difficult to determine their separate effects, let alone their interactions. Nevertheless, there is in existence a certain amount of relevant information both on the separate factors and on the overall effect of workmanship which we may review in that order.

2.0 INCORRECT PROPORTIONING AND MIXING OF MORTAR

The effect of mortar strength on the strength of brickwork has been studied in considerable detail on an empirical basis[1,2] and more recently in relation to the biaxial strength of the component materials[3]. The results of these investigations indicate that mortar strength is not a very critical factor in brickwork strength – for example with bricks of crushing strength 5000 lbf/in² (35 MN/m²) a halving of mortar cube strength from 2000 lbf/in² to 1,000 lbf/in² (14 MN/m² to 7 MN/m²) may be expected to reduce the compressive strength of brickwork from about 2300 lbf/in² (16 MN/m²) to 2000 lbf/in² (14 MN/m²). This corresponds roughly to a change in mortar mix from 1 : 3 cement to sand to 1 : 4½ or, say 30 % too little cement in the mix. A similar reduction in mortar strength could of course be brought about by an excess of water – moving from a water/cement ratio of about 0.6 to 0.8 in a typical case. McIntosh[4] has stated that 'the cement content of a 1 : 1 : 6 mix composed of cement gauged with ready-mixed lime: sand for mortar, could vary from about 13½ to 19 % of the weight of dry lime plus dry sand; the corresponding range in water: cement ratio required to produce mortar of standard consistence was from about 1.6 to 1.1 resulting in a change of 7 day strength from 420 to 940 lbf/in² (2.90 to 6.50 MN/m²). Greater differences might be expected if all the materials are batched separately on site.'

It is clear, therefore, that a 2 : 1 variation in site produced mortar to the same nominal specification is possible but fortunately the effect on brickwork strength is proportionately very much less than this and it may be expected that with reasonable care this factor will not be very serious. This conclusion must, however, be qualified in the light of the comments made in paragraph 3.

3.0 INCORRECT ADJUSTMENT OF BRICK SUCTION RATE

In order to achieve optimum brickwork strength it has long been realised that the suction rate of bricks should be controlled to prevent excessive removal of water from the mortar. It seems probable that the water absorbed by the bricks leaves cavities in the mortar[11] which fill with air and result in a weakened material on setting. On the other hand, brickwork built with saturated bricks develops poor adhesion between bricks and mortar[12] and is of course susceptible to frost damage and other troubles. Some specifications recommend a limiting suction[13] or alternatively the use of a high retentivity mortar to control the extraction of water from mortar.

Insofar as water extraction affects the final strength of mortar, one would not expect it to result in serious weakening of brickwork in compression. However, Haller has demonstrated[2] that suction rate has a considerable effect on brickwork strength because de-watered mortar tends to form a rounded joint during building owing to loss of elasticity. It would appear that with eccentric loading, an increase in the suction rate from 2 kg/m²/min. (the limit suggested in BCRA SP56) to 4 kg/m²/min. could halve the strength of the brickwork* (opposite page). It is clear, therefore, that this is a factor to be taken very seriously especially in the case of slender walls built in relatively low strength bricks.

Figure 1

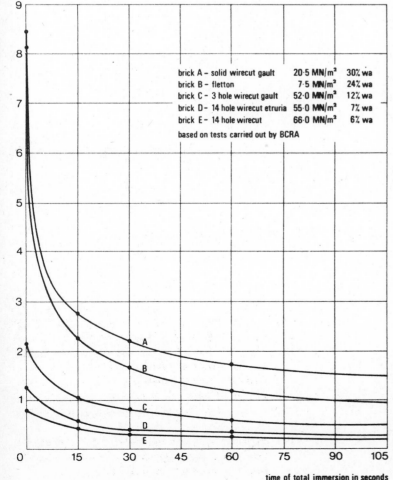

brick A – solid wirecut gault	20·5 MN/m²	30% wa
brick B – fletton	7·5 MN/m²	24% wa
brick C – 3 hole wirecut gault	52·0 MN/m²	12% wa
brick D – 14 hole wirecut etruria	55·0 MN/m²	7% wa
brick E – 14 hole wirecut	66·0 MN/m²	6% wa

based on tests carried out by BCRA

time of total immersion in seconds

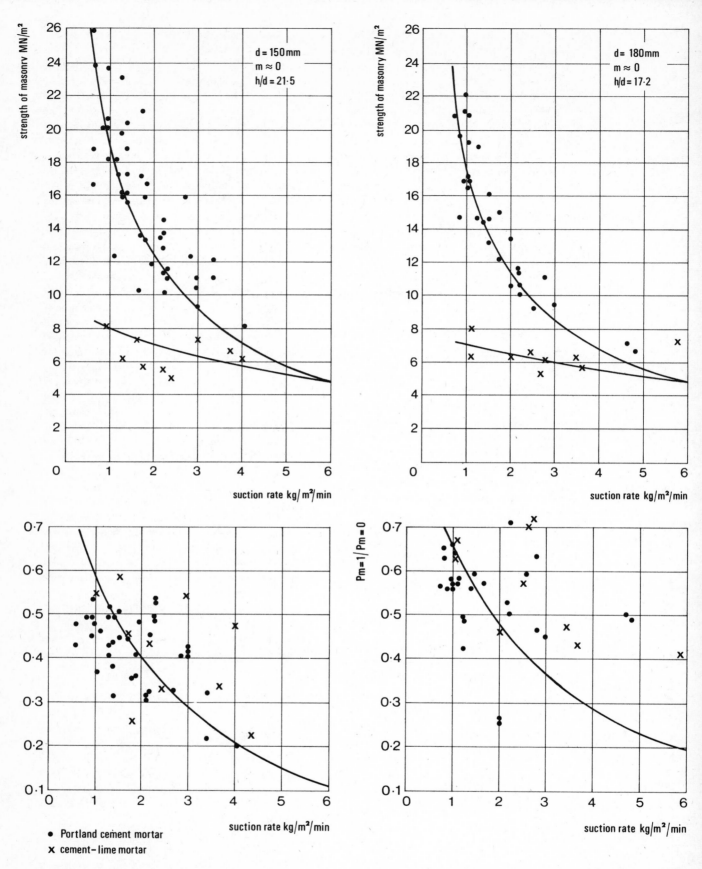

- ● Portland cement mortar
- ✕ cement-lime mortar

All suction rates should be divided by 10 to convert to kg/m²/min

All crushing strengths should be divided by 10 to convert to MN/m².

* That this could happen through failure to wet certain low strength bricks is clear from fig. 1 which is based on results supplied by BCRA. Refer also to Haller's results shown in Fig. 2.

Figure 2 *Masonry made of bricks from various brickworks. Wall thickness 5⁷/₈in (150mm) and 7¹/₁₀in (180mm). Cement mortar and lime cement mortar. Bearing strength as a function of absorptive capacity of bricks in g/min dm² (g/min 30sq in = 1,94 g/min dm²). Loading: Centric and at kern edge. (the large scatter is due to differences in brick strength, perforation etc).*

From Haller **Load Capacity of Brick Masonry. Designing, Engineering and Constructing with Masonry Products.** Ed F B Johnson.

4.0 INCORRECT JOINTING PROCEDURES

A variety of defects can arise from incomplete filling of joints and some evidence is available on the structural effects of these. The effect of incomplete filling of perpend joints has been investigated by the British Ceramic Research Association[5] and by the Building Development Laboratories of Australia[6]. A total of thirty walls was tested at BCRA with unfilled vertical joints, using two types of brick and mortar with the results shown in Table 1. Statistical examination of the results showed that there was no significant difference between corresponding sets of walls with joints filled and unfilled. The Australian tests confirmed that unfilled vertical joints had no significant effect on the strength of walls. There are also theoretical reasons for expecting that this would be the case although, of course, careless filling of vertical joints may be indicative of poor workmanship in other respects and may be unacceptable for other reasons than strength, for example, loss of sound insulation or possibility of rain penetration.

Incomplete filling of bed joints is, however, quite another matter and has been investigated by the Structural Clay Products Institute[7] in the USA and by the Building Development Laboratories[6], Australia. In the SCPI tests, the results of which are summarised in Table 2, the 'uninspected' workmanship included unfilled vertical joints as well as deeply furrowed bed joints and resulted in a reduction of strength of about 33 %. As it is known that unfilled vertical joints do not affect strength significantly, it may be assumed that most of this reduction arose from the furrowed joints. The Australian tests (Table 3) show a reduction of similar magnitude from this cause.

The third factor in brickwork jointing is that of thickness. This has also been investigated by the Building Development Laboratories, Australia, at the Universities of Edinburgh[8] and Melbourne[9] and elsewhere[10]. It is difficult to compare the results of all these investigations but it has been shown beyond doubt that excessively thick bed joints – say $5/8$–$3/6$in (16–19mm) – may be expected to reduce the strength of brickwork by something of the order of 30 % as compared with normal $3/8$in (10mm) thick joints. This is of the same order of magnitude as the reduction caused by deep furrowing but excessively thick joints are at least easily seen.

Another laying defect arises from the practice of spreading too long a bed of mortar – only sufficient mortar should be spread as will permit bricks being set in plastic mortar. There is, however, no quantitative data on the effect of this defect on brickwork strength.

5.0 DISTURBANCE OF BRICKS AFTER LAYING

Any disturbance of bricks after they have been placed will result in the bond between bricks and mortar being broken with possible adverse effects on strength. This commonly happens at corners when the bricklayer attempts to correct plumbing errors by hammering bricks into a true plumb position. There is no quantitative data available on the effect of disturbance on the strength of brickwork but it could be significant in situations where bond tension and shear strengths are critical.

6.0 FAILURE TO BUILD WALL 'PLUMB AND TRUE TO LINE AND LEVEL'

This type of defect can give rise to eccentric loading in a wall under compression and thus to reduced strength. Information on this is available from tests carried out at Edinburgh University[14] and at Building Development Laboratories[6], Australia. A summary of the Edinburgh results is shown in Table 4. In these tests $4^1/_2$in (113mm) walls were tested in compression between reinforced concrete slabs to give realistic end conditions. Two storey-height walls were tested in which the applied load was $3/_4$in (19mm) eccentric with respect to the axis of the wall and three walls were built $3/_4$in (19mm) off plumb. Comparing the strength of the walls with eccentrically applied loads with corresponding axially loaded walls indicates a reduction in strength of the order of 12 %. The reduction for those built off-plumb is about 20 %. In the Australian test (c.f. Table 3) similar walls were built with a $1/_2$in bow resulting in a 13 % strength reduction as compared with a truly plumb wall.

A recent survey by the Building Research Station[15] suggests the following permissible deviations in load bearing brickwork construction:

Wall plumb over a storey height $\pm$ 15 mm (approx $5/_8$in)

Vertical alignment between top and bottom of walls of successive storeys $\pm$ 20mm (approx $3/_4$in.)

These figures are similar in magnitude to those used in the tests which therefore give a reasonable indication of the maximum probable reduction in strength arising from lack of plumb and vertical alignment.

7.0 FAILURE TO PROTECT WORK FROM THE WEATHER

Newly completed brickwork can be adversely affected by exposure to unfavourable weather conditions, including curing under very hot conditions, frost damage and damage by rain. Some information is available on the effect of the first two of these conditions. The Building Development Laboratories report a series of tests on walls which were in temperatures between 78°F and 100°F (25.6°C to 37.8°C) and cured in the sun for five–six days. These walls showed about a 10 % reduction in strength as compared with walls cured in the shade under polythene.

At the other end of the climatic scale tests have been carried out in Norway and Finland to examine the effect on brickwork properties of laying and curing at low temperatures[16]. Masonry piers of 1m height were built in various mortars at room temperature and in cold rooms at temperatures down to —15°C; curing of the piers built at low temperatures were carried out at —15°C; The results showed, perhaps surprisingly, no deterioration in strength as between the walls built at room temperature and at in cold conditions (Table 4). On the other hand the liability of masonry built under freezing conditions to develop undesirable deformations is pointed out and one would suspect that this could give rise to indirect reduction of strength as a result of bowing or lack of plumb.

Under British conditions and with obvious and well known precautions it would appear that the curing factor is probably not of first order importance.

No information is known about the effects of damage by rain.

8.0 OVERALL EFFECTS OF WORKMANSHIP ON BRICKWORK STRENGTH

In the foregoing paragraphs the separate effect of a number of workmanship factors has been discussed. In any particular case, these defects will be present in varying degrees and the overall strength of the brickwork will reflect their combined effect. Various efforts have been made to assess the overall effect of workmanship on the strength of brickwork, the most systematic being the programme carried out by the Building Development Laboratories of Australia already referred to[6], in which controlled defects were introduced separately and in combination. The combined effect of outside curing, deep bed furrowing, unfilled perpends, $5/_8$in. (16mm) bed joints and $1/_2$in (12.7mm) bow was to reduce the wall strength from 3100 lbf/in² to 1200 lbf/in² (21.4 MN/m² to 8.3 MN/m²) (ie 61% reduction). This is generally consistent with experiments at the National Bureau of Standards, USA, where unsupervised site brickwork was from 55 to 62% of the strength of supervised brickwork.*

The Australian report makes the following assessments of the relative importance of the various defects in terms of the probable reduction in strength of a wall built under laboratory conditions:

Outside cure (warm conditions)	10%
Furrowed bed	25%
$5/_8$in (16mm) thick bed joints	25%
Perpend joints unfilled	Nil
$1/_2$in (12.7mm) bow	15%

It was concluded that these are not interactive and that the separate factors are additive.

A comprehensive formula to quantify the variables in brickwork construction has been put forward by Grimm[17] as follows:

$$f'_m = 0.087\,\omega\delta\varepsilon \left\{ 4590 + f'_b \left[1 + 1.29a\beta\gamma \left(1 + 1.46 \log C/L \right) \right] \right\}$$

where f'_m ultimate compressive strength of the brickwork in lbf/in².

ω – workmanship coefficient (associated with furrowed joints)

δ – slenderness coefficient of brickwork test specimen = $0.0178 \left\{ 57.3 - [(h_s/t_s - 6]^2 \right\}$ where $5 > (h_s/t_s) > 2$

ε – joint thickness coefficient = $1.23 (1.19 - t_j)$

t_j – mortar bed joint thickness in inches $(0.75 > t_j > 0.25)$

f'_b – compressive strength of brick in lbf/in²

─────────

* *A further study of the relative strength of supervised and unsupervised brickwork will be found in reference 18.*

a – specimen age coefficient = $0.3 (1 + 1.59 \log d)$ where a is the age in days.

β – water cement coefficient = $15.67/7^{1.3}$ (w/c)

W/C – ratio of water to cement by weight $(0.7 > W/C > 0.4)$

γ – air content coefficient = $0.021 (57.3 - A)$

A – percentage air by volume in the mortar $(30 \geqslant A)$

C/L – ratio of cement to lime by volume $4 > (C/L) > 1$

The probable extreme range of variables in Grimm's equation are as follows:
ω between 1 and 0.67

ε between 0.54 and 1.16 ($3/_4$in to $1/_4$in) (10 to 6mm) joint thickness)

β between 2.3 and 0.79 where w/c = 0.45 to 0.70

γ between 1.2 and 0.68 where A = 0 to 25%

The coefficient δ may be taken as unity for test specimens for which h/t = 5 and $a = 1$ for age of specimen $d = 28$ days.

The interest in this formula in the present context lies in its use as a basis of comparison of the relative importance of various factors involved. This has been done in reference 16 by calculating a 'significant index' for each variable equal to the difference between the highest and the lowest values of f'_m produced by substituting the high and low extremes of a particular variable in the formula divided by 10% of the average value of the range of f'_m obtained by substituting all the highest and all the lowest values. For example, for an 8,000 lbf/in² (55 MN/m²) brick the average f'_m on substitution of the low and high extremes is 2425 lbf/in² (17 MN/m²). If now the normal values of all the variables except, say, workmanship are substituted the following values are obtained:

For $\omega = 1$ $f'_m = 2386$ lbf/in² (16.5 MN/m²)
For $\omega = 0.67$ $f'_m = 1590$ lbf/in² (11.0 MN/m²)
Difference: 796 lbf/m² (5.5 MN/m²)
Significant index = $796/0.1 \times 2425 = 3.3$.

Comparative values of significant indexes calculated in this way for several variables are as follows:

Mortar	
Air content	2.7
Cement-lime ratio	3.3
Water-cement ratio	8.1

Mason	
Workmanship (filling of joints)	3.3
Joint thickness	7.9

It is difficult to compare these figures directly with results quoted in earlier paragraphs but it is clear that as far as the mortar is concerned the water-cement ratio is the important factor. This, however, is in turn affected by the suction rate of the brick and probably affects the brickwork strength in the manner described by Haller (para. 3) rather than by its direct influence on mortar strength. Grimm's analysis seems to place rather less weight on joint-filling than the Australian results which rate joint-filling and joint thickness equally.

9.0 CONCLUSIONS

This is quite clearly not a problem which is susceptible to precise quantification but enough is known to be able to make a reasonable appreciation of the effect of the various workmanship factors considered. In order of relative importance they might be noted as follows:

Incorrect adjustment of suction rate of low strength bricks, especially in slender walls.

Failure to fill bed joints.

Bed joints of excessive thickness.

Deviation from vertical plane or alignment.

Unfavourable curing conditions.

The combined effect of all these factors could under the most unfavourable circumstances result in a wall which was perhaps only half of its intended or possible strength. However, site supervision and control procedures would seem to be comparatively straight-forward and compliance with the BCRA Model Specification[13] will result in brickwork strengths close to those to be expected from walls built in the laboratory.

It remains to suggest that there should be differentiation in CP.111 as between brickwork built in accordance with the Model Specification[13], or similar, with adequate supervision and brickwork built without such control☐

REFERENCES

1 Thomas F. G. 'The Strength of Brickwork', Structural Engineer, Vol. 31, No. 2, 1953, p. 35.

2 Haller, P. 'The Physics of the Fired Brick Part 1. Strength Properties', Trans. G. L. Cairns, 1960, BRS Library Communication, 929.

3 Khoo C. L. and Hendry A. W. 'Strength Tests on Brick and Mortar under Complex Stresses for the Development of a Failure Criterion for Brickwork in Compression', Proc. BCeramSoc. (In press).

4 McIntosh, J. D. Specifying the Quality of Bedding Mortars', Proc. BCeramSoc. No. 17, Feb. 1970 pp 65–82.

5 'Investigation of the Effect on Brickwork of not Filling Vertical Mortar Joints', BCRA, Internal Report.

6 Investigation of the Effect of Workmanship and Curing Conditions on the Strength of Brickwork, Report No. W/Work/1, 1971, Building Development Laboratories Pty. Ltd., Morley, Australia.

7 Gross, J. G. Dikkers, R. D., and Grogan, J. C. 'Recommended Practice for Engineered Brick Masonry', SCPI, 1969.

8 Hendry, A. W., Bradshaw, R. E., and Rutherford, D. J. 'Tests on Cavity Walls and the Effect of Concentrated Loads and Joint Thickness on the Strength of Brickwork' CPTB, Research Note Vol. 1, No. 2, 1968.

9 Francis A. J., Horman, C. B., and Jerrens, L. E. 'The Effect of Joint Thickness and other Factors on the Compressive Strength of Brickwork, SIBMAC Proc., BCRA, 1971.

10 Monk, C. B. 'An Historical Survey and Analysis of the Compressive Strength of Clay Masonry', Structural Clay Products Research Foundation, 1965.

11 Plowman J. M. 'The Effect of Suction Rate of Bricks on the Properties of Mortar', Proc. BCeramSoc. (In press.)

12 Sinha, B. P. 'Model Studies Related to Load Bearing Brickwork' Ph.D. thesis. Univ. of Edinburgh, 1968, Ch. 5.

13 'Model Specification for Load Bearing Brickwork', BCRA Special Publication 56, Rev. ed., 1971.

14 Bradshaw, R. E. and Hendry, A. W., 'Further Crushing-tests on Storey-height Walls $4\frac{1}{2}$in Thick', Proc. BCeramSoc. No. 11, 1968.

15 Milner, R. M. and Thorogood, R. P., 'Accuracy and its Structural Implications for Load Bearing Construction', Proc. BCeramSoc. (In press.)

16 Svendsen, S. D. and Waldum, A. 'Some Remarks on Winter Masonry', Murmesteren 1–2; 3–17, 1966. Trans. D. A. Sinclair NRC Canada, TT – 1456, 1971.

17 Howard J. W., Hockaday R. B. and Soderstrum, W. K., 'Effect of Manufacturing and Construction Variables on Durability and Compressive Strength of Brick Masonry', Designing, Engineering and Constructing with Masonry Products, ed. F. B. Johnson, Gulf Pub. Co., 1969. pp. 310–316.

18 McDowall, I. C., McNeilly, T. H. and Ryan, W. G., 'The Strength of Brick Walls and Wallettes', Brick Development Research Institute, Melbourne, Special Report No. 1, 1966.

Table 1. **Effect of unfilled perpend joints on strength of brick walls.** *Tests carried out by British Ceramic Research Association*

Mean brick strength and WA		Mortar	Mean mortar cube strength		Wall thickness and bond	Cross-joints filled or unfilled	Wall strength			
MN/m²	lbf/in²		MN/m²	lbf/in²			MN	Tons	MN/m²	lbf/in²
92	13,380	1 : ¼ : 3	19.30	2809	102.5 (4½in)	Filled	2.65	265	18.40	2670
7.9%			15.30	2233	Stretcher		3.05	305	21.16	3070
			19.65	2850			3.02	302	20.89	3030
			18.13	av.2631					20.15	av.2923
			16.89	2455		Unfilled	2.98	298	20.75	3010
			16.27	2368			2.03	203	13.99	2035
			20.54	2985			3.37	337	23.37	3390
			17.92	av.2603					19.37	av.2812
92	13,380	1 : ¼ : 3	15.65	2276	215 (9in)	Filled	7.12	712	23.64	3430
			22.27	3232	English		5.10	510	16.75	2430
7.9%			19.03	2763			7.20	720	24.06	3490
			18.96	av.2757					21.44	av.3117
			20.54	2980		Unfilled	5.43	543	17.99	2610
			19.30	2805			5.85	585	19.37	2815
			18.34	2665			4.91	491	16.41	2380
			19.37	av.2817					17.92	av.2602
92	13,380	1 : ¼ : 3	15.72	2280	215 (9in)	Filled	5.49	549	18.27	2650
			17.03	2470	Flemish		7.20	720	23.85	3460
7.9%			15.10	2198			5.08	508	16.89	2450
			15.92	av.2316					19.65	av.2853
92	13,380	1 : ¼ : 3	16.68	2423	275 (11in)	Filled	6.17	617	21.37	3100
			13.85	2018	Cavity		5.98	598	21.09	3060
7.9%			13.37	1943			5.44	544	19.92	2890
			14.61	av.2128					20.75	av.3017
46	6,690	1 : ¼ : 3	13.79	2009	102.5 (4½in)	Filled	2.22	222	15.65	2270
			16.27	2368	Stretcher		2.28	228	16.68	2420
14.5%			17.51	2548			1.94	194	13.72	1990
			15.85	av.2308					15.30	av.2227
			15.44	2240		Unfilled	1.90	190	13.79	2000
			18.05	2625			1.81	181	12.75	1850
			10.41	1518			1.52	152	10.75	1560
			14.61	av.2128					12.41	av.1803
46	6,690	1 : 1 : 6	5.94	862	102.5 (4½in)	Filled	1.50	150	10.48	1520
			4.27	620	Stretcher		1.39	139	9.72	1417
14.5%			4.37	635			1.34	134	9.44	1371
			4.86	av.706					9.85	av.1436
			5.37	780		Unfilled	1.18	118	8.27	1208
			4.88	708			1.45	145	10.20	1481
			4.56	659			1.04	104	7.30	1062
			4.96	av.716					8.61	av.1250

Table 2. Effect of workmanship on the compressive strength of non-reinforced brick walls (a). *From – Recommended Practice for Engineered Brick Masonry by Gross, Dikkers, and Grogan.*

thickness type	Mortar type	h/t	Workman-ship(b)	Ultimate load kips MN/m²		Average ultimate stress psi MN/m²		Relative effect of workmanship	Average allowable stress psi(c) MN/m²		Ultimate stress allowable stress
91mm (3.6in) Single Wythe(d)	M	22.7	I	1.05	236	18.4	2675	1.00	2.30	334	8.0
			U	0.69	156	12.2	1770	0.66	1.54	224	7.9
	N	22.7	I	0.62	139	10.8	1575	1.00	1.61	234	6.7
			U	0.39	88	6.84	993	0.63	1.08	157	6.3
203mm (8.0in) Multi Wythe (metal-tied)(d)	M	20.5	I	2.03	455	16.0	2330	1.00	2.49	362	6.4
			U	1.38	310	11.0	1590	0.68	1.65	241	6.6
	S	20.5	I	1.54	346	12.2	1775	1.00	2.12	308	5.8
			U	1.00	226	7.92	1155	0.65	1.42	206	5.6
	N	20.5	I	1.12	253	8.82	1285	1.00	1.75	254	5.1
			U	0.75	170	6.02	874	0.68	1.17	170	5.1

Data from unpublished SCPI tests. Walls were tested with hinged ends; eccentricity at top = t/6 and eccentricity at bottom = 0; e_1/e_2 = 0.
Walls tested at age of 14 days. Metal-tied walls contained one 4.8mm ($^3/_{16}$in) steel tie for each 2.7 sq ft of wall area. Masonry-bonded walls contained headers comprising 10.7 per cent of wall area.
I = Inspected; U = Uninspected.
Ce C_s (0.20f'm). Values of f'm were assumed.
Brick compressive strength = 81 MN/m² 11,760 psi.
Brick compressive strength = 143 MN/m² 20,660 psi.

Table 3. Summary of wall & specimen strengths & their relationship to the 'no faults' wall & specimen strengths. F
investigation of the effect of workmanship and curing conditions on the strength of brickwork. Building Development Laboratories Pty
Melbourne

Wall type	Wall strength		Average f'm'δ prisms		Average bond-piers		Strength relationship to no faults specimens		
							Walls	f m prisms	Bond piers
	MN/m²	lbf/sqin	MN/m²	lbf/sqin	MN/m²	lbf/sqin			
No faults	21.2	3070	18.1	2620	0.524	76			
	21.5	3120	18.9	2740	0.613	89			
Average	21.4	3100	18.5	2680	0.565	82	1.00	1.00	1.00
Outside cure	19.0	2760	15.0	2180	0.310	45			
	20.8	3010	18.6	2700	0.351	52			
Average	19.9	2880	16.8	2440	0.330	48	0.93	0.91	0.59
Furrowed	16.1	2330	15.0	2170	0.841	122			
Bed	16.1	2340	14.3	2070	0.792	115			
Average	16.1	2340	14.6	2120	0.813	118	0.76	0.79	1.44
No perpends	21.9	3180	19.2	2780	0.717	104			
	21.7	3140	21.9	3170	0.579	84			
Average	21.8	3160	20.5	2980	0.648	94	1.02	1.11	1.15
5/8 inch bed joints	16.6	2410	14.1	2040	0.448	65			
	15.2	2210	14.8	2140	0.482	70			
Average	15.9	2310	14.4	2090	0.468	68	0.75	0.78	0.83
1/2 inch bow	19.8	2870	19.0	2760	0.620	90			
	17.5	2540	18.8	2730	0.565	82			
Average	18.6	2700	18.9	2740	0.592	86	0.87	1.02	1.05
All faults	8.27	1200	6.75	980	0.158	23			
	8.20	1190	8.13	1180	0.186	27			
Average	8.27	1200	7.44	1080	0.172	25	0.39	0.40	0.30

all .*	Brick strength (MN/m²)/ (lbf/in²)	Mortar strength (MN/m²)/ (lbf/in²)	Ultimate load (MN/m²)/ (tons)	Average compressive stress (MN/m²)/ (lbf/in²)	Permissible design stress (CP 111: 1964) (MN/m²)/ (lbf/in²)	Load factor	Duration of loading (min)	Strength ratio brick-work brick	Loading	Age (days)	Remarks
1)	43.0 6235	13.9 2015	1.2 120	11.2 1630	1.49 217	7.5	90 to 92 tons (12) 90 to 120 tons (18)		Axial (t/17)	46	
1)	43.0 6235	10.9 1585	1.15 114.5	12.3 1790	1.49 217	8.2	105 to 114.5 tons (17)	0.29	Axial (t/55)	18	Joint Thickness: 3/16 in
1)	43.0 6235	10.6 1540	1.35 135	14.6 2110	1.49 217	9.7	90 to 135 tons (23) 15 to 135 tons (141)	0.34	Axial (t/40)	18	
1	43.0 6235	16.9 2450	1.57 157	16.9 2450	1.49 217	11.3	30 (82)	0.39	Axial (t/62)	78	
1/w)	43.0 6235	5.96 865	1.50 150	16.1 2330	1.49 217	10.7	30 to 90 tons (48) 8 to 150 tons (291)	0.37	Axial (t/67)	2	
1/w)	43.0 6235	16.1 2335	1.72 172	18.4 2670	1.49 217	12.3	20 to 22.5 tons (17) 60 to 172 tons (44)	0.43	Axial (t/24	18	

Table 5. Compression tests on piers. *From 'Some Remarks on Winter Masonry' by Svendsen and Waldum*
Mortar 1, KC 50/50/610

Pier no.	Laid in	Brick	Breaking load		Average		Breaking strength		Cracking load
			MN	t	MN	t	MN/m²	kp/cm²	Breaking load
1	S	t	0.53	52.7	0.53	53.4	9.88	98.8	0.57
2			0.54	54.1					0.53
3	S	v	0.74	74.3	0.77	76.6	14.2	142.1	0.79
4			0.79	79.0					0.73
5	K2	t	0.58	57.6	0.56	56.0	10.3	103.2	0.59
6			0.54	54.4					0.56
7	K2	v	0.71	71.1	0.73	73.1	13.5	135.1	0.65
8			0.75	75.0					0.72
9	K1	t	0.59	58.6	0.60	60.2	11.1	111.3	0.64
10			0.62	61.8					0.63
11	K1	v	0.74	74.3	0.74	73.5	13.6	136.1	0.70
12			0.79	79.0					0.73

Mortar 2, KC 35/65/520

Pier no.	Laid in	Brick	Breaking load		Average		Breaking strength		Cracking load
13	S	t	0.60	60.0	0.60	59.9	11.1	110.9	0.55
14			0.60	59.7					0.62
15	S	v	1.12	112.2	1.03	102.7	19.1	190.6	0.87
16			0.93	93.2					0.91
17	K2	t	0.65	65.4	0.66	66.3	12.3	122.6	0.63
18			0.67	67.1					0.60
19	K2	v	0.99	98.6	0.97	97.0	18.0	179.5	0.77
20			0.95	95.4					0.87
21	K1	t	0.60	59.7	0.63	63.1	11.7	116.5	0.57
22			0.67	66.5					0.63
23	K1	v	0.86	85.5	0.84	83.7	15.5	154.6	0.79
24			0.82	81.9					0.70
25	K1	vv	0.80	80.4	0.83	83.2	15.4	154.1	0.81
26			0.86	86.0					0.69

Mortar 3, KC 20/80/440

Pier no.	Laid in	Brick	Breaking load		Average		Breaking strength		Cracking load
27	S	t	0.69	69.3	0.70	69.7	12.9	129.2	0.72
28			0.70	70.0					0.70
29	S	v	1.24	124.2	1.15	114.5	21.3	212.7	0.89
30			1.05	104.8					0.90
31	K2	t	0.79	79.0	0.75	74.9	13.9	138.5	0.78
32			0.71	70.7					0.62
33	K2	v	1.23	122.5	1.18	118.2	21.9	219.0	0.87
34			1.14	113.9					0.72
35	K1	t	0.70	69.8	0.75	74.5	13.8	137.9	0.64
36			0.80	80.1					0.85
37	K1	v	1.17	117.4	1.17	116.5	21.6	215.6	0.82
38			1.16	115.6					0.90

S = Laboratory
vv = warm, wet bri[c]
v = wet brick
t = dry brick

K1 = Room at —15°
K2 = Room at + 6—
during laying and the
reduced slowly to —

Brick floors and brick paving

K Thomas MSc CEng MIStructE FIOB ARTC
Director of Research, Liverpool Polytechnic, School of Architecture
and
L Bevis AIPHE
Technical Officer, Brick Development Association

Figure 1 *Brick paving with inset slabs*

Figure 2 *Circular paving*

INTRODUCTION

For many years architects have shown great enth
siasum for brick paving and a regular flow of enquiri
has been received by the technical staff of the vario
bodies representing the brick industry. The autho
while colleagues at the Brick Development Associ
tion received many such enquiries and felt that
might be a useful exercise to summarise the advi
given in the form of a Technical Note. The followi
is therefore a brief summary of information curren
available on this subject.

GENERAL

The use of bricks for paving and flooring has a histo
stretching back to Roman times when considerab
use was made of thin bricks or tiles for internal floo
ing and paving externally; a number of excellent e
amples still exist.

Extensive use of the brick for domestic flooring a
particularly for flooring in stables and other agric
tural buildings went on throughout the industr
development and many dwellings erected during th
period still have durable brick floors.

Abroad, notably in Belgium and Holland, brick h
specially been brought into use for paving industr
ways and for general road work. In this country, th
use has been on a more conservative scale, althou
large areas of pavings are still proving the durability
bricks in this situation and particularly fine examp
can be seen in some public buildings, such as fi
stations and hospitals, and on the domestic scene
stables and similar situations.

For the protection of industrial processing pla
extensive use is made of specialised types of pavin
(Engineering variety); these paviors are genera
smooth red or smooth blue produced in the Staffor
shire areas and achieving a surface which makes
easy cleaning of the flooring area.

More recently, brick has been more in demand
landscaping in cities and there is a growing use of bri
for domestic and commercial interiors. The value
brick in this context lies in the great variety of colo
and textures which are available throughout t
country, flexibility of the brick module and the ea
with which replacement can be completed.

PHYSICAL PROPERTIES

The physical properties of bricks vary widely, and
cannot be emphasised too strongly that *all bricks*
not suitable for all purposes. Correct selection of un
and design details should ensure a high standard
performance which will be difficult to achieve w
alternative materials, and under these circumstanc
the following functional properties should be sustai
ed:

An attractive appearance which mellows with age.

Good resistance to abrasion.

Good skid resistance

Durability or resistance to decay, particularly fr
resistance.

Economy.

Structural strength and good resistance to impact.

Fire resistance, sound insulation.

Satisfactory absorption and porosity, when used
industrial flooring.

Low thermal and moisture movement.

Flexibility in application.

Little or no maintenance.

APPEARANCE

Appearance is dependent on the colour and size of the bricks used and upon the type and colour of jointing specified. Engineering bricks and paviors used in industrial situations are usually smooth faced and in red, blue, brindled or brown colours. For the more sophisticated process, floor paviors meeting the requirements of BS 3679[1] for acid-resistance may be in dimple pattern for slip resistance or chequered or in 2, 4, 6, 8 or 32 panels and are more usually blue or red in colour.

For non-industrial and domestic situations a wide range of colours and textures are available. The majority of well-burnt bricks are suitable for use if an 'on-edge' application is adopted, and colours may range from: red to varying browns, blues to purple, yellows to dark ochre and intermingled colours, and in textures: smooth, sand-faced, sand-creased, combed, stippled or dragged. Bricks, particularly of the solid wire-cut type, can be used in the 'on flat' position, which is more economic (providing they are not specials as discussed under dimensions); a thickness which can ease level problems and increase the range of colour and texture availability. The soft mud-type brick manufactured in clamp or kiln frequently produces a rough hard-burnt stock which is eminently suitable for flooring or paving work and can be obtained in a splendid range of reds, ochres or blues.

PATTERNS OF LAYING

There is a wide variety of patterns which can be used as illustrated. Alternatively, brick flooring or paving can be used in conjunction with paving slabs. Bricks can also be used to delineate car parking spaces in tarmac or mass concrete.*

Circular brick paving can often be used to effect around young trees, but to ensure that rain-water percolates to the roots of the tree, all joints to such areas should initially be left open and later filled with chippings. As the tree grows and the trunk thickens, the inner ring of bricks can then be easily removed.

DIMENSIONS

British Standard 3921:Part 2 : 1969[2] quotes the metric brick format as 225 mm × 112.5 mm × 75 mm. From this format, if a 10 mm joint is used, the work size of the unit becomes 215 mm × 102.5 mm × 65 mm.

Special brick sizes are available from some manufacturers on request and in some instances imperial sizes are still obtainable based on the old British Standard 3921 : 1965[3] ie 9in × 4½in × 3in nominal dimensions, which deducting the ⅜in joint dimension gives a work size of 8⅝in × 4⅛in × 2⅝in.

One company produces special paviour bricks having a work size of 215 × 130 × 33 mm. In some instances standard bricks are used exposing the normal bed face to view. However, if the brick is manufactured by the extrusion process, specials may need

In Holland buff coloured bricks are used to emphasize minor road junctions with major roads and the use of two colours of bricks can be effective in defining parking areas.

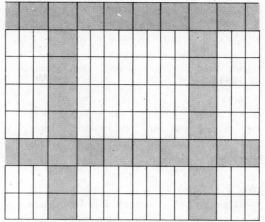

Figure 3 *Paving patterns showing bricks laid on bed with 9 × 9 in (228 × 228 mm) slabs*

Figure 4 *An alternative to basket weave with (228 × 228 mm) slabs or concrete in between*

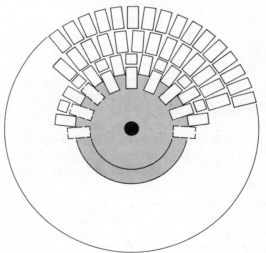

Figure 5 *Bricks set in paving around young tree. Joints are unsealed and filled with chippings to allow rainwater to percolate. The inner ring of bricks can be removed as the tree grows.*

.to be produced if the normal method of production includes perforations. When bricks are used for paving and laid in an unusual manner the manufacturer should always be consulted for his opinion as to the suitability of his material in this situation. It should also be remembered that the manufacture of special units interrupts the highly mechanised production

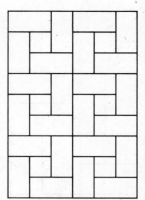

Figure 6 *Standard brick on flat*

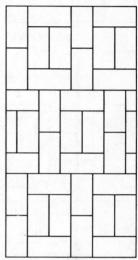

Figure 7 *Standard brick on flat*

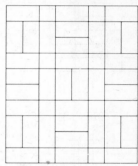

Figure 8 *Standard brick on flat*

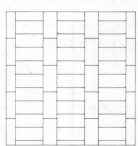

Figure 9 *Standard brick on flat and edge*

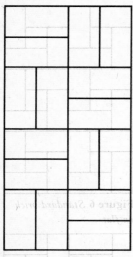

Figure 10 *Standard brick on flat. Basket weave*

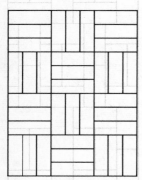

Figure 11 *Standard brick on edge*

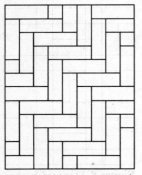

Figure 12 *Standard brick on edge. Herring bone*

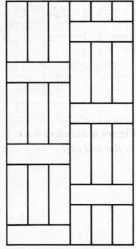

Figure 13 *Standard brick on edge*

techniques used and inevitably increases costs.

Facing brick slips are made by a number of manufacturers, the metric work sizes being 215 mm × 65 mm × 25, 33, 40 or 50 mm.

A number of manufacturers produce metric modular sizes as follows:

Format	Work size
300 × 100 × 75 mm	290 × 90 × 65 mm
300 × 100 × 100 mm	290 × 90 × 90 mm
200 × 100 × 75 mm	190 × 90 × 65 mm
200 × 100 × 100 mm	190 × 90 × 90 mm

WEIGHT

Brick paving laid on a prepared bed weighs approximately 99 to 130 kilos per square metre.

SELECTION OF SUITABLE BRICKS AND BRICK SLIPS

Not all bricks are suitable as paviors and selection should not merely be based on colour and texture although these properties are important from the aesthetic view point. Durability is of prime importance, particularly for external paving. As this note relates to clay bricks or slips, these units should be well burnt and may be of special quality* as defined in BS 3921[2] ie 'Bricks that are durable even when used in situations of extreme exposure where the structure may become saturated and be frozen, eg sewerage plants, retaining walls or pavings'. Such bricks have clearly defined limits for soluble salts content. Engineering bricks from which paviors are supplied normally attain this standard of durability. Likewise, do some facing and common bricks, but it cannot be assumed that this is so without confirmation from the manufacturer.

Crushing strength is *not necessarily* an index of durability and may be misleading if used as such. According to the standard[2]: 'it is true that bricks having a crushing strength in excess of 48.5 MN/m² are usually durable, but there are also bricks approaching this limit which decay rapidly if exposed to frost in wet conditions and others, very much weaker, which are durable'. In the latter category, the soft mud-type brick fired in clamp or kiln frequently produces a rough hard-burnt stock which is eminently suitable for flooring or paving work and can be obtained in a splendid range of red, ochre or blue.

Paviors used for industrial purposes in a corrosive environment should be highly vitrified having a low porosity (not the same property as absorption), in some cases less than 1 per cent and usually no higher than 3%, see BS 3679[1].

The direction of bedding of paviors; type of mortar; drainage; method of laying and the surface finish or lack of it are all contributory factors which can influence the durability of brick floors and these should each receive careful consideration at the design stage.

---

* *It is not essential that all pavior bricks be of special quality. Other bricks which fail to meet some of the requirements of 'special quality' (as defined in BS 3921) may give excellent service and if designers have any doubts about their suitability they should consult the manufacturer requesting advice and/or evidence that the bricks have given satisfactory service under conditions similar to those proposed.*

Calcium silicate bricks, conforming to the requirements of BS 187(7), being substantially similar in application to clay bricks are used for an equally wide range of purposes. The British Standard is most useful in this respect as it contains in Appendix E comprehensive guidance on the choice of the class of brick for a given constructional purpose, degree of exposure and also recommendations on the appropriate grades of mortar. The text of the appendix is too long to quote, but the following is an example of its usefulness:

'Work below ground or within 150 mm of the ground – recommended class of brick, not less than Class 3 (unless requirements dictate a higher strength)'.

It is therefore not unreasonable to suppose that Classes 7, 5 and 4 can be used in situations similar to those for clay brick flooring. Because of the sharp arrisses of calcium silicate bricks, joints should be maintained flush with the wearing surface, and situations of high impact avoided.

Installation

When planning a brick floor or external paving the principal elements to be considered are:
1 the base; 2 the bedding surface; 3 the wearing surface and 4 the joints.

For interior floors, the base should consist of a concrete slab which is laid to falls where drainage is an essential requirement or, alternatively, the slab should be screeded to maintain the required level. A fall of 1 in 60 should be taken as the absolute minimum. The base should generally be 1 : 2 : 4 concrete and of a depth of 100 mm to 150 mm (see Fig. 14). The base should include a separating layer which may consist of building paper, tar felt, bitumen felt, polythene film, or a layer of sand well consolidated to a depth of approximately 20 mm. Sand should be soft building sand, slightly damp, and be enclosed in wooden fillets of workable areas to maintain the true finished floor level.

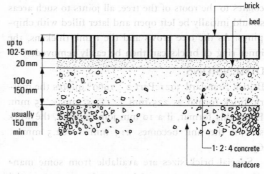

Figure 14 *Typical section, disturbed ground and wheeled traffic, concrete sub base to be designed to suit soil and intensity of traffic*

The sheet separating layers should be laid on an accurately formed base having a true and smooth surface so that the bedding material may be constant in thickness throughout its entire area. The separating material should be laid, not stuck down, and should be lap-jointed to approximately 100 mm.

For industrial floors where wet processes are carried out, or where there is a possibility of spillage, the treatment of the base is far more sophisticated as described under 'special aspects' of corrosion-resistant flooring'.

In designing industrial floors, first consideration should be given to the layout of plant, manufacturing processes and disposal of pipelines, etc. Pipes, if passing through floors, should, where possible, be grouped so that an effective seal may be provided to the least area. Equally, where such floors are to be designed to take excess spillage and will be continuously wet, plant and machinery should be raised above the general floor level to avoid ponding and be so arranged that the flow to drainage channels is not impeded.

When bricks are used for industrial floors, it will often be necessary to use a variety of specials, such as channel bricks and cove skirtings.

Exterior paved areas, including paths, also present drainage problems.

Drainage

Good drainage is essential for external pavements and for internal flooring when spillage of liquids is likely. This not only requires provision for a suitable number of drainage points and adequate falls, but also the correct siting and design of any obstacles that will be located on the floor.

The slope of the falls on the floor is an important issue and often a compromise has to be reached to accommodate aesthetic as opposed to functional requirements. A slope of 1 in 40 may be desirable from the point of view of the rapid discharge of fluids but, on the other hand, the physical effort involved in walking up such a slope is considerable.

A slope of 1 in 60 should be taken as the absolute minimum that can be tolerated, if adequate drainage is to be achieved and pooling avoided. Whenever possible, this fall should be increased to 1 in 50 or even 1 in 40 bearing in mind the earlier comments.

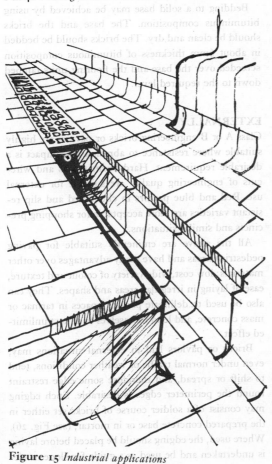

Figure 15 *Industrial applications*

Some brick pavements may prove troublesome where they are subjected to a considerable degree of saturation and, the bonding of the units should be considered in conjunction with the fall. It is desirable to have long, unbroken mortar joints running in the direction of the flow of the rainwater. These joints may be less durable than the actual paviors, and any concentration of water at these points should be avoided. In addition, it is preferable to arrange traffic across the courses of bricks rather than parallel to them as illustrated in Fig. 16 & 17.

Figure 16 *Stretcher bond*

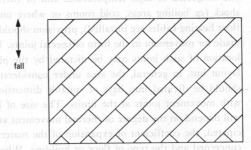

Figure 17 *Diagonal stretcher bond*

Lack of provision for drainage in external situations can be responsible for causing problems with excessive moisture, such as efflorescence, staining, fungi and mould growth, or disintegration due to frost attack. Thus adequate drainage must be provided.

For pathways to buildings, garden paths etc. drainage may be effected by sloping the paving on one side, or both sides by forming a crown. Patios and surrounds to housing estates or paving to shopping precincts should be sloped away from buildings, retaining structures, columns etc. and the slope must be adequate as already discussed. Drainage to large areas may require slopes which discharge into specially designed channels and subsequently to a main drain. Pathways close to buildings may be drained by means of edge gutters, so assisting the flow of run-off from adjacent buildings.

Although a slope will provide adequate surface drainage, it may also be necessary to allow for sub-surface drainage for large paved areas as impervious soils will retain surface water unless means of drainage are provided. This can usually be achieved where a concrete base is used, by bedding the brick floor units over a layer of fine gravel or sand laid over the sub-base. This will not be necessary over sandy or silty soils where the effect will be attained by seepage.

Internal floors will usually be bedded over a concrete base and separating layer and will not usually require a fall. Where, however, industrial processes are carried out, then careful planning of a drainage system is essential. Such floors should be sloped a minimum of 1 in 60 and drainage channels provided

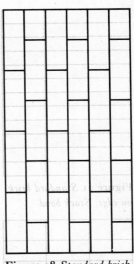

Figure 18 *Standard brick on edge. Stretcher bond (Longitudinal)*

Figure 19 *Standard brick on edge. Stretcher bond, (across)*

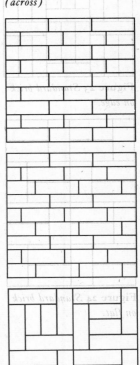

Figure 20 *Metric modular bricks, 290 × 90 × 90 mm and 190 × 90 × 90 mm, also 90 × 90 × 90 mm specials or cut bricks*

191

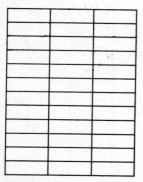

Figure 21 *Standard brick on edge. Stack bond*

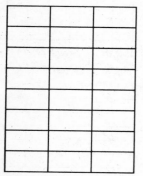

Figure 22 *Standard brick on flat. Stack bond*

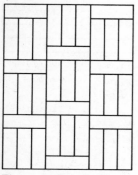

Figure 23 *Standard brick on edge.*

Figure 24 *Standard brick on flat.*

to defined areas or at the extremities of the paved area and be connected to a main drain. Where acids, inflammable fluids and other dangerous liquids are likely to spill on the paving, special arrangements, if not indeed, independent drainage may be necessary to ensure that dangerous chemicals do not enter the normal drainage system.

MOVEMENT JOINTS

Movement joints should always be provided where the floor finish abutts a wall. Such joints may be formed by inserting a batten during laying which can be subsequently withdrawn to allow the space provided to be filled with a non-rigid material ie polyurethane or polyethylene, or by using a suitable prefabricated strip. If the former materials are used it is important to seal the joint with a mastic such as polysulphide.

It is important to make provision for movement in large areas of flooring and the use of movement joints should be considered when the horizontal dimensions exceed 6m if the jointing between paviors is based on a cement mortar.

When brick paving is used for industrial purposes and subjected to high temperatures and or thermal shock (eg boiling areas, cold rooms or where under floor heating cables are installed), provision should be made for movement in the form of special joints. The location of such joints may be dictated by the plant layout but, in general, the area under consideration should be divided into bays of suitable dimensions, with movement joints at the limits. The size of bay will depend on the degree of thermal movement anticipated, the coefficient of expansion of the materials concerned and the type of floor or building. When a floor is required to be acid resisting it is vital that the movement joints are filled with a compound having equal chemical resistance to that of the floor in general.

The most common materials used for movement joints may be summarised as follows:

Extruded polyvinylchloride sections (plasticized PVC resin);

Butyl rubber putties; Silicone rubbers; Polysulphide rubbers and PVC strip elastomer compounds.

If soft materials are used without protection they can give rise to trouble on floors subjected to wheeled traffic, as a result of chipping of the edges of the bricks. The system illustrated in (Fig. 25) is sometimes used to overcome this difficulty.

An alternative is to leave an open joint in the paving and set PVC strip in these joints using one of the elastomer types.

LAYING A BRICK FLOOR

Class A Engineering bricks and paviors, are generally

used for industrial premises where resistance to acid, alkalis and oil, together with freedom from self-generating dust, are essential requirements. They are also highly suitable where resistance to abrasion and impact is a necessary feature.

Internal floors for normal traffic (pedestrian) may be Class B Engineering, or solid wire cut types for use in the 'on flat' position. The majority of well-burnt bricks are equally suitable when used in the 'on edge' position which can provide a wide range of colours and textures.

Bricks or paviors should be bedded on a separating layer over a screeded base. The flooring units should be bedded in mortar approximately 10 mm thick and be tamped down to the required thickness. The mortar mix used should be one part Portland cement; half part hydrated lime: four or five parts sand by volume with only a minimum amount of water to achieve workability. A joint space of 10 mm is generally used but this is up to the designer.

For bedding – in situations where the base is subject to high temperatures, eg around boilers and heating installations, bituminous bedding may be used. The bituminous bedding should consist of one part of aqueous bitumen emulsion and two-and-a-half parts of soft dry sand by volume. This should be laid only on a dry screeded surface which has been primed by brushing a coating of emulsion over it. A minimum of water should be added to the composition to improve workability.

For industrial floors where wet or special processes are carried out, there are a number of specialist bedding techniques using one of the following bedding compounds; supersulphated cement, high alumina cement, silicate cement, sulphur cement, cashew nut resin or phenol formaldehyde resin cement and these are discussed in BRS Digest No. 120 [4].

Bedding to a solid base may be achieved by using bituminous composition. The base and the bricks should be clean and dry. The bricks should be bedded in about 6mm thickness of bituminous composition screeded over the base and the flooring units tamped down to the required level.

EXTERNALLY

Class A or B engineering bricks or paviors are highly suitable where resistance to abrasion and impact is a desirable requirement. Hard-burnt stocks and wire-cuts of engineering quality are suitable for external use. Red and blue paviors in chequered and slip resistant varieties are more acceptable for shopping precincts and similar situations.

All the above are eminently suitable for paving pedestrian areas and have many advantages over other materials: low cost, wide variety of colour and texture, ease of laying in irregular areas and shapes. They can also be used to delineate parking spaces in tarmac or mass concrete, and lend themselves to almost unlimited effect.

Bricks or paviors laid in external situations may, even under normal traffic or weather conditions, tend to shift or spread, and therefore some edge restraint round the perimeter edges is desirable. Such edging may consist of a soldier course of bricks set either in the prepared concrete base or in mortar, (see Fig. 26). Where used, the edging should be placed before laying is undertaken and be used as a guide for slope or ele-

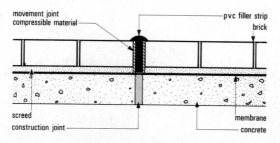

Figure 25

vation. It may in some instances be advisable to leave even-spaced open joints as weep holes in the edging to assist drainage.

Where a concrete base is not used then the paving units may be laid on a well consolidated foundation, rolled to the required fall (minimum 1 in 60). The bricks being laid on a bed of sand or lime : sand (1 : 4) or ash with 6 mm to 10 mm joints. The joints can be filled with sand or lime : sand (1 : 4) or run in grout. Care should be taken not to disturb the bedding which should be not less than 50 mm thick.

The appearance of the area can be considerably enhanced by the use of suitable colouring of the joints, but care should be taken to ensure that the pigment manufacturers instructions are carefully followed as over – prescription can seriously affect bond.

Alternatively, the paving may be laid over a 50 mm layer of sand placed on the sub-base and carefully screeded to the required level.

Having decided upon the bond pattern, the bricks are then laid or 'dropped' onto the sand bed leaving a 10 mm to 15 mm joint between each unit. The joints may be made by grouting carefully poured into the joints; the mix required should be 1 part Portland cement to $\frac{1}{4}$ part lime to 3 parts sand by volume.

Another method of jointing is to mix the proportions of cement : lime : sand dry and sweep it into the joints, taking care to remove all surplus from the surface of the paving. When complete the whole should be dampened using a fine spray garden hose. In order to effect curing of the joints the paving should be kept damp by intermittently spraying for two to three days.

REDUCTION OF SUCTION RATE

Clay bricks are sometimes wetted to reduce the initial suction by hosing or by immersion in a tank of water for a short period – the latter method being preferable due to the more uniform wetting. However, bricks should not receive this treatment unless the suction is excessive. SCAG[5] quotes a maximum suction rate for structural brickwork of $2 Kg/m^2/min$. Indeed, some manufacturers do not recommend wetting as 'iron staining' sometimes appears on the mortar joints as a result of a constituent within the brick being taken into solution and then being absorbed by the joints.

It is not necessary to wet highly vitrified bricks such as engineering bricks; indeed, to do so would make them extremely difficult to lay due to a floating action on the mortar. Such practices create a film of water between the bricks and the mortar and in addition to difficulties in laying tend to reduce the desirable tensile bond between brick and bedding material.

Highly porous bricks such as London Stocks tend to have a high suction rate and it may be desirable to wet them before laying, particularly during hot dry

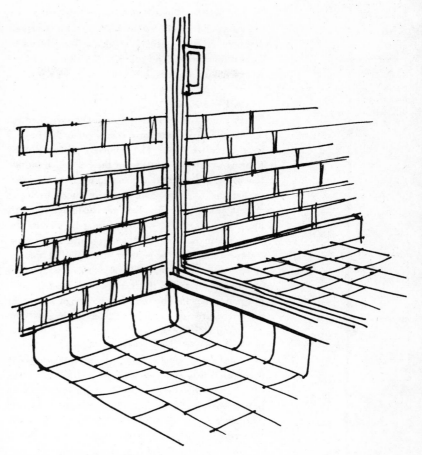

Figure 27 *Outside/inside use of paving*

weather, otherwise there may be a tendency to extract too much moisture from the bedding material, the fine particles of cement being drawn to the surface with insufficient moisture to effect complete hydration and hence, once again, poor bond.

MATURING

All floors should be allowed to mature undisturbed and protected for at least three days, after which they may be used for light pedestrian traffic.

MAINTENANCE

Brick floors, although perhaps noisy and cold to the tread, are very abrasion resistant and hard wearing. They have a particularly high resistance to compression and impact shock, especially when laid on edge. Where chemical resistance is an important requirement, then the critical factor is the grout or joint between them and, accordingly, this aspect needs periodic examination. Brick floors and pavings can become slippery when wet.

Normal maintenance should be carried out by sweeping all loose dirt and soilage, followed by washing with a solution of a neutral detergent in water. If the floor is particularly dirty, it can be scrubbed with a scouring powder and warm water. Soap should not be used, as it tends to form a scum which could cause slippery conditions. This is particularly the case in hard water areas.

Sealing is not generally recommended, but should a seal be necessary, one or two coats of a water-based seal of the acrylic type might be applied. While brick floors are not normally waxed, if necessary a water emulsion floor wax can be used to improve the appearance and facilitate maintenance. Solvent-based waxes should not be used as they might cause the

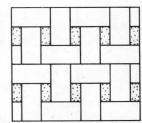

Figure 28 *Standard brick on flat, quetta bond with concrete infill*

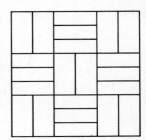

Figure 29 *Standard brick on edge and flat*

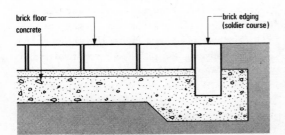

Figure 26

floor to become slippery, particularly when wet.

Ribbed and chequered brick surfaces may harbour dirt unless they are cleaned regularly. This should be carried out with a floor polishing/scrubbing machine or deck scrub, or hose and water.

SLIPPERINESS

There is no such condition as non-slip. Slipperiness on a brick floor or paved area is only partly determined by the surface itself. The tendency to slip on the flooring or paving depends on the resistance to motion along the surface of contact between the flooring/paving and footwear, and can be reduced by using ribbed, diamond, dimpled or chequered pattern units, or with smooth surfaces either by dishing the joints round each unit or incorporating an abrasive (carborumdum) in the jointing material. The amount of resistance or friction will vary for footwear made of different materials (leather, rubber, plastic, composition or steel). The presence of foreign matter on the paved area or footwear is likely to increase the amount of slipping that could occur. The kind of use to which a paved area is put is very important. Slipping on a paved floor is often due to an unexpected decrease in friction, such as may be caused by water or grease on the surface. In practice, it is difficult or near impossible to ensure absolute safety without sacrificing appearance and ease of cleaning, and a compromise has usually to be made which balances the use made of the paved area and appearance.

BRICK SLIP OR TILES FOR FLOORING

Brick slips are usually a slice of fired clay approximately 25 mm thick either cut from a brick in the 'green' state or sawn from a fired brick. In some instances brick slips are specifically manufactured for flooring and for prefabricated brickwork panels.

If brick slip floors are not prefabricated, they should be laid in a similar manner to quarry tile floors. The slips should be bedded in Portland cement and sand mortar. If the traditional method is used, this should be in the proportions 1 part Portland cement to 3 of sand approximately 10mm thick. If the semi-dry method is used, the proportions should be 1 to 4 or 1 to 5, the thickness of the bedding to be a minimum of 25 mm. By this method, it is possible to increase the thickness of the bedding to a depth of 75 mm, thus avoiding the use of a screed, where this would normally be used to raise the height of the flooring.

It is vital that a separating layer be provided to accommodate differential movement using one of the following techniques:

Sand method

A layer of soft building sand, slightly damp, should be put on to the base concrete, compacted by tamping to a maximum thickness of 20 mm; temporary wooden fillets are necessary to enclose the sand. The brick slips are then bedded in the usual way, using about 10mm thickness of bedding mortar not richer than 1 part Portland cement to 3 of sand by volumne. (Clean, sharp, washed sand must be used).

Before placing the bedding mix, the sand should be checked to ensure that the moisture content is still adequate.

Felt, building paper or polythene film method

The concrete should be finished with a cement/sand screed in the usual way, care being taken to provide a true and smooth finish. The felt, building paper or film should be laid flat as a continuous layer and should not be stuck down; adjacent sheets should be lap-jointed. The brick slips should be laid with a bed of about 10mm thickness.

SPECIAL ASPECTS OF CORROSION RESISTANT FLOORING

The design of corrosion resistant flooring is a subject in itself but, as semi-vitrified clay bricks and tiles are widely used in this field, it is considered appropriate to include some general information on this aspect of flooring.

The materials which are suitable for this type of work and the techniques employed require very careful consideration. Not only must thought be given to the materials which are available, but attention must be paid to the design of the floor also, to services passing through it, machine fixings etc. No one flooring composite is resistant to all corrosive agents and the weak link in brick or tile flooring is the jointing or the supporting concrete.

Good practice demands that all flooring subjected to spillage of chemicals be built up on a double layer basis. The two layers comprise the actual wearing surface, which is resistant to the chemical and mechanical conditions prevailing, and a membrane or 'defence line'. The purpose of this membrane is to protect the sub-floor if the top surface is damaged and to act as a slip plane between the fired clay material which tends to expand and the concrete base or support which tends to shrink.

The requirements of a membrane are:

(a) The material must be resistant to corrosion by the chemicals with which it may come in contact.

(b) It must be completely impervious to all liquids which may be spilled on the floor.

(c) The interlayer must be completely continuous. ie if a sheet material, all joints must be sealed or welded.

(d) The material should, whenever possible, possess flexible properties.

(e) The material should be sufficiently strong to support the loads it will be required to carry. In addition it should be sufficiently robust to permit the topping layer to be broken out without damaging the membrane. This point is of particular importance when repairs have to be carried out to the topping.

Suitable membranes are asphaltic and bituminous materials; plastic foil and sheet materials; and sometimes lead sheeting or rubber sheeting. In some instances latex cement compounds are used, but there may require an additional building paper layer to provide the slip-plane.

When asphalt membranes are specified, they are normally located immediately beneath the paviors. When plastic sheet membranes are used (eg polyvinylchloride film, polyethylene film, polyisobutylene, etc.) the material may be laid directly beneath a brick paviour providing the thickness or depth is a minimum of 50mm. Should the wearing surface be less than 50mm thickness, the membrane should be located between the base slab and the screed. The reason for this restriction is that the sheet membranes normally provide a 'floating' surface and the dead weight of the topping must be sufficient to ensure

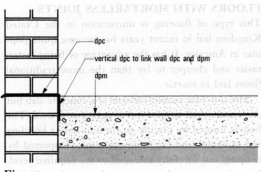

Figure 30

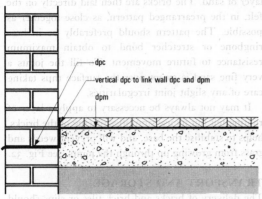

Figure 31

that the surfacing does not move or break up under traffic.

The purpose of the membrane is to provide a tanking to the floor area and it is essential that it be turned up and tucked into any wall or plinth with which the floor abuts Fig. 30 & 31.

The techniques for installing this type of flooring vary, but several general requirements apply:

(1) The work should be carried out by a specialist contractor.

(2) The mortar manufacturers recommendations should be strictly adhered to.

(3) Concrete sub-floors should be dry and free from laitance, oils, grease etc.

(4) If the concrete sub-floor is impregnated with oil and/or grease, the supplier of the corrosion resistant mortar should be consulted for a suitable cleaning agent. Failure to do this could result in using a chemical which may affect the setting, hardening or final properties of the mortar. The only satisfactory solution to the problem may be to replace the affected concrete.

(5) The thickness of the mortar bed and the joint width should be restricted to 5mm. If the accuracy of the units permit even slimmer vertical joints this is preferable. Thin joints are desirable since the chemical resistance of the mortar is usually of a lower order than that of the bricks. Thus, the surface area of exposed mortar is kept to a minimum. In addition, the cost of the mortars is usually high in relation to the cost of the bricks.

(6) The cutting of bricks should be kept to a minimum and, when necessary, should be carried out with a masonry saw or carborundum disc. This technique results in neat edges, and obviates possible damage to the bricks during cutting.

(7) When laying the bricks, the mortar should be buttered on to the bricks, and then placed on the floor and pressed home so that any excess mortar exudes from the joints thus ensuring full and tight joints.

(8) When bricks are laid over plastic sheet membranes, or over rubber or latex/cement membranes, it is essential for the person laying the floor to wear soft (eg rubber soled) shoes in order to avoid damaging the membrane. It is also essential that the working area be kept clean to avoid treading grit into the membrane.

REMOVAL OF STAINS
Staining with Portland cement

It is good practice to remove any mortar droppings as the work proceeds. If they are not removed they will harden and bond to the surface of the paving creating an unsightly appearance. Most mortar droppings can be removed by brushing or by wiping with a piece of sacking. Coating the exposed face of individual paviours with a silicone solution prior to laying may help to reduce the bond of mortar droppings, but this technique if not carefully executed may result in a poor bond between the units and the jointing material where the solution has been absorbed on faces other than the proposed wearing surface. This practice may be extremely time consuming and hence costly and may not be justified except on major projects.

When staining with cementitious products occurs and the mortar or grout has been allowed to harden, removal of the resulting disfigurement may be difficult. To remove Portland cement stains rather aggressive agents are usually necessary if they are to be capable of dissolving hardened cement. Many proprietary cleaners are available, alternatively a simple solution based on hydrochloric acid (usually a dilute solution 1 part hydrochloric acid to 10 parts water) can be used. The latter may be perfectly satisfactory for vertical areas if correctly applied and subsequently washed down but, may be disastrous if used in the horizontal plane without caution. There are inherent dangers in using acid, some of which are obvious. Acid solutions are harmful to human skin, especially dangerous to the eyes, and are capable of attacking clothing.

Such solutions can kill grass and shrubbery and, since acid is used to dissolve Portland cement, will attack mortar joints as well as unwanted mortar stains. Acid solutions do not dissolve fired clay products but may react with some elements in the clays to produce stains. Such discolourations are not usually noticeable on darker bricks but may be unsightly on the lighter coloured units. Should acid cleaning be attempted it is highly desirable to experiment on test specimens initially rather than risk disastrous results. The obvious danger in using such aggressive chemicals for cleaning floors and pavements is that the surface tends to hold the cleaning solutions and makes rinsing down more difficult.

For internal floors, additional precautions are necessary as strong acid fumes must not be inhaled. They also tend to cause rapid corrosion of numerous metals, particularly aluminium and galvanised steel. It is therefore vital that, when cleaning takes place, adequate ventilation is provided and all metal objects are removed or protected during the operation. Proprietary cleaners are available and those described as inhibited acids (ie somewhat self-neutralising) although rather expensive may be well worth trying as they tend to be somewhat less harmful to metal, clothing and skin etc.

Before applying any cleaning solution, the initial suction rate of the surface should be reduced by applying a fine spray of water. Thus, cleaning agents applied after wetting will not be absorbed into the construction. This simple precaution will reduce the risk of discolouration due to 'acid burns' and will minimise the dissolving action on the mortar joints. The cleaning solution should be applied with a stiff fibre brush and only small areas dealt with at a time (approx. 3m square), the floor should then be rinsed immediately with clean water. It is always advisable to try a small test area initially in an inconspicuous position to ensure that a satisfactory result can be obtained.

In the UK there are several organisations who specialise in cleaning buildings and, if staining is severe, it may well be advisable to seek their advice and/or service.

Disfigurement due to organic materials

The growth of organic materials such as lichens and mosses etc can cause dangerous situations which result in permanently wet and slippery surface to the paving. Such growths can readily be removed by the application of a proprietary weed killer. Alternatively, they can be killed by applying a $2\frac{1}{2}$% solution (4oz per gallon) of zinc or magnesium silicofluoride, repeated if necessary. However, regardless of the method of removal of the growth, the important issue is to remove the source of the complaint ie provide adequate drainage and hence remove the stagnant water which encourages plant life.

PREFABRICATED PANELS

Prefabricated panels of reinforced concrete with a brick slip finish have been produced in the UK and have the advantage of rapid placing on a prepared bedding surface. The disadvantages are of course high haulage costs for transporting large panels and a higher initial panel cost although the speed of laying and uniformity of panel may well outweigh these items.

UNDER-FLOOR HEATING

The detailed design of under-floor heating is beyond the scope of this technical note. Hot air ducts within the floor thickness was a system favoured by the Romans but contemporary methods tend to be based on hot water pipes under the brick finishing layer, electric heating using resistance wires embedded in the mortar joints or alternatively, cast into the actual supporting slab/screed. Some designers prefer heating by radiation from a warm floor, whilst others favour background heating from this source only which can be augmented by local radiators or even open fires. The high thermal capacity of brick floors makes the electrical storage heating system based on off-peak rates economical as described elsewhere[6]. Whatever the source of floor heating, the effect is to increase comfort by taking the chill from the floor and reduce heat losses.

FLOORS WITH MORTARLESS JOINTS

This type of flooring is uncommon in the United Kingdom but in recent years has become quite popular in America. It has the advantage of being faster, easier and cheaper to lay than the more traditional floors laid in mortar.

The sub-base usually consists of a concrete slab but on occasion well tamped soil or aggregate is used. The base is formed by placing on the sub-base a levelling bed of damp sand (25 mm – 50 mm), compacted by tamping and carefully screeded to the required level. In order to provide a smooth bed surface for the bricks a layer of roofing felt is placed over the levelling layer of sand. The bricks are then laid directly on the felt, in the prearranged pattern, as close together as possible. The pattern should preferably be in herringbone or stretcher bond to obtain maximum resistance to future movement. To fill the joints a very fine sand is swept over the surface thus taking care of any slight joint irregularities.

It may not always be necessary to apply a layer of roofing felt but its use simplifies placing of the bricks, helps to stabilise the base, and will prevent weeds and grass from growing up through the joints, see Fig. 32.

TRANSPORT AND STORAGE

The delivery of bricks and brick tiles or slips should be so arranged that the minimum of handling is necessary, and adequate precautions are taken to guard against possible accidental damage. Clean, dry storage should be provided on site for all materials; flooring units should be stacked in such a manner that excessive handling and accidental damage are avoided.

WORKMANSHIP

Good workmanship and efficient supervision are essential to the laying of brick and brick tile flooring, necessitating the employment of skilled operatives.

PROTECTION

Generally, flooring should not be laid until the heavy work in a building has been completed. When laid, it should be protected until the bedding has set covering with sheets of hardboard or similar material. After the set has taken place, the flooring may be used for light pedestrian traffic.

It should be kept clean and free from cement droppings, plaster droppings and all materials likely to cause staining. If plant, such as trestles, ladders, steps, etc have to be used for electrical and similar light work, parts in contact with the flooring should be padded, and sliding of plant on the finished floor should not be allowed.

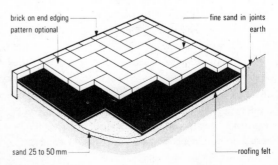

brick on end edging pattern optional — fine sand in joints — earth

sand 25 to 50 mm — roofing felt

Figure 32

ACKNOWLEDGEMENTS

The authors are indebted to numerous organisations for providing useful information for inclusion in this Technical Note. Information has been extracted from several sources and the authors wish to make special acknowledgement to the following organisations:

Redland Bricks Limited.

The Clay Brick Institute of America (formerly Structural Clay Products Institute).

Brick Development Research Institute, Melbourne, Australia.

The Engineering Equipment Users' Association □

REFERENCES

1 BS 3679 : 1963 **Specification for Acid-resisting Bricks and Tiles.** British Standards Institution.

2 BS 3921 : Part 2 : 1969 (Metric Units) **Specification for Bricks and Blocks of Fired Brickearth, Clay or Shale.** British Standards Institution.

3 BS 3921 : 1965 **Specification for Bricks and Blocks of Fired Brickearth, Clay or Shale.** British Standards Institution.

4 BRS Digest No. 120 (Second Series) August 1970. **Corrosion Resisting Floors.** HMSO.

5 **Model Specification for Loadbearing Clay Brickwork,** Special Publication No. 56, by the Structural Ceramics Advisory Group of the Building Science Committee BCeram RA, (Revised 1971).

6 **Notes on the Construction and Finish of Floors that are to be Electrically Warmed.** Electrical Development Association.

7 BS 187 : Part 2 : 1970. (Metric Units) **Specification for Calcium Silicate (Sandlime and Flintlime) Bricks.** British Standards Institution.

Brickwork and the enforcement officer

F. D. Entwisle FRICS CEng FIStructEPPIAAS
Surveyor, Department of Planning and Architecture, Sheffield

BACKGROUND TO ENFORCEMENT

Legislation relating to the making and enforcement of Building Regulations is contained in the Public Health Acts, 1936 and 1961. Sections 61, 62, 64 & 65 of the 1936 Act cover the scope of the powers and the legal machinery of enforcement, and Sections 4, 5 and 6 of the Public Health Act of 1961 introduced the present Building Regulations, making the enforcement of regulations the function of local authorities and amending and extending the relaxation procedure.

It is important to remember that the authority is the local District Council, not a particular officer, and decisions relating to the Building Regulations, although made by an officer technically, are legally given by the council.

The regulations require the submission of plans, indicating the proposal, which must clearly show compliance with these regulations, and a local authority must give a decision within a prescribed period of 5 weeks, which may be extended to a maximum of 2 months, only with the permission, in writing, of the developer.

It is obvious that, for projects of any magnitude, all the necessary particulars are rarely available at the time a decision is sought and local authorities have introduced a considerable amount of flexibility in the application of the law relating to the approval of proposals. For example, the majority of authorities accept undertakings as to later submission of structural and other details, not readily available at the time the normal plans are submitted, to allow developers to submit such particulars at appropriate stages.

In addition, there is a considerable amount of latitude allowed to the officers responsible to local authorities for the enforcement of Building Regulations, so far as the choice of materials and detailed design and construction are concerned. Normally, reports of officers to their authorities are in the form of a simple recommendation to approve, and this is always followed. Only in cases of proposals to reject or to relax regulations is more detailed information given to the council members forming the Local Authority Committee set up to deal with Building Regulation matters.

The regulations require notices to be submitted, by the builder to the local authority twenty-four hours before foundations, damp proof courses, site concrete and drainage works are covered, and before comple-

tion and occupation of buildings in order that compliance with regulation standards can be confirmed.

Throughout the country there is a considerable variation in the quantity and quality of staffing for Building Regulation enforcement, resulting in a variation of approach and flexibility. The situation should become more uniform after the 1st April 1974, when the number of Building Regulation authorities will be reduced and most of sufficient size to employ the correct professional staff required for an objective enforcement unit. On the whole, it can be claimed with some confidence that local authorities apply a reasonable flexibility, and have regard for the problems of the developer.

So far as the staff of an enforcement unit is concerned, the work can be divided broadly into three sections:

1 Dealing with applications and the submission of plans for approval in the first instance.

2 Dealing with proposals that involve specialist structural design problems.

3 Enforcement of the actual regulations on the site.

All three sections are co-ordinated, so that there is a general agreement on basic principles, yet each section is provided, by the Regulations, with considerable flexibility, and the means to deal with innovation and departures from the norm.

The regulations themselves provide a number of options, to the developer and the local authority, as to how they can be satisfied. Firstly, there is the specific requirement, which is rarely used in the present regulations. An example is the requirement that the surround to chimneys shall comprise 100 mm thick non-combustible material. There is no flexibility here, related to the insulation quality of the surrounding material, and any variation from this thickness must be by way of relaxation of the regulations. Secondly, there is the functional requirement, which is the predominant approach, and the structural aspect is perhaps the best example. All foundations and the superstructure must be designed in such a way as to satisfy the mandatory requirements of the regulations, that they shall safely withstand all the loads to which they will be subjected, without undue settlement or deformation which could be hazardous or cause damage to the building.

Thirdly, there are the 'deemed to satisfy' provisions. These are either specified methods of construction set out in italics in the Regulations, which normally deal with such matters as provision of damp proof courses and types of flue construction, or Codes of Practice and British Standard Specifications listed in the regulations. Any design or method of construction to one of the deemed to satisfy provisions must be accepted by the local authority, although the developer is free to use any other method or material, provided he can prove that it satisfies the functional requirement. It is important to note that the only British Standards and Codes deemed to satisfy the regulations are those actually listed, any later amendments are not deemed to satisfy until included in the regulations. Nevertheless, a developer is free to quote a new code or to use a new approach arising from recent research and prove that this would satisfy the functional requirements. Similarly, the local authority is free to accept such amending documents, not as deemed to satisfy provisions, but as a means of complying with the regulations.

Fourthly, there is the performance standard, where the regulations specify a particular requirement in performance; fire resistance and thermal insulation are examples.

Flexibility is provided by the powers of relaxation, which rest with the Secretary of State for the Environment, for Part A the application of the Regulations, Part D structural stability, and some of Part E structural fire precautions, and with the local authority for the remaining parts of the regulations. However, local authorities cannot relax any of the regulations with regard to their own developments, all of which, except for school designs approved by the Department of Education, are subject to Building Regulations, and must always be referred to the Secretary of State for the Environment.

It is clear that there is ample scope for innovation in design and the introduction of new materials and methods; the Building Regulations cannot be claimed to be restrictive. It is again emphasised, however, that the departure from deemed to satisfy provisions places the onus upon the developer to prove his case. Uniformity throughout the country was one of the objectives of a National Code of Building Regulations, but it must be accepted that Regulations based primarily on functional requirements give rise to variations in the use of materials and types of construction. A considerable contribution could be made towards a good combination of flexibility and unformity if there could be a faster circulation of acceptable changes, by the Department of the Environment to local authorities, with broad guidelines as to how these could be achieved. Whilst this would never remove the freedom of local authorities, to make their own decisions based upon their skill and knowledge, nevertheless it would make a considerable contribution to uniformity and eliminate many frustrations experienced at present.

GENERAL SCOPE OF REGULATIONS

When considering the use of any material or mode of construction, in relation to Building Regulations, it is essential to bear in mind that the mandatory requirements are almost entirely functional. Designers are left with great freedom of choice, provided they can show that the functional requirements of strength, stability, weather resistance, fire resistance, sound and thermal insulation are satisfied, and the enforcement officer must also judge the design on this basis.

In considering the use of brickwork in relation to Building Regulations, there are variations in the type of approach for different parts of the regulations. For example, the requirements on structural stability, whilst flexible are nevertheless definite – stability must be provided; a standard of performance is specified for thermal insulation and fire resistance; a thickness specified for chimney construction; adequate sound insulation is required, with a standard quoted as a yard-stick only and specific weights and flanking arrangements given as deemed to satisfy provisions.

However, brickwork is frequently used in situations such as internal non-loadbearing partitions, inspection chambers and manholes, and non-loadbearing external panels, where the regulations make no definite demands. Here, the performance must be considered in relation to the scope of the enabling powers under which Building Regulations are made; in the interest of public health and safety. The correct functioning of drainage inspection chambers and the stability of non-loadbearing panels, for example, can both reasonably be regarded as matters of health and safety. Nevertheless, there are a great many points of design detail and workmanship that cannot be so regarded.

The regulations are part of a penal code and demands must be enforceable within the framework of that code. Enforcement officers may, and indeed should, aim at the highest standards by persuasion, but make clear what is recommended as good practice and what is to be enforced.

APPLICATION OF REGULATIONS

The use of brickwork will be discussed for appropriate parts of the regulations.

PART B Materials – refers to suitability of materials, and although those satisfying the British Standard Specifications quoted in the Regulations, and used in appropriate positions, must be accepted, all others must be judged on merit; provided a material can be shown to fulfill the appropriate function to the satisfaction of the local authority, it can be accepted.

The question of durability of materials poses a problem for the Enforcement Officer, but Regulation B1 should be used to guide developers in the proper use of materials appropriate for any particular purpose, and this surely includes the choice of bricks.

British Standard specifications and BS Code of Practice CP121, Part 1 discuss the appropriate types of bricks for various conditions of exposure. BS.187: 1967 classifies the strengths of calcium silicate bricks suitable for use in different locations. BS.3921: Part 2: 1969 refers to three broad classifications and defines the different quality of clay bricks as follows:

1 **Internal quality.** Suitable for internal use only.

2 **Ordinary quality.** Less durable than special quality, but normally durable in the external face of a building.

3 **Special quality.** Durable even when used in situations of extreme exposure where the structure may become saturated and be frozen, eg retaining walls, parapet walls, positions below damp proof

courses and sewage plants.

Special quality bricks are suitable for conditions of extreme exposure and may even be used as a damp proof course, if the average water absorption does not exceed 4.5 % of dry weight. Ordinary quality load-bearing clay bricks, ranging in strength from 7 MN/m² to 103 MN/m², are usually durable in external use, but not all are suitable for wet conditions; for example, below ground or for inadequately protected parapet walls. Although crushing strength is not a complete guide, ordinary quality bricks stronger than 48 MN/m² are frequently durable; nevertheless, it should not be assumed that all bricks of ordinary quality are suitable for all conditions of exposure, even in the external face of a building and, in the case of doubt, information should be sought from the manufacturer. Perforated bricks are now well established and acceptable with certain reservations as to weather protection and fire resistance which will be discussed later.

The metric modular format has resulted in a 90 mm wide brick, an acceptable dimension under appropriate circumstances although below the usual brick width.

Clay bricks fresh from the kiln should not be used, as they are subject to expansion in such an unweathered condition. Calcium silicate bricks are subject to longer term drying shrinkage.

Undesirable detailing

A further problem relative to durability in use arises from the unfortunate, yet fashionable tendency to set brickwork forward of structural beams and to rake out the joints for architectural effect, without proper regard for the extent of exposure or the suitability of the bricks. Such arrangements provide very testing conditions and special quality bricks should normally be used.

Perforated bricks are rather more vulnerable when the perforations are relatively close to the face of the brick. A very deep recessed joint, or excessive projection beyond the face of a structural member above, could expose the perforations and render the brick liable to damage when any moisture in the holes is subjected to very low temperatures.

Opinions differ on the desirability of this type of detail, but this is a situation where the enforcement officer can only think in terms of Regulation B1 and the proper use of materials. In heavily stressed structural brickwork, such a reduction of area of the bed joint might not be permissible, but in the majority of cases it is unlikely that there would be a structural reason for refusing such an arrangement.

Whilst the writer considers this is poor detailing in any type of brickwork, and should be discouraged, only in the case of grossly exaggerated recesses or projections could such proposals be refused, within the terms of the Building Regulations.

Brick edges exposed by setting back a member above should be protected by a suitable flashing, and in the writer's view, all unprotected recesses should be limited to 5 mm, and should be discouraged in very exposed positions unless special quality bricks are used. See Figure 1.

Volume changes in materials

The fact that materials used in building change

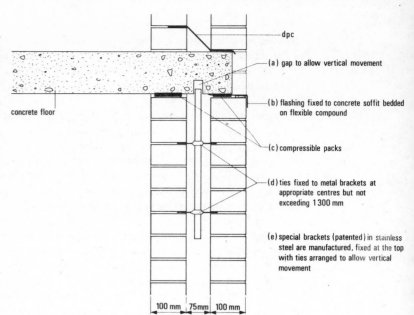

concrete floor

- (a) gap to allow vertical movement
- (b) flashing fixed to concrete soffit bedded on flexible compound
- (c) compressible packs
- (d) ties fixed to metal brackets at appropriate centres but not exceeding 1 300 mm
- (e) special brackets (patented) in stainless steel are manufactured, fixed at the top with ties arranged to allow vertical movement

dpc

100 mm   75mm   100 mm

Figure 1

volume with variation of moisture content and temperature, as well as when subjected to stress, is not sufficiently appreciated. In spite of increased knowledge of the behaviour of materials and more accurate structural analysis, modern buildings in brickwork, and indeed other materials, suffer far greater disfiguration from minor cracking than older buildings in masonry.

The major reason probably is the greater mass and the weaker flexible mortar used in the past, compared with the thinner walls and high strength cement mortars prevalent today, although much of the cracking could be avoided if designers paid greater attention to the provision of movement joints, as discussed below, and made a more careful selection of mortars.

Cracking does not, however, always arise from brickwork movement but from movement of other elements. For example, in-situ concrete floors will generally shrink, but a concrete roof subjected to sunshine will expand, and, if built tight to surrounding brickwork, will displace the latter, causing cracking. A space or easily compressible movement joint should be incorporated between the edges of the roof and the surrounding brickwork. See Figure 2.

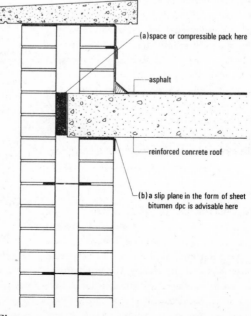

- (a) space or compressible pack here
- asphalt
- reinforced concrete roof
- (b) a slip plane in the form of sheet bitumen dpc is advisable here

Figure 2

PART C: Preparation of site and resistance to moisture. This part is concerned with damp penetration and weather resistance.

External walls

Whilst the cavity wall is the normal and best method of providing a barrier to driving rain, solid brickwork of good design and construction can keep out the weather in reasonably sheltered areas and, if properly rendered, also in more exposed positions, provided there is adequate eaves and gable protection by way of generous overhang. Tile hung or similar clad solid walls will resist driving rain even in positions of extreme exposure. Dense solid walls may require a vapour barrier as near the internal surface as possible, and may need additional treatment to improve the thermal insulation in dwellings.

The enforcement officer will always show preference, quite rightly, for cavity walls, but he should consider carefully any proposals to use solid construction in relation to the extent of exposure and the design details.

Prevention of rising damp

The suitability of high quality engineering bricks for damp proof courses to walls is well known, and the British Standard 3921 deals with these (not all, by the way, blue bricks). The use of bricks in lieu of a bitumen membrane avoids the problem of damage which often occurs when the membrane is set back and the joint pointed with a strong cement mortar. The squeezing out of the bitumen often forces out this pointing which, because of its strength, takes small pieces of brickwork with it, leaving an unsightly spalled area. See Figure 3.

Figure 3 *Shows damage to brickwork by the squeezing out of a bitumen DPC forcing away dense mortar pointing*

This is an example of a defect not within the scope of Building Regulations, where the only object is to prevent dampness rising through the brickwork, but nevertheless worth bringing to the attention of developers.

Plastic damp proof membranes are now available which do not squeeze out under load and so do not produce this defect.

PART D: Structural stability. Stress analysis for brickwork, as far as the legal code (outside Inner London) is concerned, was first introduced in the Bye-laws made under the Public Health Act 1936, which allowed designers either to use a schedule of rules for brick walls based on the height and length, or to prove that the loads to be supported did not induce stresses in the brickwork greater than those acceptable. The available data on allowable stresses in brickwork was rather limited until the issue of the first British Standard Code of Practice CP.111 in 1948. This code was included in the Bye-laws of 1953, as a deemed to satisfy method of complying with the mandatory structural requirements of the Bye-laws. CP.111 was completely revised in 1964, and later metricated as CP.111, Part 2, 1970, which forms the basis of the deemed to satisfy provisions in Regulation D15.

Structural design

The only mandatory requirement is contained in Regulation D.8, namely: that structures shall carry the appropriate loads without such movement as would cause damage to the building or part of the building. This is the only demand upon a designer, and so far as brickwork is concerned, it can be satisfied in four ways:

1 By using the rules set out in Schedule 7, provided that the proposal is within the range of such rules.

2 By using CP.111 as a basis for design.

3 By using other authoritative data based on sound structural principles.

4 By relating the structure in question to full-scale testing.

The range is such that the regulations cannot be accused of inhibiting new techniques, and enforcement officers certainly cannot, and indeed should not, restrict designers to the rules laid down in Schedule 7. They must examine all designs and their decisions should be based on the merit, or otherwise, of the designs.

While there are no restrictions on the designer, other than to produce a sound design, the enforcement officer is restricted in that he *must* accept designs to CP.111 and Schedule 7, even if he considers them unsuitable to some degree for the building in question, since they are 'deemed to satisfy'. Although it would be rare indeed for a design to CP.111 to be unsuitable for a particular case, there are cases where proposals to Schedule 7 are unsuitable and even dangerous. This is one of the hazards of deemed to satisfy provisions, and in the writer's view it is high time that these were qualified. This could easily be done by introducing the following clause immediately before the deemed to satisfy provisions:

'*without prejudice to the generality of Regulation D.8, the designs to the following rules or Codes of Practice are deemed to satisfy the Regulations.*'

This would allow the use of any documents quoted as 'deemed to satisfy' and give developers ample freedom, yet empower the enforcement officer to require adjustment when appropriate to suit any special case.

Criticism of Schedule 7 and Regulation D.17

The approach in Schedule 7 is limited to buildings up to 12 m in height (not including basement storeys) and, in one case (Clause 7, Schedule 7), to floor loadings not exceeding 3 kN/m². The specification for bricks is limited to minimum strength requirements. There are special rules for cavity walls, with concessions for limited use of a 75 mm inner leaf for single storeys. However, there is no requirement for floors or roofs to be tied to walls to provide lateral restraint, neither is the effect of openings in walls or local loading taken into proper consideration. Whilst for most types of buildings, the schedule is over-conservative, it should be used with care. See Figure 4.

Figure 4

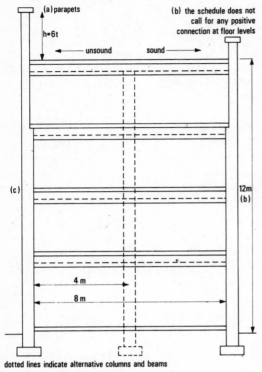

dotted lines indicate alternative columns and beams

(c) Superload w Kn/m²
Without column load on wall 4w per metre concentrated at beam bearing
With column load on wall 2w per metre
(Note: w could be any superload acceptable for use)

The susceptibility of the top storey of normal cavity brickwork to damage by wind forces, due to low vertical load and lack of effective lateral support, is not sufficiently recognised, and many dwelling house designs need investigation to ensure effective lateral support from roof, ceiling or cross walls, which is a matter not covered by Schedule 7.

The requirements of Clause 14 of Schedule 7 for parapet walls and Regulation D.17 as to maximum height of chimneys are also equally unrealistic. See Figure 4.

The permissible height above a roof is not truly related to the point of lateral structural support to the parapet or chimney stack, and, because the Schedule is applicable in all situations, the pressure due to wind can be anything from 480 N/m², which is quite safe, to 1440 N/m², which can be quite dangerous.

CP.111 deals with all loadbearing brickwork designs adequately, and Schedule 7 can be neglected; in fact, in the writer's view, it should no longer form part of the Regulations, except in a very much modified form.

CP.111 method

In CP.111, Part 2, 1970, walls and individual piers are designed on a stress basis, varying with the strength of brick and mortar, and factored relative to slenderness ratio and eccentricity of loading.

The vertical load capacity of brickwork depends upon the strength of the bricks and mortar, although the contribution of the latter is limited. Guidance as to suitable mortars can be obtained from several sources (2), (3). Brickwork is strong in compression

and relatively weak in tension, hence the object is to keep structures in compression. By reason of their low ratio of height to thickness, stocky walls can develop maximum vertical strength, which falls as this ratio increases. Slenderness ratio is an important factor in design because it is a pass/fail requirement; recently, the maximum ratio in CP.111 has been increased to 27, based on the effective height or length of a wall, and the actual thickness of the brickwork.

Examples of lateral support are given in the Code covering simple cases, but where lateral forces are applied these should be investigated, and the supports and anchorages must be capable of resisting such forces plus $2\frac{1}{2}$ % of the total vertical load which the wall is designed to carry at this point. The latter requirement is to accommodate any lateral force that might occur due to the wall being out of plumb.

Loads applied eccentrically cause increased stresses, but over only a small area, and in consequence a permissible working stress increase of 25 % is allowed for such local effects, in CP.111. Similarly, concentrated loads are only applied over a small area, with the brickwork fully contained by surrounding brickwork, and so a 50 % overstress is allowed. This concession should not be applied at the end of a wall because the brickwork is not fully contained on all sides; normal stresses should apply at such points.

The use of a boot lintel bearing only on a short length of the inner leaf of a cavity wall is to be deplored, because the eccentric bearing, due to the tortional load on the toe of the lintel, can cause sufficient rotation of the lintel to produce cracks in the brickwork. If such a method is adopted, a long bearing on a good padstone on the inner leaf is essential. Padstones and spreaders should be designed to spread concentrated loads over a sufficient area of brickwork, but should be of sufficient depth relative to length to ensure that lines at an angle of 45°, drawn from the edges of the beam bearing, intersect the base of the padstone at the extreme edges. A longer padstone will tend to deflect under load and fail to spread the load uniformly over the brickwork beneath.

Cavity walls

In cavity work, both leaves can be used for slenderness ratio calculation purposes, as required by CP.111, the effective thickness being two-thirds of the combined thickness of the two leaves. A similar approach can be applied to piers in cavity work. The load can either be shared between both leaves or taken by one leaf only, but with cavity widths more than 75 mm it is advisable to take the load on one leaf. With such wide cavities, special ties of the fishtail type are needed, in twice the normal quantity.

Careful consideration should be given to the choice of materials to ensure their proper use; for example, normal brickwork for one leaf and lightweight blocks for the other, because, in addition to variation of movement due to moisture and thermal effects, the materials have different Young's Modulus values and, therefore, different stiffnesses. Under no circumstances should these dissimilar materials be rigidly bonded together.

Since departure from these basic facts can lead to serious defects and to failure to satisfy the Building Regulations, the Enforcement Officer should have regard to these matters.

There are differences of opinion as to whether, in loadbearing cavity walls to high buildings, the ties should be of stainless steel or non-ferrous metal. There is no evidence that the normal galvanised fish-tail ties are lacking in durability or that there is any height at which a change from galvanised ties to stainless steel or non-ferrous metal should occur. Ties cannot be seen, and replacement would be a major operation, so it cannot be denied that a more durable tie for higher buildings has advantages, but in the writer's opinion it is doubtful whether these can be enforced under Building Regulations, except possibly where a corrosive atmosphere exists, for example, near the coast or in heavy industrial areas.

When applying CP.111 to cavity walls on a stress basis, the cavity width can be reduced below 50 mm, subject to proper control to ensure an effective cavity, although this is not an advisable reduction in exposed positions, and increased to 150 mm, provided the length and spacing of ties are varied in relation to leaf thickness and cavity width. Where the cavity is more than 75 mm, only the standard strip ties or their equivalent should be used. The cavity width is limited to 75 mm, if either leaf is less than 100 mm, but the use of a 75 mm leaf is not limited to single storey buildings.

Allowance for horizontal and vertical movement in structures

For cavity walls, CP.111 recommends horizontal movement joints at every fourth storey or 12 m to allow for the differential movement between the external and internal leaves.

Vertical movement joints are also required in long lengths of both clay and calcium silicate bricks, the latter are subject to movement due to the alternate wetting and drying experienced in all external positions. These joints should occur at about 10–15 metre intervals (3). Lack of horizontal movement joints can lead to dangerous defects in brickwork and these come within the scope of the enforcement officer's powers, but lack of vertical movement joints normally only causes unsightly cracks in brickwork, and so they are in the advisory category.

In the normal framed building, non-loadbearing external infill panel walls are widely used, but because of the definitions of loadbearing in the present CP.111, in relation to Building Regulation definition of loading, these do not come within its scope (5).

There have been unfortunate experiences of failures in brick panels, mostly in the region of the slip bricks used to cover the concrete floors or edge beams. These failures are primarily due to a combination of the long term shrinkage of reinforced concrete, and creep under load, causing excessive compression in the brickwork, emphasised by the eccentric loading on the concrete corbel, resulting in fracture of the brickwork in this area and in some cases bowing of the wall. This condition sometimes has been further aggravated by the use of kiln fresh bricks. See Figure 5.

In such cases it is essential to provide a movement joint at the top of each panel to allow for the shortening of the reinforced concrete frame. The effective tying of the brickwork at this top edge to allow such movement yet provide lateral support to the top of the panel is a matter for careful detailing (4). See Figure No. 1.

Reliance upon pure adhesion of slip bricks and tiles at high levels cannot be recommended but, if adopted, the bedding should be an epoxy resin mortar, carried out as a separate operation from the bricklaying.

Non-loadbearing internal walls can pose stability problems, yet 100 mm and 75 mm thick partitions, effectively bonded or tied at the ends, and nominally pinned at the top, behave remarkably well when subjected to lateral loads; and better still when all internal surfaces are plastered, because the sides and the top edge are further restrained by the plaster. Edge support is essential for stability, and with a free top edge, the length of the wall becomes more critical and should not exceed three times the height.

The only possible legal control, for these two situations at the moment is by way of Regulation B.1, as to suitability in the use of materials. This is a doubtful, but nevertheless reasonable, argument to guard against unsafe conditions.

Exceptional loading

The Fifth Amendment to the Regulations, introduced after the Ronan Point collapse, laid down rather stringent rules, for buildings of 5 storeys or over (including basement storey) requiring either the main structural members to resist a force of 34 KN/m², or, in the event of one main member failing, the surrounding structure to be designed to limit the extent of collapse. This is now contained in Regulation D.19 of the consolidated version of 1972, and it has been shown that brickwork can satisfy these requirements without undue difficulty. Cantilever and composite action can easily be provided by suitable reinforced areas and the use of reinforced concrete floors and beams. Tension ties laterally and vertically are also easily introduced within brick walls (7).

Metric modular brickwork

Mention must be made of the structural problems associated with the 90 mm wide modular brick.

CP.111, Part 2, 1970 provides for the use of 90 mm bricks and, of course, the mandatory requirements can be shown to be satisfied by other methods, but such approaches must be supported by suitable calculations

Figure 5 *Shows the squeezing out of the bitumen tray DPC due to compression in the brickwork from the shortening of the RC frame. The movement joint provided at the top of the panels has not been fully effective, hence the DP course has acted as a movement joint.*

acceptable to the local authority. Schedule 7, a rough rule of thumb method, dispenses with the need for calculation, but does not include reference to 90 mm bricks, hence they do not strictly qualify for use with this Schedule.

The difference in width of 12 mm between the standard and modular bricks, is marginal and will only affect slenderness ratio and bedding area to a limited extent. The difficulty can be overcome by a simply check on the slenderness ratio and bedding area relative to the standard brick, and from this it will be easy for the enforcement officer to accept the use of 90 mm brickwork with reference to the principles of CP.111 rather than Schedule 7.

The note on metric modular bricks, issued by Structural Clay Products (13), gives calculations of designs using 90 mm bricks, and it could be claimed that circulation of similar guidance documents, to all local authorities by the DOE Building Regulations Division, whenever a basic change such as this occurs, would be a most useful practice (14).

PART E: Structural fire precautions. The high fire rating of fired clay brickwork is well known, but its ability to retain strength on cooling compared with other materials is not sufficiently appreciated; only the mortar generally receives permanent damage and then only to a shallow depth beyond the surface, needing only hacking out and pointing for complete restoration. Calcium silicate bricks perform similarly.

Schedule 8 in the regulations sets out fire resistant periods deemed to satisfy the regulations for structural elements. The normal 102.5 mm solid brickwork with 12.5 mm vermiculite/gypsum plaster will provide the Regulation maximum fire resistance of 4 hours, and 2 hours with normal plastering. Cellular clay blocks, not less than 50 % solid, have less fire resistance. Many modern bricks are perforated, yet are not classed as cellular, but research work indicates that, in loadbearing brickwork, they do not have equivalent fire resistance to the truly solid brick at the upper limits. A period of $1\frac{1}{2}$ hours can be obtained with 102.5 mm perforated bricks and vermiculite/gypsum plaster 12.5 mm thick, under the standard test (15). Without plaster, or with normal plaster, only $\frac{1}{2}$ hour resistance is obtained. A non-loadbearing 102.5 mm unplastered or nominally plastered wall will provide $1\frac{1}{2}$ hours resistance, and with a vermiculite/gypsum plaster 3 hours.

Schedule 8 of the Regulations is not particularly helpful in dealing with these variations, but the figures quoted above are based on recent authoritative research and can be accepted by the Enforcement Officer.

In loadbearing conditions requiring more than a $\frac{1}{2}$ hour fire resistance, a vermiculite/gypsum plaster 12.5 mm thick will be required, a change in the usual specification that will require special consideration. It must, of course, be acknowledged that all clay bricks behave remarkably well when subjected to fire, and most have ample margins for the lower periods stated.

The 90 mm wide modular brick is not listed in Schedule 8, yet a solid loadbearing modular brick can be accepted, for a period of one hour, or 2 hours if the finish is 12.5 mm vermiculite/gypsum plaster, and for $1\frac{1}{2}$ and 3 hours respectively if non-loadbearing. If perforated, a vermiculite/gypsum plaster finish is required for $\frac{1}{2}$ hour resistance in loadbearing conditions, and for $1\frac{1}{2}$ hours in non-loadbearing conditions.

The mandatory requirement of the Regulations is for a period of fire resistance and Schedule 8 provides but one guide to methods of construction deemed to satisfy. Other authoritative information from research sources can, and should be, used and the Codes of Practice now under revision will, no doubt, include the results of the most recent research (6).

The term 'loadbearing' should be interpreted realistically for, under test, the load will be very near the allowable design load. For fire resistance purposes it would be acceptable to allow say, a 90 mm untreated wall in bricks to be used for a $1\frac{1}{2}$-hour fire resistance, in lieu of 102.5 mm, if the wall is only subject to a nominal load from a light roof or ceiling.

PART F: Thermal insulation. Standards of thermal insulation for walls can only be enforced for dwellings, although of course they are desirable in all heated buildings for reasons of economy as well as comfort.

Schedule 11 sets out a variety of combinations of materials and methods of construction, providing the appropriate insulation standards to satisfy the Regulations. The normal 275 mm cavity wall, plastered internally, meets the requirements for dwellings, with a thermal transmittance coefficient 'U' of 1.70 W/m² deg.C, assuming the surface resistance at 0.18. The reduction of the thickness of a leaf, or the omission of plaster, can reduce the insulation below the allowable, but other measures are available to raise the standard of insulation.

Pumping a foamed urea-formaldehyde resin into the cavity is now an acceptable method of obtaining a high quality insulation for cavity walls, and a 'U' value of 0.51 can be achieved. Another method is to use fibre glass sheets inside the cavity, close to the inner leaf, with a cavity of sufficient width to leave an air space between the sheets and the inside face of the external leaf. A 'U' value of 0.67 can be achieved by this method.

Other applications, for example, dry lining, can also improve the standard of insulation, achieving 'U' values of between 0.65 and 1.00.

Part G: Sound insulation. Since density is the easiest way to achieve sound insulation, bricks with their natural weight and easily filled joints present an effective sound barrier, and the efficient bond at angles provides good flanking conditions. When using frogged bricks, it is important that the large frogs are laid upwards, and filled to ensure maximum density, for test results have shown that the required standard insulation can be achieved only if this is done.

The comparatively recent introduction into Building Regulations of sound insulation requirements no doubt arose from the use of walls of much lighter construction than the traditional brick wall, which would be capable of satisfying the existing legal requirements regarding fire resistance and stability, but having poor sound insulation. There is now an acceptance that minimum standards of sound insulation are in the interest of public health. It is unfortunately true that, without these regulations, walls and floors of a very low sound insulation value would be used to separate dwellings.

The mandatory requirement for walls is contained in Regulation G.1: simply that walls separating dwellings, and also separating habitable rooms from other parts of the same building such as common corridors and machinery rooms, shall, in conjunction with the associated structure of external walls, floors and roof, provide *adequate* resistance to the transmission of airborne sound. It rests entirely with local authorities as to what is adequate, although they will be guided by the specification set out in Regulation G.2, the 'deemed to satisfy' clause. This consists of: (a) walls subject to field testing to produce results comparable with a table from the Code of Practice on Sound Insulation (12), which sets out the required sound reduction at various frequencies, and (b) reference to Schedule 12, which sets out detailed specifications of acceptable types of walls, together with the types of associated structure for flanking conditions, specified in Regulation G.2 itself.

The basic need is density, but there is a considerable concession for cavity construction. For solid construction, a mass of 415 Kg/m² is specified for brickwork and blockwork (including 12.5 mm of plaster each side). The same density is required for a cavity wall with a 50 mm cavity, but with a cavity of 75 mm the mass is reduced to 250 Kg/m², with a specific requirement that the leaves be of lightweight concrete or (it could be assumed) any lightweight brick or block. This reduced density requirement is no doubt due to the absorption properties of the lightweight leaf, plus the increased effectiveness of the wider cavity.

Ties must in all cases be of the butterfly type, although further advantage may well be gained by omitting ties entirely, widening the cavity to 100 mm and designing each leaf as structurally independent, using joist hangers if the wall supports a timber floor or roof. Such an approach is new and will need careful design and good construction. The associated structure to the party wall includes flanking construction which may be one of the following arrangements:
(a) The continuation of the party-separating wall beyond the external wall for a distance of 460 mm, **or**

(b) The party separating wall tied or bonded to one leaf of an external flanking wall of an average mass of not less than 120 Kg/m², *provided* that the distance horizontally between any door or window opening is not less than 690 mm, *unless* the height of each such opening does not exceed two-thirds of the height of the storey, and the external wall above and below the openings extends for 3 m, measured horizontally, on both sides of the separating wall, **or**

(c) The party-separating wall extends to the *outer* face of a flanking wall of *timber* or *light construction, (other than tile hanging)* and at the top and bottom the party wall is tied or bonded into either –

1 a solid floor next to the ground, **or**

2 a suspended concrete floor of average mass of not less than 220 Kg/m², **or**

3 a concrete roof of a mass of not less than 145 Kg/m².

Method (a) is the simplest but not always acceptable, Method (b) requires an inner leaf of specified weight in a considerable area of external wall with only a short window, and method (c) allows lightweight infill panels but only with a concrete floor and roof,

and also with the party-separating wall carried to the face which presents weathering problems. See Fig. 6.

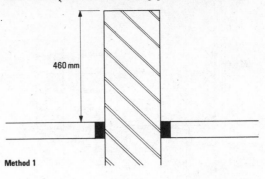

Method 1

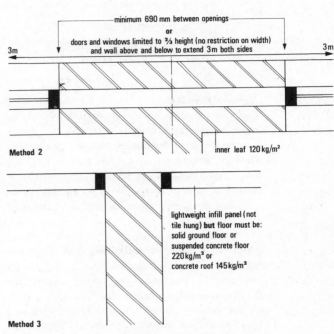

minimum 690 mm between openings
or
doors and windows limited to ⅔ height (no restriction on width) and wall above and below to extend 3m both sides

3m — 3m

Method 2

inner leaf 120 kg/m²

lightweight infill panel (not tile hung) **but** floor must be: solid ground floor or suspended concrete floor 220 kg/m² or concrete roof 145 kg/m²

Method 3

Figure 6 *Regulation G2. Deemed to satisfy requirements for associated structure (walls).*

Regulation G.3 refers to sound insulation of floors and in Regulation G.4 the deemed to satisfy requirements for floors also refers to associated structures. Concrete floors must seat fully on the inner leaf of external walls, and must be bonded or tied, not only to party-separating walls but also to any other wall which supports the floor. This means that the inner leaf must not pass the floor slabs but be built on them. Timber separating floors must be bounded on at least three sides by walls of a mass of 415 Kg/m², and every external flanking wall must extend not less than 600 mm above the underside of the floor, the only openings that are allowed within this height are over a balcony or similar construction. This means that a lightweight wall can only be used on one side of a timber separating floor, and a low window (other than one opening onto a balcony) is not allowed. See Figure 7 (back cover).

It is emphasised that these specifications are 'deemed to satisfy' provisions only, other methods of construction can be used, provided adequacy to the satisfaction of the local authority is proved.

PART J: Refuse chutes. Brickwork surrounds to fired clay refuse chutes provide not only stability and fire resistance but very useful sound insulation, so desirable in these situations.

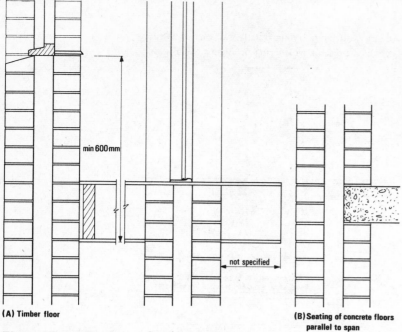

(A) Timber floor

(B) Seating of concrete floors parallel to span

min 600 mm

not specified

Figure 7 *Regulation G3. Sound insulation-floors Associated structures and openings.*

Continued from page 8

The impervious surface of engineering quality bricks makes them eminently suitable for refuse chambers where easily cleaned surfaces are essential. The stiff structure provided by brick-built refuse chutes and chambers can serve a useful structural purpose in resisting lateral forces on buildings from wind and accidental forces covered by Regulation D.19.

PART L: Chimneys and flues. Clay brickwork responds well to high temperatures and is so well established as a material for chimney construction that Regulation L.6 requires traditional chimneys to be constructed of non-combustible material 100 mm thick, which is a specific requirement. There is no functional requirement for the material surrounding flues in chimneys, hence materials of a thickness less than 100 mm, even if possessing better insulation, are not acceptable, except by way of relaxation.

The introduction of the perforated brick has caused some unnecessary confusion and some comment is called for. The perforated brick is regarded as 'solid' for structural purposes when over 75% solid, and responds very much as a truly solid brick except when subjected to very high temperatures for long periods.

Whilst it can be argued that the perforated brick does not satisfy the letter of the Regulation for chimney construction, it is nevertheless satisfactory and can be used with perfect safety. The relaxation procedure could be adopted if considered necessary, although this would be taking the term 'solid' in relation to brickwork to rather ridiculous limits. It is difficult to visualise a situation where the perforations could present a problem in flue construction, because they are always perpendicular to the bed joint and hence sealed.

A rather different situation arises with the 90 mm brick, which does not satisfy the regulations, but in suitable circumstances could be used via the relaxation procedure, and could be justified quite easily, either by the use of a special liner or by rendering the flue with gypsum/vermiculite plaster.

PART N: Drainage. Inspection chambers and manholes can be subject to harsh treatment, depending on ground conditions, and most fired clay bricks are suitable for use in these structures, but in very

wet and chemically aggressive environments it is necessary to specially select suitable bricks.

The choice of mortar is important and should suit the type of brick, and the environment, for example, sulphate-resisting cement should be used in sulphate-bearing soils.

For small inspection chambers up to 1 m deep, in areas free from vehicular traffic, particularly in non-cohesive soils, a thickness of 90 or 102.5 mm brickwork should be acceptable but, otherwise, a thickness of 215 mm should be used, with increased thickness for greater depths (9).

GENERAL

Although brickwork is a building element with a long tradition, it has not been sufficiently appreciated, particularly as a loadbearing element. The selection of bricks and mortars to suit a particular need deserves the same care as do other structural materials, and efforts to provide high quality work using all the latest research material can result in a first class, durable and pleasing structure. □

REFERENCES AND BIBLIOGRAPHY

1 BS Code of Practice CP.111: Part 2: 1970, **Structural Recommendations for Loadbearing Walls'.**

2 **'Model Specification for Loadbearing Clay Brickwork',** Special Publication No 56. The British Ceramic Research Association 1971.

3 Thomas, K. **'Movement Joints in Brickwork',** CPTB Technical Note, Vol. 1, No. 10.

4 Foster, D. **'Some Observations on the Design of Brickwork Cladding to Multi-storey RC Framed Structures',** BDA Technical Note, Vol. 1, No. 4.

5 Bradshaw, R.E. and Entwisle, F.D. **'Wind Forces on Non-loadbearing Brickwork Panels',** CPTB Technical Note, Vol. 1, No. 6.

6 BS Code of Practice CP.121: **Walling: Part 1: Brick and Block Masonry: 1973.**

7 Haseltine, B.A. and Thomas, K. **'Loadbearing Brickwork – Design for the Fifth Amendment',** BDA Technical Note, Vol. 1, No. 3.

8 BS3921: Part 2: 1969, **'Bricks and Blocks of Fired Brickearth, Clay or Shale'.**

9 BS Code of Practice, CP.301: 1971, **'Building Drainage'.**

10 **'The BDA Metric Brick Format',** CPTB Technical Note, Vol. 2, No. 7.

11 BS.187: Part 2: 1970, **'Calcium Silicate (Sandlime and Flintlime) Bricks'.**

12 BS Code of Practice CP.3: Chapter 111: 1960, **'Sound Insulation and Noise Reduction'.**

13 Foster, D. **'Metric Modular Bricks and Building Regulations',** Structural Clay Products Ltd., Report No. SCP6, August 1972.

14 **'Housing on a Dimensional Framework – Brickwork Design (90 mm wide bricks) with reference to the Building Regulations'.** Housing Development Notes, Department of the Environment, December 1972.

15 BS 476 : 1972 **'Fire tests on building materials and structures. Part 8. Test methods and criteria for the fire resistance of elements of building construction.'**

Note: *Reference 2,3, 4, 5, 7 & 10 are available free from The Brick Development Association.*

Performance characteristics of perforated and solid bricks

Comparative testing of perforated and solid bricks

This represents a summary of *The performance of walls built of wirecut bricks with and without perforations* published by the British Ceramic Research Association as Special Publication No 60, and describes experiments carried out to determine the load bearing and weather resisting properties of perforated bricks, compared with traditional solid units. Throughout, the term 'perforated brick' is used in the literal sense, meaning a brick with holes passing completely through it, rather than in the BS3921 usage, which only applies the term 'perforated' to a brick whose hollow volume is greater than 25% of its total volume.

Advantages of perforated bricks

Since the perforations facilitate a more even heat distribution throughout the brick during the firing process, a more homogeneously vitrified product results. This is of considerable importance in view of the increasing number of structures being built in calculated loadbearing brickwork, requiring therefore, as uniform a building unit as possible. The perforations also provide improved thermal insulation properties, compared with solid bricks, and facilitate on-site handling.

Previous research

Despite the fact that perforated bricks are used very widely throughout Europe and the United States, little extensive work on the comparison of solid and perforated brick performance characteristics has been done. A certain amount of comparative testing has been carried out on the Continent, the results of which indicated that walls built of perforated bricks were approximately as strong as those using solids, but the number of specimen walls built was insufficient for any statistically significant interpretation to be made.

Tests carried out by the Building Research Station on solid and perforated brick walls have also indicated that there was little difference in the average value of the load factor (ultimate failure load/working load) between the different wall types, when using the working load calculated from Table 3 of CP111, but here again, the number of samples built and tested precluded statistical analysis.

Aims

The following experiments set out to establish the relationship between wall strength and brick strength for different types of solid and perforated bricks, to verify the validity of the basic stress table of CP111, to determine the resistance to rain penetration of walls built of solids and perforateds, and to determine the stress distribution in perforated bricks under load, by photo-elastic analysis techniques, and hence to establish guide lines for optimum perforation patterns.

MATERIALS
1 Bricks

Extruded wirecut solid and perforated bricks were obtained from eight different works. Bricks were designated by a letter (A – H) indicating the works, and a number indicating the number of perforations. Solid units were designated by the works letter followed by S (eg AS).

2 Sand

Sieve analysis of sand used throughout tests:

(1)

BS Sieve	% Passing
Above 3/16 in	Nil
3/16 in – 7's	Nil
7's – 14's	1.9
14's – 25's	4.4
25's – 52's	25.2
52's – 100's	49.9
100's – 200's	14.8
Below 200's	3.7

3 Lime

Limbux conforming to BS890 (1966).

4 Cement

Standard production cement obtained from local supplier for walls A – G (except G3, B11 and all F series). Since this material proved to be more variable than was hoped, another grade, specifically supplied for experimental purposes was obtained direct from the Cement Marketing Board. This cement, described as 'typical', was used for the remainder of the tests and yielded a maximum coefficient of variation of compressive strength of 2.7% (1:3 mortar cube).

PREPARATION OF MATERIALS
1 Mortar

Two different mixes were employed for comparison purposes, $1:\frac{1}{4}:3$ (x) and $1:1:6$ (y). The bricklayer was allowed to choose the preferred consistency of mortar for the first walls, subsequent water proportions being kept constant. All mixes were adjusted to a consistence of 10 ± 3mm by the dropping ball method. Proportioning was by volume and the mixing procedure was in accordance with CP121:101:1951.

2 Walls

All walls were single leaf, 34 courses high, and 4ft 6in (1.372m) long. Bricklaying technique conformed to the requirements of Special Publication 56. All joints were pointed flush and both faces of each wall were rubbed down. Where necessary the suction rate of the bricks was adjusted by wetting so as not to exceed 20g/dm²min.

EXPERIMENT 1
Wall performance under axial load

Method Two sets of test walls were built of each brick category, using mortar mixes x and y. The walls were built on to a concrete plinth set on a steel plate, in order to facilitate loading into the testing machine. Walls were allowed to cure for 14 days before testing. A grano-fondu capping was cast on to the top of each wall 24 hours before testing.

All walls were tested to destruction at a rate of loading of 100lbf/in²/min.

Results For the purposes of this article typical mean results from the three main brick strength categories will be quoted below (2). The examples are taken from the middle range of each category.

(2)

Brick type	Ultimate compressive strength (lbf/in²)	Ultimate wall loading (lbf/in²)		Ratio of wall strength to brick strength	
		Mix x	Mix y	Mix x	Mix y
HS	12,500	4,637	3,527	0.372	0.283
A3	12,000	3,090	2,260	0.259	0.190
C7	8,000	2,593	1,760	0.320	0.217
ES	6,000	2,770	2,257	0.458	0.373
F5	6,000	2,277	1,763	0.391	0.302

Figure 1

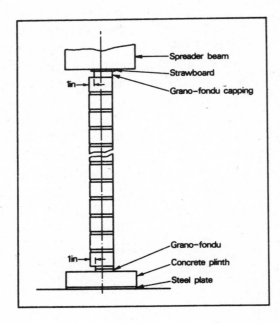

- Spreader beam
- Strawboard
- Grano-fondu capping
1in
- Grano-fondu
- Concrete plinth
1in
- Steel plate

Figure 2

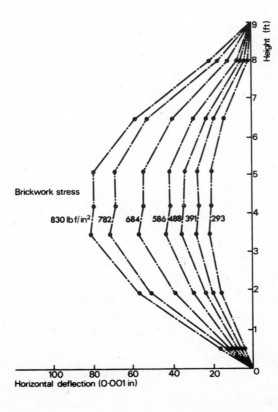

Brickwork stress

830 lbf/in² 782 684 586 488 391 293

Height (ft)

Horizontal deflection (0·001 in)

EXPERIMENT 2

Wall performance under eccentric load

This experiment seeks to simulate the situation where a brick infill panel is used in a concrete framed structure where no provision has been made for shrinkage and/or differential expansion of frame and panel. The infill panel construction method usually calls for the brickwork to overhang the concrete floor slab a distance of one third of the brick's thickness. This allows slip tiles to be positioned over the edge of the floor slab, resulting in an overall appearance of continuous brickwork. Under such conditions the loading on the infill panel is eccentric, with the result that in situations of excessive loading the brickwork will tend to spall off, and ultimately the whole wall will fail.

Method Two sets of test walls were built of each brick category, using mortar mixes x and y, on a 1in grano-fondu plinth, set on to a concrete plinth, similar to that described in experiment 1. The grano-fondu plinth was set one inch back from the facing edge of the wall (see fig 1). Walls were allowed to cure for 14 days, and 24 hours before testing were topped with a grano-fondu capping. The capping was also set back one inch from the facing edge of the wall.

All walls were tested to destruction at a rate of loading of 100 lbf/in²/min.

Results As can be seen from the graph (fig 2), there was considerable buckling of the wall before failure, the central section of the wall being about 0.08in out of true at 30–40% maximum load.

As for experiment 1, the results listed below are representative, not complete.

(3)

Brick type	Ultimate compressive strength (lbf/in²)	Ultimate wall loading (lbf/in²)		Ratio of wall strength to brick strength	
		Mix x	Mix y	Mix x	Mix y
HS	12,500	2,500	1,656	0.201	0.133
A3	12,000	1,677	995	0.141	0.084
C7	8,000	1,638	1,012	0.203	0.125
ES	6,000	1,144	814	0.189	0.134
F5	6,000	1,147	839	0.197	0.144

Discussion of results of experiments 1 and 2

Generally a perforated brick has a lower apparent strength than an equivalent solid. This however is mainly a computational effect, since brick testing in accordance with

Figure 3

Figure 4a

Figures 4b and 4c

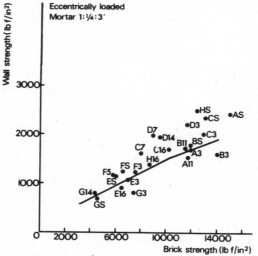
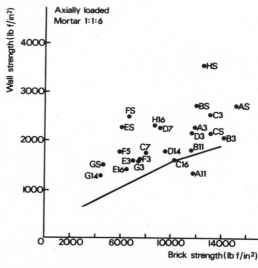

Figure 4d

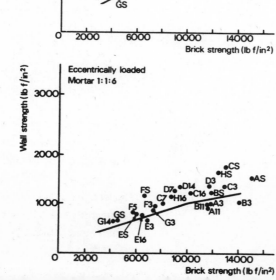

BS3921 ignores the fact that the actual bearing surface of the perforated unit is less, by the area of the perforations, than the bearing surface of the solid (see fig 3). The question of the relative strengths of individual units however, is not the main point. What has to be established is whether walls built of perforated bricks are weaker, by any significant amount, than those built of solid units.

Comparison of the results of mortar cube tests and wall test results revealed that while variations in mortar cube strength for individual walls had little or no significant effect on wall strength, the variations in mortar mix did, thus necessitating separate consideration of the results for the different mixes (see fig 4).

CP111 allows loadbearing walls to be designed either by reference to the brick strength and mortar type, or by the result of a wall test on the actual bricks and mortar to be used. In the present case the slenderness ratio of the test walls is 18 so the basic stress in table 3 of CP111 has to be multiplied by 0.5 to give the permissible stress in the wall. Since the wall under test is less than 500in² a correction factor of $0.75 + \dfrac{216}{2000}$ has to be applied.

The permissible stress in axially loaded walls is thus 0.43 times the basic stress in table 3 of CP111. For the eccentric case the factor is 0.28. The following table (table 4) compares the actual compressive stress in the wall at failure with the permissible stress for each of the brick categories. The load factor, that is the ratio of the ultimate stress to the permissible stress, is also given.

Table 10 of CP111 stipulated by interpolation a load factor of 8 for these conditions of wall testing. This, when multiplied by the

Wall type	Axial loading			Eccentric loading (e = t/8)		
	Permissible stress lbf/in²	Ultimate wall strength lbf/in²	Load factor	Permissible stress lbf/in²	Ultimate wall strength lbf/in²	Load factor
HSx	333	4,637	13.9	217	2,500	11.6
HSy	220	3,527	16.0	143	1,656	11.6
A3x	323	3,090	9.6	210	1,677	8.0
A3y	214	2,260	10.5	139	995	7.2
C7x	235	2,593	11.0	153	1,638	10.7
C7y	164	1,760	10.7	107	1,012	9.5
ESx	182	2,770	15.2	118	1,144	9.7
ESy	132	2,257	17.1	86	814	9.4
F5x	177	2,277	12.8	115	1,147	10.0
F5y	129	1,763	13.7	84	839	10.0

basic stress, and applying the correction factors referred to in the above paragraph (0.43 axial, eccentric 0.28) provides the minimum value at which the walls of various brick strengths should be crushed. These values are shown as solid lines on the graphs left, (fig 4) and from these it can be seen that since the majority of the individual wall strength values fall well above the lines, the figures given in CP111 are, on the whole, conservative.

These results indicate that while the statistical relationship between brick strength and wall strength is significantly different for solid and perforated units, the actual wall strengths, when compared with the requirements of CP111, cannot be readily differentiated. There is no significant difference for either of the mortars or either of the loading conditions between solid and perforated bricks.

EXPERIMENT 3
Resistance to rain penetration
Method Two single leaf 6ft x 6ft test walls, built into frames, were constructed using mortar mixes x and y. The frames were suspended from a steel yard so that records of weight changes could be made. All walls were whitewashed on the back in order to make photographic records of moisture penetration.

Walls were subjected to an air pressure of 2in water gauge (equivalent to a wind velocity of 64mph) and at 30 minute intervals were sprayed with one gallon of water in one minute (equivalent to 2.5in of rain over the whole face of the wall in a 24 hour period).

The experiment continued for 48 hours except in cases where the rate of penetration appeared to be abnormally slow, when the time was extended to 168 hours.

After testing, walls were demolished course by course in order to determine whether any water had collected in the perforations.

Results As in experiments 1 and 2, representative selection of the results obtained is shown. Graphs (figs 5a and 5b) show the relationship between time and weight of water absorbed by the walls for the different mortar mixes.

Table 5 relates the wetted area of the rear face of the wall after a 24 hour period, to the percentage perforation area of the bricks concerned.

(5)

Brick category	% area of perforation	% area wet after 24 hours	
		Mix x	Mix y
HS	Nil	38	Nil
A3	13.8	3	8
C7	19.8	3	3
ES	Nil	87	100
F5	17.1	44	94

Discussion of results
Within any particular brick category the rate of absorption of the wall appears to be related in a general way to the water absorption and suction rate of the bricks, the presence of perforation having no seeming relevance.

No water was found in perforations when the walls were demolished.

From Table 5 it can be seen that the wetted area of the rear of the walls is completely unrelated to the degree of perforation of the bricks, although it varies with brick type and is more frequently higher in the case of mortar mix y.

Workmanship is obviously an important factor in the resistance to rain penetration of any wall, and whatever the rain resistance of an individual brick, the wall it is built into will fail to give good results if it is not constructed in accordance with CP121:101: 1951.

EXPERIMENT 4
Determination of stress distribution in perforated sections
Introduction Photo–elastic analysis is the technique of examining the stress distribution in materials by photographing the stress patterns set up when a model, made usually of a plastics material, is subjected to various loadings under controlled conditions.
Theory When brickwork is subjected to compressive loading, eventual failure occurs in the form of vertical splitting, since the quasi-fluid properties of the mortar lead to a tensile strain in the bedding plane of the brick. A thin horizontal section of the brick therefore, behaves as if it were subjected to horizontal tensile forces, hence a brick can be represented experimentally by a thin section of plastics material, subjected to appropriate bi-axial tensile forces.

In a solid section under bi-axial load, all stress is symmetrical about every point in the section. As soon as perforations are introduced however, the stress is no longer symmetrical, and stress concentrations are set up. The object of the following experiments was to investigate these concentrations for various patterns of perforations, to compare the stress concentration factor (SCF) for the various patterns, and to attempt to establish suitable perforation patterns that will give rise to minimum SCF's.
(*Note SCF is used, in relation to a perforation, to describe for the same applied load the number of times the stress is greater, due to the perforation, than the applied stress. The higher the factor, the lower the stress required to cause the material to fail.*)
Apparatus An accurate loading frame was constructed which enabled measured loads of up to 100lbf and 200lbf to be applied to the short and long sides respectively of the test section. The test sections were ⅛in thick, Araldite CT200 plastic. The fringe patterns were photographed with a camera fixed to the photo–elastic bench.
Method The first condition to be investigated was that of a basic plate with no perforations, thus checking the model and loading frame design. The second condition was the case of a single circular perforation. As can be seen from fig 6 this gave rise to a circular fringe concentric with the hole. On increasing the diameter of the hole the diameter of the fringe correspondingly increased, but its SCF remained constant.

In order to see what effect on the stress pattern neighbouring holes had, the next case to be examined was that of three circular holes equispaced down the centre line of the section, two different hole diameters being examined separately. A similar concentric circular fringe effect to that in the previous example was obtained, but in addition the fringes showed a tendency to join together along the centreline of the holes (see figs 7 & 8).

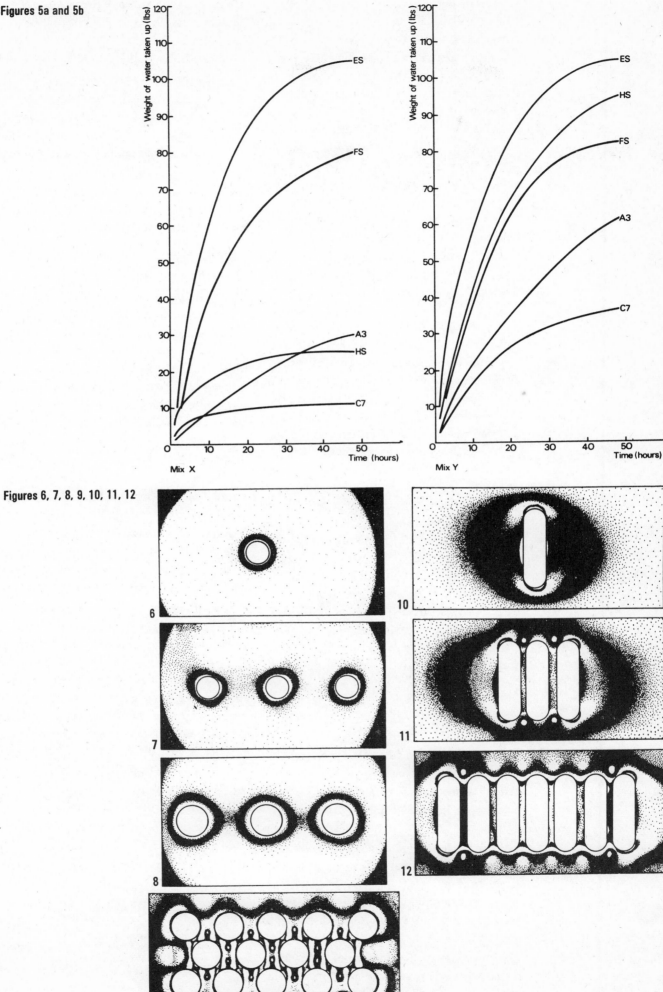

Figures 5a and 5b

Mix X

Mix Y

Figures 6, 7, 8, 9, 10, 11, 12

213

Examination of a system consisting of two equal circular holes showed that a maximum tensile stress would occur at the centre of the web separating the two holes. The SCF would vary inversely with the distance between the two holes, reaching a maximum as the holes just touched.

The analysis, as a whole, indicated that the effect of the holes was confined to a relatively small region around them, distant holes giving rise to only small percentage SCF changes.

Results of tests on four, eight and fourteen hole perforation patterns showed that as the number of holes increased the actual stress increased slightly, both at the edge of the holes and in the webs, as can be seen from fig 9.

Fig 10 shows the fringe pattern for a single parallel sided slot. As can be seen the stress is greater at the semicircular ends of the slot since stress varies inversely with radius of curvature, and this stress is also greater than that around a circular hole of the same diameter as the width of the slot. In the case of three slots (fig 11) it was noted that the stress concentration at the ends of the outer slots appeared to be slightly greater than that at the ends of the centre one, also the webs were subjected to a slight tensile stress. In the case of the seven slot configuration (fig 12) the tension in the centre webs was greater than in the outside ones, but nevertheless the tension obtained was still very small compared with the stress at the ends of the slots. As in the three slot configuration, the stress at the ends of outer slots was significantly greater than that at the ends of the intermediate slots.

Conclusions

1 Since the stress field without perforations is symmetrical, it was found, as expected, that a symmetrical perforation (ie a circular hole) gave a lower SCF than other shapes (ie slots).

2 When two or more circular holes are present, the maximum tensile stress will occur at the mid-point of the web between the holes, the magnitude of the SCF at this poitn varying inversely with the distance between centres.

3 An increase in the number of holes does not appear to cause any significant increase in the SCF.

From this it follows that:

1 The size and pattern of perforations should be such as to keep stress concentrations to a minimum.

2 Webs between holes should be maintained at least equal to the hole radius.

3 Since the undisturbed stress field is symmetrical it would probably be advantageous to keep the perforation pattern symmetrical, with hole radius and web thickness constant.

4 In the case of slots, it is a better policy to avoid long narrow slots because of the high SCF's at their rounded ends.

GENERAL CONCLUSIONS

The following general conclusions can be drawn from this series of experiments:

1 The indications are that although CP111 (1964) is slightly conservative, the validity of its theory is not in question, and when using the appropriate brick strengths as specified in Table 3 of CP111, both solid and perforated units will satisfy the requirements of this table equally well.

2 The stringent rain penetration tests established that perforated bricks do not behave differently from solid units, hence under normal weathering conditions perforations should not become full of water.

3 From (2) it follows that quality of workmanship and mortar performance are the most important determinants of rain penetration through brickwork.

4 Certain anomalies in the results would suggest that the ceramic properties of the bricks are important and overshadow the more obvious geometrical differences, hence in the absence of a wall test the best estimate of brick performance can be made using the average ultimate compressive strength obtained from tests on a representative sample of bricks.

5 The major factors determining satisfactory performance are proper design and good workmanship. Whether the bricks are solid or perforated is of secondary importance □

References
For further information:
The performance of walls built of wirecut bricks with and without perforations.
by H. W. H. West, H. R. Hodgkinson and S. T. E. Davenport.

The strength, function and other properties of wall-ties

by K. Thomas

ABSTRACT

This paper discusses the properties of wall-ties and their suitability regarding differential movement, fire resistance, corrosive conditions and frost. Research on the shear and pull-out values of wall-ties for use as peripheral fixing for brickwork panels is described and a table of recommended safe loads included. The tensile strength of wall-ties with varying degrees of embedment and precompression due to wall deadload are also included together with a discussion on the suitability of polypropylene ties. Finally the paper summarizes work on the strength of wall-ties shot-fired to concrete members.

1. THE FUNCTION OF WALL-TIES

Cavity-wall construction offers advantages over solid construction as far as resistance to water penetration and thermal insulation are concerned, but from the structural point of view, as discussed elsewhere,[1] the solid wall is generally to be preferred.

When cavity walls are subjected to bending due to eccentricity of loading a tendency towards differential lateral deflection occurs between the leaves. Wall-ties play an important function in equalizing deflections of the two leaves, particularly when only one leaf is loaded. When ties are capable of transmitting the tensile and compressive forces without appreciable stretching or buckling, the bending stresses will be the same in the two leaves, providing they are of equal stiffness.

Davey and Thomas[1] state that the tendency towards lateral deflection is greater for the cavity wall, and in CP 111[2] allowance is made for this, by taking the effective thickness of a cavity wall (for slenderness ratio purposes) as equal to two-thirds of the sum of the thickness of the two leaves. Tests at the Building Research Station have shown that for cavity walls approximately 9 ft high and 4 ft 6 in long, loaded eccentrically on one leaf, the unloaded leaf followed approximately the curvature of the loaded leaf regardless of whether the ties were in compression or tension. In the first series of tests Davey and Thomas record that a 1:3 cement-mortar was used, and the ties were of the 'butterfly' type made from 9 SWG steel wire. In the second series a 1:1:6 mortar was adopted, the ties being of the same form but made from 12 SWG hard-drawn copper wire. The ties in both series were spaced at 3 ft centres horizontally and 18 in vertically. It was noted in the second series of tests that lateral movement of the two leaves was approximately the same when the ties were in tension but rather less when in compression, probably due to bending of the wall-ties.

Although wall-ties may ensure that both leaves of a cavity wall have the same radius of curvature, it does not follow that the ultimate strength of the wall will be greater, indeed, the reverse may be the case. In the worst case, with the load eccentrically applied to one leaf and the other leaf unloaded, the flexural stress induced in the unloaded leaf may, at high loads, cause it to crack so that its contribution to the loaded leaf will disappear. Even in such conditions, however, ties are normally required, in order to stabilize the unloaded leaf.

1.1 Differential movement

CP 111[2] recommends that where the outer leaf of an external cavity wall is not greater than half a brick, the uninterrupted height and length should be limited so as to avoid undue loosening of the wall-ties due to differential movements between the two leaves. It is recommended by the code therefore that the outer leaf be supported at intervals of not more than every third storey or every 30 ft, whichever is less. However, for buildings not exceeding four storeys or 40 ft, in height, whichever is less, the outer leaf may be uninterrupted for its full height.

In view of the current availability of double-triangle type stainless-steel wall-ties perhaps the recommendation in the Model Specification for Load-bearing Clay Brickwork[3] that the vertical twist type tie be used might be reconsidered. The more rigid ties will almost certainly cause more distress when differential movements occur and the more flexible stainless-steel ties, although more expensive, will accommodate greater movement. Some research is necessary to determine the optimum tie spacing necessary and no doubt fewer more efficiently positioned ties would result in an overall reduction in cost, ie it might be shown that ties are only necessary at mid-height and floor levels.

Foster[4] quotes one example of Swiss practice and describes how the outer leaf of the structure is restrained at each floor level connected at 300 mm intervals to a ring of Stahlton clay units (Figure 1). It is not necessary to use Stahlton planks and perhaps a precast concrete beam might be considered, no doubt incorporating brick slips to give the appearance of a monolithic structure. This is yet another possibility not generally adopted in the UK.

Figure 1 *Detail of wall construction at Biel (after Foster[4]).*

1.2 Fire resistance

Little information is available on the fire resistance of wall-ties or indeed of the effect when ties may be damaged after major fires.

It is suggested that when cavity walls are subjected to major fires, some check on the ties is necessary before passing a wall as sound, based merely on external visual inspection. The reason for this recommendation is that in at least one instance corroded wall-ties have been responsible for structural distress. The building concerned consisted of three storeys of 11 in cavity brickwork and was used as a warehouse for inflammable materials. On two separate occasions severe fires gutted the building leaving the walls in perfect condition from the visual point of view, and rebuilding of the damaged areas of the structure commissioned. Some five years after the second fire the walls began to buckle and twist to an alarming extent and on opening up the cavity wall, it was noted that galvanizing from the vertical twist-type ties had been removed (almost certainly by one of the fires, zinc vaporizing at approximately 400°C) and subsequent corrosion accompanied by considerable expansion of the ties had occurred.

The fire resistance of polypropylene wall-ties can only be determined by actual test but in a private communication from the Fire Research Station in May 1966, the following opinion was given: 'It is necessary to differentiate between load-bearing walls and non-load-bearing walls. A cavity wall of two leaves of 4¼ in brick, gives a fire resistance of 6 h when load-bearing with galvanized steel ties, and 6 h for a similar wall without ties tested as non-load-bearing. A single-leaf wall when load-bearing, can give a fire resistance of 2 h'. From this statement it was suggested that a cavity brick wall tied together with polypropylene wall-ties can give a fire resistance of 6 h when non-load-bearing and at least 2 h when load-bearing.

Such a statement by the Fire Research Station is no doubt encouraging to manufacturers of prefabricated brickwork panels who may wish to provide ties between leaves of brickwork at each floor or roof level only.

1.3 Corrosion

The life of a steel wall-tie is largely dependent on the weight and quality of the zinc protective coating. This varies according to the type of tie, which should be selected according to the conditions to which it will be exposed during use. Metal thickness in excess of that recommended in the standard[5] is of little value, particularly if corrosion takes place in the wall cavity.

Corrosion of galvanized steel wall-ties when used in conjunction with black ash mortar is well known. The standard[5] recommends that when used in such situations, special precautions should be taken such as coating with bitumen, preferably on site. Figures 2 and 3 illustrate the corrosive effects on galvanized steel and the subsequent cracking of mortar joints due to the accompanying expansion.

It is important that marine sands are thoroughly washed, before use, as mortar containing chlorides may be responsible for accelerated corrosion of metal ties. Sulphate-bearing materials may also aggravate the situation.

The use of lime in mortar is said to inhibit corrosion of wall-ties, although some authorities maintain that there is sufficient free lime in cement for this purpose.

Brass or manganese bronze wall-ties are not recommended in areas where nitrates or ammonia are likely to be present, as the ties may suffer from stress corrosion cracking due to an embrittling effect upon the metal, a phenomena which may occur in a matter of days in some cases and years in others. Organic-based foaming or wetting agents used in mortar mixes are another possible source of danger; also some detergents. The latter method of entraining air in mortar mixes is not recommended.

1.4 Frost

The effect of frost on traditional wall-ties appears to have little effect but this may not be true of poly-

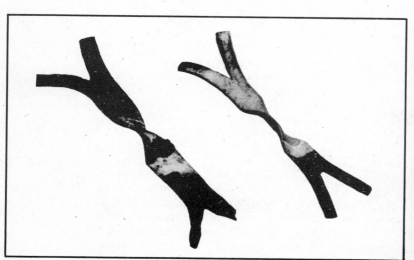

Figure 2 *Galvanized-steel wall-ties after embedment in black ash mortar. (Courtesy: Building Research Station.)*

Figure 3 *Expansion of black ash mortar joints. (Courtesy: Building Research Station.)*

propylene ties. It is well known that certain plastics become embrittled when subjected to low temperatures and the author has witnessed the shattering of certain plastic objects when given a slight blow. This occurred after storage out of doors for only one winter. Some research is necessary on this topic over an extended period of time.

Calcium chloride or additives based on this salt are occasionally used as frost inhibitors. The use of such admixtures have no significant advantage in mortar and may lead to subsequent dampness and corrosion of wall-ties. It is therefore recommended that the use of such admixtures be discouraged.

1.5 BS 1243:1964

Amendment No 2 (PD 5974), published January 28, 1967, permits the use of stainless steel for wall-ties. Under 'Materials' the same amendment specifies that the steel shall comply with BS 1554, 'Rust-, acid- and heat-resisting steel wire', thus effectively excluding the use of stainless-steel vertical twist-type wall-ties. No doubt this is an error as it seems illogical to exclude such ties.

2. SUITABILITY OF NORMAL CAVITY TIES FOR USE AS PERIPHERAL FIXING OF BRICKWORK PANELS

The Brick Development Association Limited sponsored an investigation at Napier Technical College, Edinburgh, to assess the possibility of using normal cavity ties for securing brick panels to structural reinforced concrete members. Whilst cavity ties are not normally designed for this purpose, it was felt that their ready availability would justify their adaptation for peripheral fixing of wall panels if sufficient holding power could be developed. Accordingly, specimens were devised for testing the load-carrying capacity of cavity ties embedded in mortar joints in brickwork having regard to the constructional details involved and the consequent shear or tensile nature of forces the ties would be required to resist.

Five types of cavity ties were tested in three different mortar mixes and, although uniformity was aimed at by having all the specimens made by the same person, under the same conditions, no attempt was made to produce a quality not easily attainable on the site.

Panels of brickwork may be located either between structural members or in continuous form with the structural members behind the panels. Ties are usually bent temporarily into a right angle and fixed to the inside of the formwork. On striking the formwork the ties could then be straightened for insertion into the mortar joints of the brickwork panels.

Wind or other lateral pressure on the panels would result in a shear loading on the ties where the brickwork came between structural members, but in a direct tensile (or compressive) loading where the brickwork was in front of them. Specimens for test were therefore devised to investigate the effects of these two types of loading.

Ties situated near the bottom of a panel undoubtedly have a greater pull-out or shear resistance owing to the effects of precompression but this effect was ignored in the tests and care was taken to ensure that the grip of the mortar joint on the tie being tested resulted only from the bond of the mortar itself.

2.1 Ties tested

The following standard cavity ties were used in both the shear and tension tests:
(a) Wire 'butterfly':
1. Zinc-coated steel, 10 SWG.
2. Copper, 12 SWG.
(b) Strip 'fishtail':

Figure 4 *Tension specimen for testing fishtail strip metal tie.*

Figure 5 *Tension specimen in position for testing.*

1. Zinc-coated steel, $\frac{1}{8}$ by $\frac{3}{4}$ in
2. Copper, $\frac{1}{8}$ by $\frac{3}{4}$ in
3. Bronze, $\frac{1}{8}$ by $\frac{3}{4}$ in

2.2 Mortar mixes

For each type of specimen three different mortar mixes were used (Table 1).

Table 1 mortars used for shear and tension tests

Mix	Proportions by volume		
	Cement	Lime	Sand
1	1	$\frac{1}{4}$	3
2	1	1	6
3	1	2	9

Note: The sand was fully dried before use.

2.3 Test specimens

The two groups of test specimens were, respectively, *tension* and *shear* and these were arranged as shown in Figures 4–9 so that, as far as possible, the types of forces induced would be those for which the tests were intended.

2.4 Testing equipment

A large rectangular steel box-frame was used for mounting the test equipment (Figure 10). The latter consisting basically of a vertical steel tension rod passing through a hollow hydraulic ram secured at the top of the frame. The specimens were anchored, without being clamped, to the bottom of the frame and attached to the lower end of the tension rod by suitably designed fittings. A proving ring inserted

Figure 6 *Shear specimen for testing fishtail strip metal tie.*

Figure 7 *Fishtail tie shear specimen in position for testing.*

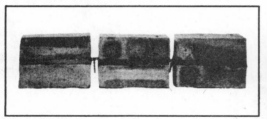

Figure 8 *Shear specimen for testing wire butterfly ties.*

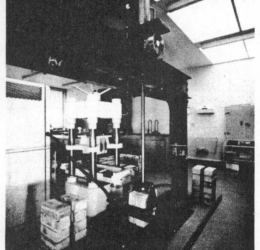

Figure 9 *Butterfly wire tie shear specimen in position for testing.*

Figure 10 *General arrangement of testing equipment.*

between the underside of the ram and the top of the frame enabled a direct measurement of the applied load to be made.

Attention was given to the requirement that no compressive force be applied to the mortar joints since this could have the effect of gripping the ends of the ties more tightly and producing an artificial increase in the failure load. At the same time, wooden packings were introduced between contact surfaces, where the load was applied, to distribute the pressure uniformly across the sometimes uneven or irregular surface of the bricks.

2.5 Test details
Tables 2 and 3 give the mean values of the loads sustained by the three specimens of each type at failure and Table 4 gives the mortar properties.

2.6 Conclusions
2.6.1 Tensile tests – butterfly ties
For the 1:¼:3 and 1:1:6 mortar mixes failure occurred in the tie itself. The mortar remained intact in all cases. Tests involving the 1:2:9 mortar produced 50% failures owing to fracture of the ties and 50% owing to the mortar joint opening and releasing the tie. Where failure was by breaking of the tie the strength of the galvanised ties was greater than that of the copper. When failure was due to opening of the mortar joint (1:2:9 mortar only) this occurred before either type of tie realised its full tensile strength so that neither had the advantage.

2.6.2 Tensile tests – fishtail ties
Failure of specimens was either by opening of the mortar joints or by pulling out of the ties leaving the joint intact. In no test did the wall-tie fracture. The galvanised fishtail ties all failed by opening of the mortar joint whatever the mortar used. The behaviour of the copper fishtail ties was slightly less efficient in that one in five of the failures was due to pulling out; the remainder being due to opening of the joints. The bronze ties produced the poorest results, just over half of the specimens being pulled out of the mortar joints when failure occurred. Figures 11 and 12 show typical 'fishtail' ties before and after testing and it will be noted that deformation of the bronze ties is greater than the others. The galvanised ties maintained their shape in all instances.

Table 2 tensile test results

Mortar mix	1:¼:3	1:1:6	1:2:9
Type of wall-tie	Ultimate load (lbf)		
Bronze fishtail	1587	1118	905
Copper fishtail	1703	1119	758
Zinc-coated fishtail	1767	1202	873
Copper butterfly	582	612	584
Zinc-coated butterfly	733	748	576

Table 3 shear test results

Mortar mix	$1:\frac{1}{4}:3$	$1:1:6$	$1:2:9$
Type of wall-tie	Ultimate load (lbf)		
Bronze fishtail	987	987	755
Copper fishtail	1610	1162	782
Zinc-coated fishtail	910	1150	494
Copper butterfly	552	560	511
Zinc-coated butterfly	605	557	482

Table 4 mortar strength for tensile and shear tests

Crushing strength of 4 in cubes (lbf/in²)			
Mix	$1:\frac{1}{4}:3$	$1:1:6$	$1:2:9$
By volume	1607	537	210

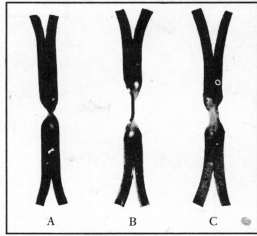

Figure 11 *Fishtail ties before testing. (A) Bronze. (B) Copper. (C) Zinc-coated steel.*

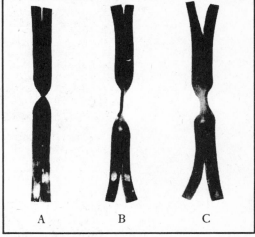

Figure 12 *Fishtail ties after tension test. (A) Bronze. (B) Copper. (C) Zinc-coated steel.*

Figure 13 *Reinforced-concrete column for testing dovetail slots and anchors.*

Figure 14 *Dovetail anchor positioned for tension test.*

Figure 15 *Dovetail anchor positioned for shear test.*

2.6.3 Shear tests – butterfly ties

Almost all of the specimens failed by breaking of the wall-tie itself whatever the mix of mortar used. The very small proportion in which opening the of joints occurred was not considered to be important although it is probable that if a mortar weaker than 1:2:9 was used, failure in this way would be much more frequent. It was found difficult to produce a pure shear failure owing to the tendency of the wire to bend across the joint width and ultimately fail in tension. A comparison of the shear and tension failing loads seems to support this.

2.6.4 Shear tests – fishtail ties

The wall-ties used in the specimens tested proved to be stronger than the mortar in all cases. Any deformation of the ties at failure was very difficult to detect and almost negligible.

2.7 Dovetail slot and anchor ties

The dovetail slot and anchor system was devised for attaching brickwork to concrete walls, columns, beams and slabs, etc. In essence, the system consists of dovetail section channels formed of metal sheet which are cast into the face of the concrete members by attaching them to the inside of the formwork before assembly. A variety of anchor plates with ends shaped to match the dovetail slots are available for building into the joints of the brickwork after being inserted into the metal channels.

This type of wall-tie offers a considerable advantage over traditional ties for peripheral attachment of brickwork panels, in that the vertical adjustment of the anchor permits alignment with the nearest bed joint in the brickwork.

Wind pressure on panels of brickwork would result in a shear loading on the anchors where the brickwork occurs between structural members, but in a direct tensile (or compressive) loading where the brickwork was in front of them.

All the ties were tested in slots cast in the 'face' of reinforced concrete columns using a standard 1:2:4 concrete mix and two different metals for the slots and anchors (Figures 13, 14 and 15).

2.8 Slots and anchors tested

The slots and anchors used in the tests were as follows:
(a) Galvanised steel anchors $\frac{1}{8}$ in thick in 18-gauge galvanised steel slots, 6 in long.
(b) Galvanised steel anchors 14-gauge thick in 18-gauge galvanised steel slots, 6 in long.
(c) Copper anchors $\frac{1}{8}$ in thick in 18-gauge copper slots, 6 in long (see Figure 16).

2.9 Testing equipment

The testing equipment was basically that used for the traditional ties (Figure 10).

Tables 5 and 6 give the mean values of the loads sustained by the six specimens of each type of anchor at failure.

2.10 Concrete

Some 6 in cubes made from the concrete used for the columns were tested at 28 days. The mean cube crushing strength was 5150 lbf/in².

Table 5 tensile test results

Type of anchor	Ultimate load (lbf)
$\frac{1}{8}$ in Galvanised steel	1185
14 g Galvanised steel	840
$\frac{1}{8}$ in Copper	970

Table 6 shear test results

Type of anchor	Ultimate load (lbf)
$\frac{1}{8}$ in Galvanised steel	1368
14 g Galvanised steel	1190
$\frac{1}{8}$ in Copper	1100

2.11 Conclusions

2.11.1 Tensile tests

In all cases failure of the specimen was due to opening of the dovetail slot and the subsequent crushing of the surrounding concrete in the immediate vicinity of the anchor. The appearance of the anchors was very little different after failure. The copper specimens suffered a very slight rounding of the corners of the dovetail but otherwise the deformation was negligible.

2.11.2 Shear tests

The failure of the shear specimens was by fracture at the narrow neck of the dovetail (Figure 16). This occurred in all cases although the lighter gauge galvanised-steel anchors had a tendency to buckle rather than fracture. Crushing of the concrete adjacent to the dovetail slot generally occurred at approximately 1200 lbf **load and** consequently very

little crushing was apparent when the copper anchors were tested as the average load at failure was 1100 lbf. Even after failure of the concrete adjacent to the slot, the specimens in galvanised steel were still capable of sustaining further load owing to the bridging effect of the slot.

Some improvement in shear strength would no doubt result from filling the slot with mortar, particularly when light-gauge galvanised-steel anchors are used, as this would have the effect of restraining the dovetail portion against lateral buckling and thus permitting it to carry its full shear load. Such practice would of course fix the tie position vertically and mean that alignment with the nearest bed joint may be more difficult.

Table 7 recommended working loads for wall-ties

Specification	Working loads per tie (lbf)					
	Tension			Shear		
(1) Abbey slot-type ties						
(a) *Galvanised steel fishtail anchors, $\frac{1}{8}$ in thick in 18 gauge galvanised-steel slots, 6 in long. Slots set in 1:2:4 concrete*	230			270		
(b) *Galvanised-steel fishtail anchors, 14 gauge thick in 14 gauge galvanised-steel slots, 6 in long. Slots set in 1:2:4 concrete*	160			240		
(c) *Copper fishtail anchors, $\frac{1}{8}$ in thick in 18 gauge copper slots, 6 in long. Slots set in 1:2:4 concrete*	190			220		
	Mix			Mix		
(2) Cavity-wall type ties	1:$\frac{1}{4}$:3	1:1:6	1:2:9	1:$\frac{1}{4}$:3	1:1:6	1:2:9
(a) *Wire butterfly* (1) *Zinc-coated, 10 SWG*	140	140	110	110	110	90
(2) *Copper, 12 SWG*	110	110	100	110	110	90
(b) *Strip fishtail* (1) *Zinc-coated, $\frac{1}{8}$ by $\frac{3}{4}$ in*	310	220	150	200	200	150
(2) *Copper, $\frac{1}{8}$ by $\frac{3}{4}$ in*	310	220	150	200	200	150
(3) *Bronze, $\frac{1}{8}$ by $\frac{3}{4}$ in*	310	220	150	200	200	150

2.11.3 Recommendations

As a result of the testing described above, Table 7 has been prepared by the author to give recommended working loads for the various wall-ties.

3. TENSILE STRENGTH OF WALL-TIES FOR LATERAL SUPPORT OF WALLS

During the summer of 1968 tests on wall-ties were carried out at Stockport College of Technology. The

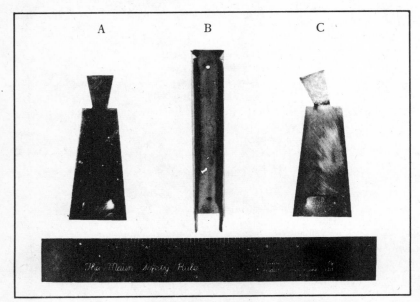

Figure 16 *(A) Anchor plate before testing. (B) Dovetail slot. (C) Anchor plate after shear test.*

reason for the research was to investigate the tensile strength and suitability of various types of wall-ties in providing lateral support to walls in accordance with Clause 304 of CP 111[2].

The investigation consisted of building storey-height walls and testing four different types of wall-tie, the degree of embedment ranging as indicated in Table 8. The free end of each wall-tie was cast in a small block of concrete and the tensile strength of the connection obtained by jacking against the wall. Comprehensive details of the tests are to be published shortly.

Table 8 tensile strength of wall-ties with varying degrees of embedment and precompression due to wall dead-load (After Thorley[6])

Type of wall-tie	Mortar mix	Embedment in joint (in)	Embedment in concrete block	Average of 10 specimens: ultimate load (lbf)
Kavi-ties (Poly-propylene)	1:1:3	$3\frac{3}{4}$	$3\frac{3}{4}$	339
	1:1:6	$2\frac{3}{4}$	$4\frac{3}{4}$	238
Butterfly ties	1:1:3	$3\frac{3}{4}$	$4\frac{1}{4}$	1236
	1:1:6	$3\frac{3}{4}$	$4\frac{1}{4}$	1000
Fishtail ties	1:1:3	$3\frac{1}{2}$	$4\frac{1}{2}$	1814
	1:1:6	$3\frac{1}{2}$	$4\frac{1}{2}$	1139
1 by $\frac{1}{8}$ in mild-steel flats (hooked)	1:1:3	$4\frac{1}{4}$ and hooked 1in over bwk.	$4\frac{1}{4}$	2517
	1:1:6	$4\frac{1}{4}$ and hooked 1in over bwk.	$4\frac{1}{4}$	1901

It is not proposed to discuss the results of the metal ties, the values being provided for comparison with the work described earlier in this paper.

Polypropylene wall-ties are relatively new and the results quoted in Table 8 are perhaps the only values currently available. The mode of failure when bedded in a 1:1:3 mortar joint was by extension of the tie (approximately 0.11 in average) and breaking of the tie; this occurred in eight of the ten specimens, the two remaining specimens did not fracture but produced a yield greater than the capacity of the jack. The polypropylene ties bedded in a 1:1:6 mortar joint all failed by breaking of the tie, fracture occurring $\frac{1}{2}$ in in the wall for nine specimens and flush with the face of the wall for the tenth.

Previous tests on polypropylene wall-ties carried out by Stanger's gave a mean tensile breaking load of 265 lbf on six specimens and an ultimate compressive load of 654 lbf being the mean of three specimens. The latter tests were all based on axial loadings, a condition which might not occur in practice.

4. STRENGTH OF WALL-TIES SHOT-FIRED TO CONCRETE MEMBERS

The technique of shot-firing high-tensile steel pins into concrete has been used for some time and the following paragraphs give an account of tests carried out in the laboratory and on site.

The laboratory test was carried out on a block of concrete, 6 by 6 by 30 in long approximately, which was designed to simulate concrete members for an actual construction. For this purpose, details of the original concrete materials were given to the research assistant. Testing was carried out when the concrete was approximately 7 weeks old.

A Hilti DX 100 hammer was used for firing the pins and consists basically of a barrel and plunger. The pins are forced into the concrete when a hammer blow on the plunger activates a cartridge. The design of the pin ensures, due to a fitted washer (Figure 17), that the flight of the pin is controlled both in direction and depth of penetration.

Three pull-out tests were carried out in the first instance and the following values obtained:

Test 1: 900 lbf.
Test 2: 1400 lbf.
Test 3: 800 lbf.

In spite of the fact that the test block was reinforced with a lightweight oblong mesh, some splitting of the concrete occurred. After the first three tests, two pull-out tests were carried out on pins approximately $\frac{1}{8}$ in shorter ($\frac{7}{8}$ in overall – type 735DX; Figure 18) to allow for the $\frac{1}{8}$ in thickness of the wall-tie. The corresponding values were as follows:

Test 4: 800 lbf.
Test 5: 800 lbf.

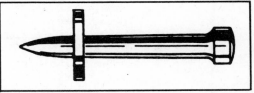

Figure 17 *Hilti nail.*

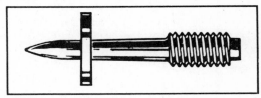

Figure 18 *Hilti screw stud.*

It can be seen from the above figures that the average value for the first three tests is just over 1000 lbf per pin. With an allowance made for the shorter embedment in the concrete due to the thickness of the tie, the pull-out value of a single pin would appear to be reduced to 800 lbf. These values are low compared with an earlier site test when the average value was approximately 2000 lbf. It is not clear why lower values should be obtained in the laboratory, but scale effect may be one reason.

Ultimate tensile strength tests on the wall-ties were carried out by shot-firing the ties into a test block (1 in overall – type NK35DX) and placing the block with the tie in a specially constructed jig in a Denison testing machine and pulling the tie out of the test block.

Test details are as follows:

Test 6: Ultimate strength 240 lbf. Some splitting of the concrete occurred around the pins. The pin pulled out of the concrete virtually undamaged.
Test 7: Ultimate strength 400 lbf. Pins pulled out with the tie virtually undamaged.
Test 8: Ultimate strength 200 lbf. Inner pin was loose after firing and before the start of the test. The pins pulled out with the tie virtually undamaged.

During the shot-firing it was noticed that the splitting of concrete usually occurred after firing the second pin, and the probable reason for this was that the two holes in the tie through which the pins were fired were too close together. It appeared as though the second pin was fired into a zone of stress concentration from the first shot, thus resulting in the splitting of concrete and lowering the pull-out value. The ties consisted of 1 in wide by $\frac{1}{8}$ in thick galvanized steel, 8 in long, plus 2 in long return, the holes being $1\frac{1}{8}$ in centres.

It was decided to use longer pins for the next two tests as it was thought that this might produce better results. Pins ($1\frac{3}{8}$ in overall – type NK35DX) were used giving 1–$1\frac{1}{8}$ in embedment into the concrete after allowing for the thickness of the tie.

Test 9: Ultimate strength 620 lbf. One pin only was through the inner hole (nearest to the tie). The tie was loose on the pin after firing and before the test commenced. The pin pulled out with the tie virtually undamaged.

Test 10: Ultimate strength 400 lbf. Two pins were used, the concrete splitting after the second pin was fired. The pins pulled out with the tie virtually undamaged.

For Tests 1–9 inclusive a 2 lbf hammer was used for firing the Hilti pin but for Test 10 a heavy block of metal was used.

Site tests were later carried out on four ties as used for the laboratory exercise.

Test 1: Pin ($1\frac{3}{16}$ in overall – type NK30DX) used through outer hole. The pin pulled out at 1480 lbf even though it was not fully home after firing.

Test 2: Pin NK30DX fired through the metal between the two holes. Pin pulled out at 1800 lbf.

Test 3: Pin (1 in overall – type NK35DX) fired through outer hole. Pin pulled out at 870 lbf.

Test 4: Two pins NK30DX were used, one through each hole. Both pins pulled out at 870 lbf. Some shattering of the concrete occurred after the second pin was fired. The last test generally confirms the

observations made with laboratory tests that the ultimate strength of a tie secured, with two pins is less than a tie fixed with one pin, when the centres of pins are no further apart than $1\frac{1}{8}$ inches.

ACKNOWLEDGMENTS

The author wishes to acknowledge the work carried out by Mr F. R. Denson, formerly Head of the Building Department at Napier Technical College, Edinburgh. Mr Denson was responsible for carrying out all the tests relating to peripheral fixing of brickwork panels.

Thanks are due to the Principal of Stockport College of Technology for permission to reproduce part of the work on the tensile strength of wall-ties for lateral support of walls. This was carried out by Mr W. Thorley.

Thanks are also due to Lowe and Rodin, Consulting Engineers, for permission to quote work sponsored by them on the strength of wall-ties shot-fired to concrete members (Section 4) □

References

1 Davey, N., and Thomas, F. G. T., 'The Structural Uses of Brickwork', Structural & Building Paper No. 24, *I.C.E.*, February, 1950.

2 British Standards Institution, 'Structural Recommendations for Load-bearing Walls', CP 111:1964.

3 British Ceramic Research Association, 'Model Specification for Load-bearing Clay Brickwork', Special Publication No. 56, 1967.

4 Foster, D., 'The Use of Structural Brickwork for Frameless High Building', *Architectural Review*, April and May, 1962.

5 British Standards Institution, 'Specification for Metal Ties for Cavity Wall Construction', BS 1243: 1964.

6 Thorley, W., **Private Communication**, Stockport College of Technology.

Wind load analysis of multi-storey brickwork structures

by Professor A W Hendry, BSc (Eng) PhD DSc MICE MIStructE FRSE
Professor of Civil Engineering,
University of Edinburgh.

Figure 3 *General view of 5-storey cross-wall structure under lateral loading test at Torphin Quarry, Edinburgh*

1 INTRODUCTION

Wind loading is frequently a serious factor in the structural design of multi-storey brickwork buildings, particularly following the introduction of the 1970 version of CP3, Chapter V. The problem has for some years been the subject of extensive investigation by the Structural Ceramics Research Unit at Edinburgh University and the purpose of this Research Note is to summarise the results of this work in relation to practical design procedures. In particular, comparisons will be shown between calculated and measured deflections and stresses in simple crosswall structures and an indication will be given as to the most appropriate methods of calculation for this and more complex structures in brickwork.

2 THEORETICAL METHODS FOR THE ESTIMATION OF WIND STRESSES AND DEFLECTIONS

With the analysis of shear wall structures has attracted a good deal of attention in recent years and various methods of calculation have been proposed. These methods have been discussed extensively in the literature and it is not proposed to examine them in any detail in this note. It may be appropriate, however, to summarise and comment briefly on those methods which have been compared with experimental results.

All these methods represent to a greater or less extent simplifications of the actual system, as indicated in Figure 1.

Simply connected cantilever approach

With this method, which is the simplest possible solution, the structure is assumed to consist of a series of vertical cantilevers and the wind moment on the structure is divided amongst the walls in proportion to their flexural rigidities.

Elementary wall and diaphragm analysis

This method has been developed by Benjamin and is based on a simple deflection compatibility analysis of the structure taking into account flexural and shear displacements. It is frequently assumed that the floor slabs are rigid but this is an approximation which may not always be valid as the deflections of the floor slabs may in some cases be of the same order as those of the walls and may therefore have to be taken into account in setting up the equations. In certain cases it may be sufficiently accurate to treat a multistorey structure as a stack of single storey structures and to analyse each storey level separately. Benjamin has discussed this approach and has indicated that it may be applied with sufficient accuracy in cases where the wall sections are continuous from one storey to the next and where the loading is regular.

Continuous interconnection shear wall theory

The continuous shear connection method has been developed for simple interconnected shear wall systems although it has been applied to a complex brickwork structure using an analogue computer to solve the governing equations. In the case of simple shear walls the method is attractive in that design curves can be prepared giving moments and shears in the system.

Wide column frame analogy

In the wide column analogy, the shear wall structure is replaced by an equivalent frame in which the columns have the same flexural rigidity as the walls and the interconnecting members are of infinite stiffness for part of their length and of the same flexural rigidity as the actual beams or slabs in the

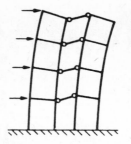

Basis
Cantilever walls caused to deflect equally by floor slabs

Simply connected cantilevers

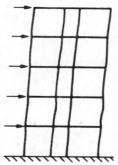

Basis
Distribution of lateral loads calculated by consideration of relative flexural and shear rigidities of component walls

Simple shear wall analysis

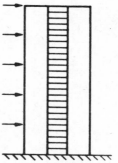

Basis
Walls interconnected by a continuous shear medium permitting formulation of differential equations for system

Continuous shear interconnection

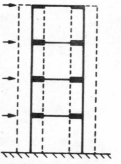

Basis
Replacement of actual structure by an equivalent frame in which interconnecting beams have infinite rigidity over part of their length.

Wide column frame analogy

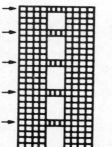

Basis
Subdivision of structure into a large number of elements connected at their nodes. Solution of this system in terms of inter-nodal forces and displacements by matrix methods

Finite elements

Figure 1 *Methods of analysis of laterally loaded brickwork structures*

central part between the walls. The system is dealt with by frame analysis methods which may take into account axial and shear deformations in the walls. As standard computer programmes are available for this kind of analysis the method can be applied quite easily provided that the equivalent frame parameters can be satisfactorily estimated. This, however, may be difficult in a structure consisting of an array of irregularly shaped walls.

Finite element analysis

The finite element method is now very well known. It consists of dividing up the walls and other elements of the structure into a large number of triangular or rectangular elements connected at their nodes and setting up matrices in terms of the forces acting at the nodes and the resulting displacements. The method provides a powerful analytical tool and in principle could deal with a structure of any description. A practical limitation, however, lies in the rapid proliferation of unknowns when the structure becomes at all extensive. It is true that the whole structure can be partitioned to reduce demands on computer storage but, even so, the work is likely to be substantial and probably excessive for the majority of cases arising in design. It is, however, of great value in research and in special cases in practice. Suitable computer programmes are readily available.

3 OUTLINE OF EXPERIMENTAL INVESTIGATION

Preliminary work served to develop a model testing technique and explored the calculation of the rigidity of an unsymmetrical structure involving torsional effects. The second stage was a detailed investigation of the behaviour of a five-storey cross-wall structure having the plan form shown in Figure 2. Experiments were first carried out on a 1/6 scale brickwork model and repeated on a similar full scale structure, details of which are also given in Figure 2. The full scale structure was built in a disused quarry and lateral load was applied by jacking at each floor level against the quarry face, which had been previously lined with concrete to give an even working surface. Figure 3 (see page 1) is an overall view of the test installation.

In both model and full scale experiments, measurements were made of strains and deflections at various loads. Figure 4 shows a comparison between model

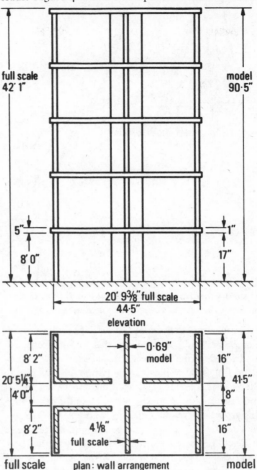

Figure 2 *Details of model and full scale experimental cross-wall structure*

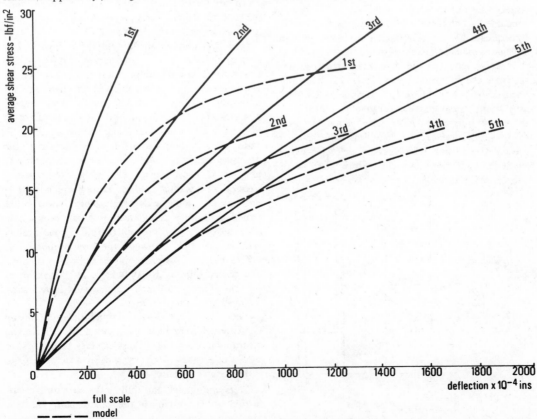

Figure 4 *Comparative model and full scale structure deflections in lateral loading test*

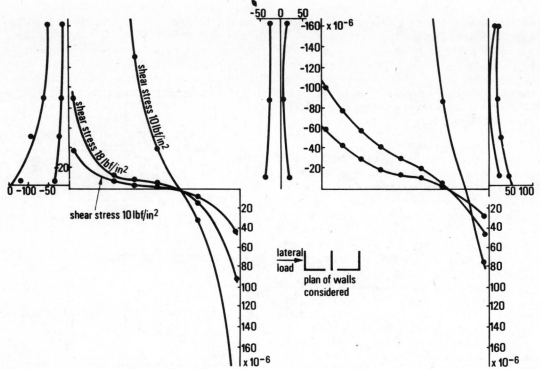

Figure 5 *Measured vertical strains in full scale cross wall structure across section near foundation level*

and full scale deflections at each floor level with increasing load. In making this comparison, the model deflections have been scaled up to give equivalent full scale values. Examination of the results shows that there is good agreement between model and full scale up to a shear stress of 10 lbf/in². Above this stress, the model deflections are proportionately very much higher than the corresponding full scale values. The reason for this lies in the fact that pre-compression due to self weight was smaller in the model than in the full scale structure and therefore horizontal tensile cracking started in the range 10–18 lbf/in² shear in the model, leading to progressively larger deflections with increase in load. It may be concluded from these results that model brickwork structures will give a reasonable representation of full scale deflections before the onset of tensile cracking.

Measurements in the full scale structure at a section close to ground level shows decidedly non-linear distribution of vertical strains in Figure 5. The strain measurements also give an indication of the non-uniform stress distribution across the width of the return walls. This point was investigated in greater detail using 1/6th scale models of a pair of single storey shear walls with return walls of various widths, acting as flanges. For this case the experimen-

tal results indicate that a maximum width of 0·35 of the storey height, may be assumed to act as a flange to the shear wall. This conclusion was confirmed by finite element analysis of the structure. A similar investigation of the effective width of floor slab inter-connecting the two shear walls indicated that a maximum of 0·5 of the bay width could be assumed to act in this manner.

With the exception of the preliminary study, all the tests so far referred to were on symmetrical models under symmetrical loading. It is frequently the case in practice that loads or wall arrays are not symmetrical and the structure is often subjected to torsional effects. A start has been made on the study of this more complex problem with experiments at 1/6th scale in a simple symmetrical structure subjected to eccentric loading, producing lateral shear and torsion, as shown in Figure 6.

It is clearly important in applying any method of structural analysis that appropriate values of the elastic constants are adopted. In the case of brickwork, the matter is complicated as there are no unique values which can be quoted for Young's modulus of elasticity (E) and modulus of rigidity (G) even for a given brick-mortar combination. Investigation both at model and full scale shows that, for example, the shear modulus increases very rapidly at pre-compressions within the range of working stresses. This is illustrated in Figure 7 which is based on tests on model scale single storey shear panel structures. Similar results from full scale tests follow the same general pattern but there is a good deal of variation in the values of the elastic moduli in different tests. This probably results from variations in workmanship, as it may be expected, for example, that the efficiency of joint filling would have a pronounced effect on deflections.

The value of E will vary in much the same way as G. Although very little information is available the assumption of a value of Poisson's ratio of 0·1, giving E = 2·2G, appears to give reasonably accurate results in deflection calculations.

Tests on model and full scale structures were carried out to establish the relationship between precompression and shear strength. The results of these tests are summarised in Figure 8 which also

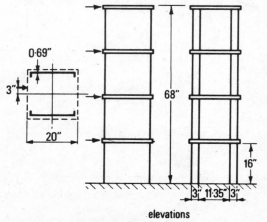

elevations

Figure 6 *Model shear wall structure with eccentric lateral loading*

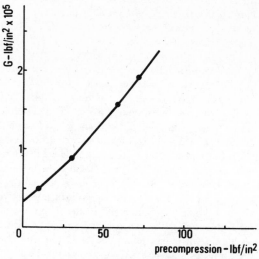

Figure 7 *Variation of shear modulus with precompression*

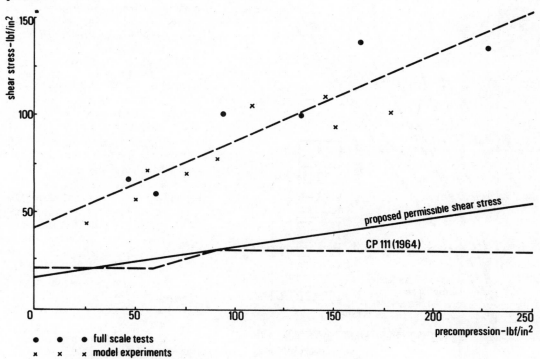

● ● ● full scale tests
x x x model experiments

Figure 8 *Variation of shear strength of brickwork with precompression*

shows the permissible stress given in CP.111 (1970) and that proposed in the recently issued Draft Revision of the Code. Tests indicate that failure of multi-storey cross-wall brickwork structures under lateral load will occur as a result of shear in the lowermost storey and that calculation of the resistance of the structure may be based on the cross-sectional area of brickwork resisting the lateral shear multiplied by the permissible shear stress appropriate to the precompression. Much less uncertainty attaches to the shear strength of brickwork than to the elastic constants upon which deflection and stress calculations must be based.

4 COMPARISONS BETWEEN CALCULATED AND EXPERIMENTAL RESULTS

A number of comparisons between measured and calculated deflections have been made for the model and full scale test structures referred to in the previous section and some of these are shown in Figures 9 – 12. Figure 9 shows that it is possible to calculate the deflection of interconnected shear walls either by finite elements or by the simple storey by storey method provided that the value of the shear modulus is varied in accordance with Figure 7.

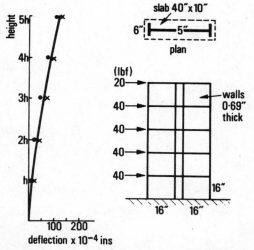

Figure 9 *Comparison of calculated and experimental deflections for a coupled shear wall model*

Figures 10 – 12 refer to tests on a five-storey cross-wall structure which is somewhat more complicated than the previous case. A variety of results can be obtained according to what assumptions are made as to the effective widths of the return walls and of the floor slabs. This point is illustrated in Figure 10 which shows the effect of five different assumptions as to effective wall and slab widths on a calculation using the shear continuum method. Further comparisons are given in Figures 11 and 12 for model and full scale tests respectively. It would appear that neither the shear continuum method nor the simply connected cantilever methods give a satisfactory representation of the structure as the deflection curves are not of the same form as the experimental ones. The wide column frame analogy gives results which are identical with the shear continuum method. However, both the finite element analysis and the simple storey-by-storey method show very good agreement with the experimental result. Similar methods give very satisfactory estimates of the deflections induced by torsion in the structure shown in Figure 7. In this investigation the deflection of the structure was calculated on the basis of the theory of torsion of thin walled sections and also by a simplified

method in which the shear is distributed amongst the individual wall elements according to their relative rigidities. As may be seen from Figure 13, both methods show close agreement with the experimental result.

Some comparisons have been made of measured and calculated stresses in model and full scale structures. Good agreement was obtained for a five storey structure consisting of a pair of interconnected shear walls using the finite element method but indifferent agreement was found between theoretical and measured stresses in the full scale structure as will be seen from Figure 14 which gives extreme values of vertical stresses on a horizontal section close to ground level. Although the lateral load stresses are apparently not very accurately predicted, it must be remembered that the experimental values are based on a somewhat uncertain value of E and also that these stresses must be superimposed on the dead load

≫→8

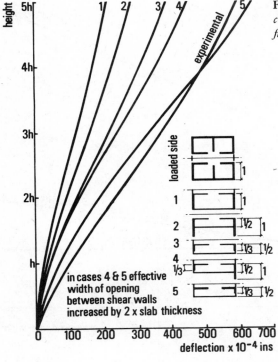

Figure 10 *Deflection of cross wall structure by the continuum method assuming various widths of slab flange and opening. Average shear stress 10 lbf/in²*

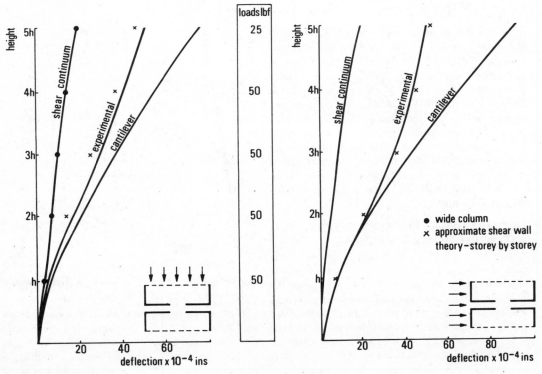

Figure 11 *Comparison of measured and calculated deflections in model five storey shear wall structure*

228

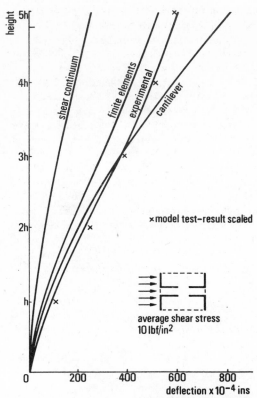

Figure 12 *Comparison of measured and calculated deflections of shear wall structure tested at full scale*

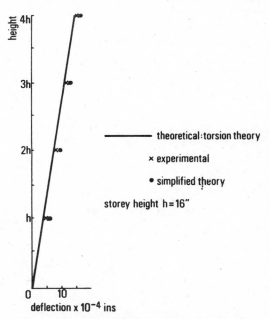

Figure 13 *Comparison of measured and calculated torsional deflections due to a torsional loading of a simple shear wall structure*

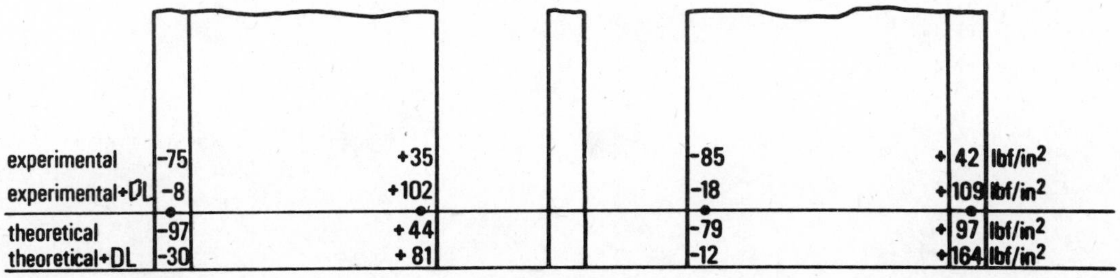

experimental	−75	+35	−85	+42	lbf/in²
experimental+DL	−8	+102	−18	+109	lbf/in²
theoretical	−97	+44	−79	+97	lbf/in²
theoretical+DL	−30	+81	−12	+164	lbf/in²

Figure 14 *Comparison between measured and calculated vertical stresses for lateral loading and for lateral loading +67 lbf/in² dead load. Average shear stress 18 lbf/in² in brickwork, corresponding approximately to design wind load under CP₃ Ch v, 1970. Assumed Value of E = 1·2 x 10⁶ lbf/in².*

compressive stress of 67 lbf/in². The figures shown would apply to a partially completed building under the full design wind load of the new CP3, ChV. The addition of the stresses due to the full dead load plus superimposed load would, of course, reduce the relative importance of inaccuracy in the lateral load stresses. Furthermore, the criterion for lateral load resistance is most likely to depend on the average shear strength of the brickwork which can be predicted with reasonable accuracy.

5 CONCLUSION

The work so far carried out on brickwork structures under lateral loads indicates that deflections of cross wall structures can be estimated with reasonable certainty by the finite element method or by the elementary wall and diaphragm analysis method, provided that the relevant elastic moduli are known.

Precise calculation of stresses is not possible but the need for refined analysis can be decided by carrying out preliminary calculations based on elementary assumptions. For example, if wall layout, thicknesses and brickwork strengths have been determined on the basis of space planning and vertical load bearing capacity, a check on lateral load stresses by the simply connected cantilever method which showed satisfactory results would indicate that no further calculation was necessary as this method will always overestimate the effect of lateral loads. If the results of this simple calculation were not satisfactory, more refined analysis might be required. The method to be adopted would depend on circumstances but present indications are that the approach developed by Benjamin would generally be adequate. Finite element analysis might be justified in special cases

and in experienced hands might be expected to yield accurate results. The continuous shear connection method does not seem to be satisfactory for brickwork structures and its use is not recommended. The wide column frame analogy appears to give the same result as the shear continuum method and would not seem to be of very wide application to brickwork structures.

The work carried out on brickwork structures so far, gives some indication of the probable level of accuracy of design calculations for wind loading on multi-storey shear wall structures. Although great accuracy in estimates of stress is not possible, the situation is probably not greatly different with that existing in the case of similar structures in other materials which have not so far been the subject of full scale testing. There is undoubtedly scope for greater refinement by further research but, for the present, designers will be able to proceed with confidence using existing knowledge and methods. □

6 Selected bibliography
Tall Structures Ed. A. Coull and B. Stafford Smith, Pergamon 1967.
Statically Indeterminate Structures J. R. Benjamin, McGraw Hill, 1959.
Designing, Engineering and Constructing with Masonry Products, Ed. F. B. Johnson, Gulf Publishing Co., 1969.
Proceedings Second International Brick Masonry Conference, 1970 British Ceramic Research Assoc., In press.

structural brickwork-
materials
and performance

Thomas

Figure 1 above *General view of 5-storey crosswall structure under lateral loading test at Torphin Quarry, Edinburgh.*

Figure 2 right *Aerial view of the BCeramRA explosions test site.*

SYNOPSIS

Calculated loadbearing brickwork is discussed and the basic requirements are specified. Materials and their effects on strength and performance are considered and recommendations made. The mechanism of brickwork failure under vertical and lateral loading is covered and the results of current research in this field are included. Factors affecting strength are discussed and information is provided on composite action with concrete beams, also quality control.

INTRODUCTION

Loadbearing brickwork is not new but calculated loadbearing brickwork is of comparatively recent origin. The first British Code of Practice *CP 111*[1] was issued in 1948 and to many architects and engineers this Code of Practice opened up new possibilities in design which were tantamount to the birth of a new structural material. In 1964 a revised edition was issued with substantially increased permissible stresses which were undoubtedly due to a great deal of research and many full-scale tests carried out at the Building Research Station.

Since that time the brick and structural ceramics industries have intensified their structural user research programme and over 2000 storey-height panel walls have been tested. In addition, a five-storey brick crosswall structure is being used for research on stability (Fig. 1) and a three and a half-storey building (Fig 2) was specially constructed to determine the resistance of brickwork to explosive forces.

It is obvious that structures designed in calculated loadbearing brickwork must be cheaper than framed buildings. The amount of savings would depend upon the layout of the building but maximum saving is normally achieved when slender crosswall construction is used.

TYPES OF LOADBEARING BRICKWORK
Crosswall construction

Brickwork is strong in compression but weak in tension when traditional mortar joints are used. It therefore follows that several requirements must be fulfilled:

(a) differential settlement must be avoided and rigid foundations should therefore be provided;

(b) the plan formation should repeat itself on each floor level, thus facilitating stress distribution to the foundations;

(c) simple crosswalls in themselves have a vast inertia and should be so positioned as to accommodate lateral thrusts due to wind forces;

(d) in the longitudinal direction, the geometrical proportions of the building often ensure that this is the weaker axis of the crosswalls, and resistance to a pack of cards type failure can easily and adequately be provided for in the form of flank walls, corridor walls or by the stiffness of stair wells and lift shafts;

(e) the arrangement of windows is important and 'hole in the wall' type of fenestration gives greater stability to the building than vertical strips of brickwork which tend to act as cantilevers from the ground when infill panels of glazed curtain walling are the only means of tying such slender elements of brickwork together; however, such details can sometimes be permitted, but a careful structural assessment should always be made in such circumstances;

(f) if large openings are required at ground level, these can quite easily be accommodated with a reinforced concrete frame, indeed, this is common practice (Fig 3).

Cellular construction

As the title implies this form of construction involves the use of cells or boxes and the requirements for crosswall construction equally apply but, as with the more traditional domestic housing, intersecting walls occur so frequently that stability is rarely a problem (Fig. 4). This is not to say that such construction should be built 'willy nilly', and these structures must always have the necessary assessment.

Figure 3 *15-storey loadbearing brickwork tower block at Beaconsfield. (Architects: Robert Matthew, Johnson-Marshall & Partners. Consulting engineers: Felix J. Samuely & Partners).*

MATERIALS

The overall stability of a building depends not only on the proportions of the structure but also on the individual units and jointing materials. It is therefore necessary to study the individual units themselves prior to any structural analysis and to understand the individual characteristics of the materials involved.

Bricks and blocks

Different varieties of bricks and blocks are available; these come within the following categories:

(a) common,

(b) facing,

(c) engineering

Common and facing are self-explanatory terms but engineering bricks have clearly defined limits for absorption and compressive strength, as indicated below:

Class	Average compressive strength MN/m² not less than	Average absorption boiling or vacuum per cent weight not greater than
A	69.0	4.5
B	48.5	7.0

Architects and engineers often assume that engineering bricks must be either Staffordshire blue or smoothfaced red bricks but many variations are possible including brown and white, and those with attractive textures and multi-colours. The term engineering tends to become associated with smoothfaced units used for sewer construction.

Unit strength

Unless a higher compressive strength is required and agreed between the purchaser and manufacturer, the

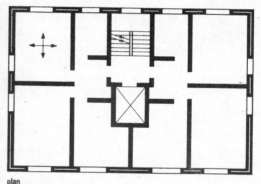

plan
Figure 4 *Example of complex cellular brick building.*

strength is not required to be more than 5.2 MN/m² for brick and 2.8 MN/m² for block.

The compressive strength of bricks and blocks is dependent on (i) the type of material used, (ii) the method of manufacture and (iii) the shape of the unit. The shape of the unit has a considerable effect upon the strength and stress modification factors which are quoted in the current *CP 111* takes account of this fact.

Another important point is that when a crushing strength is quoted by a manufacturer this refers to the strength as laid on the normal bed face, and if it is intended to lay the unit in any other way it may be necessary to carry out additional tests.

When frogged bricks are used for structural brickwork it is important to make sure that the units are laid frog up and that the frogs are completely filled with mortar. Simms[2] carried out a series of tests to determine the relative strengths of frog-up and frog-down construction, and his results indicate that considerable differences in strength can be achieved but that for normal domestic construction frog-down construction is satisfactory.

Effects of special cements and admixtures on mortar wall performance

The 1948 edition of *CP 111* was based largely on research carried out at the Building Research Station, and the walls and piers tested were constructed in traditional mortars based on mixtures of lime : sand, cement : lime : sand or straight cement : sand. In all cases where cement was used this was of the ordinary Portland variety. Subsequent research elsewhere has involved the use of other types of cement and admixtures, and as these can affect the properties of mortars and performance of walls it is considered germane to discuss them under this heading.

Cements—general

Portland cement, ordinary or rapid-hardening, Portland blast-furnace cement or sulphate-resisting cement to the appropriate British Standards can be used without special requirement unless specifically stated by the manufacturers, although it should be recognized that considerable strength differences can be achieved using the same type of cement when obtained from different sources.

Masonry cement

Manufacturers of this type of cement are normally reluctant to publish its constituents but it is understood that in the UK it consists of approximately 75 per cent Portland cement, the remainder being made up of a fine filler such as ground limestone and an air-entraining agent. The S.C.A.G.[3] specification limits the use of mortars based on this cement to equivalents of 1 : 1 : 6 ordinary Portland cement : lime : sand.* This is an excellent material for normal two-storey domestic construction and, in some instances, for perhaps more adventurous building but, unfortunately, it is grossly misused and often confused with ordinary Portland cement. Plasticizers should not be added to mixes gauged with masonry cement as over-aeration is likely with subsequent loss of bond and compressive strength.

Super-sulphate cement

This is not a Portland cement and advice on mix proportions should be sought from the manufacturers. It is susceptible to extremes of temperature and great care should be taken in the UK in frosty weather as low temperatures tend to delay its setting time and serious difficulties may result.

High-alumina cement

A recent model specification prohibited the use of this type of cement for loadbearing brickwork. The main reason for this was undoubtedly the danger of a flash set occurring if it is accidentally mixed with Portland cement and the inconvenience of having to remove all trace of ordinary Portland cement from mixers and tools. High-alumina cement can be successfully used for loadbearing brickwork and recent research work has made use of its excellent properties, but if specified it is important to remember that richer mix proportions may be necessary, a low water/cement ratio must be used and lime must not be used as a plasticizer. This cement has excellent sulphate resisting properties.

Pigments

Pigments should be used strictly in accordance with the manufacturer's instructions. Over-prescription can have disastrous results as, due to their non-cementitious character and relatively high surface area, they tend to adulterate the mix reducing bond strength and in some instances, compressive strength. Where carbon black is used as a colouring agent, quantities greater than 3 per cent by weight of the cement may affect these properties. Recent research[5] has shown that 10 per cent black pigment (containing carbon black) can reduce drastically the compressive strength of mortar (Fig 5).

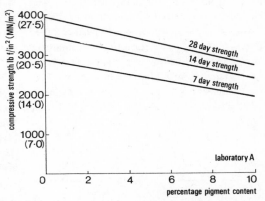

Figure 5 *Compressive strength of 4in mortar cubes with varying pigment content–cement/lime/sand mortar.*

Plasticizers

Plasticizers of the air-entraining type improve frost resistance during laying and aid workability, but over-specification, particularly with high suction brick if not correctly treated, may be the cause of poor bond resulting in lower compressive strength in the wall and total loss of tensile bond strength.

Frost inhibitors

Additions of frost inhibitors based on calcium chloride should never be used in mortar joints as, apart from being ineffective (ie there is no evidence as far as the author is aware that sufficient heat can be generated in a normal mortar joint to depress the freezing point), they cause deliquescence with the subsequent danger of corrosion of embedded steel and an increased possibility of efflorescence.

Sand

The type of sand used for mortar can have a considerable effect on its working properties, durability and performance in the wall and upon its ultimate strength. Ideally from the mortar strength and durability points of view a good sharp, clean sand should be used but difficulties may arise in brick-laying unless a lime or plasticizer is added to the mix.

Most natural sands contain a proportion of 'fines', consisting of two types of particles. The fraction known as 'silt' usually consists of quartz, whilst minerals such as felspars break up more easily and are partially decomposed by water to form clay. A

---

* *A recent draft Australian Specification[4] supports the point and requires that 1 :3 masonry cement : sand (by volume) 2-in mortar cubes should have a minimum strength at seven days of 500 lbf/in² and 900 lbf/in² at twenty-eight days. Comparing these values with those suggested by SCAG (Table 1 of this paper), it will be noted that the difference is substantial. Perhaps Australian masonry cements have a lower percentage of actual Portland or equivalent cement.*

233

Grade	Description	Mix (Vol) Cement : lime : sand			Minimum compressive strength MN/m² 7 days	Minimum compressive strength MN/m² 28 days
I	Cement : sand	1	0–¼	3	11.0	16.0
II	Cement : lime : sand	1	½	4½	5.5	8.0
III	Cement : lime : sand	1	1	5–6	2.8	4.0
	Cement : sand with plasticizer	1	—	5–6		
	Masonry cement : sand	1	—	4½		
IV	Cement : lime : sand	1	2	8–9	1.0	1.5
	Cement : sand with plasticizer	1	—	7–8		
	Masonry cement : sand	1	—	6		
	Hydraulic lime : sand	—	1	2		

After SCAG 1907

Table 1 *Minimum average compressive strengths of laboratory specimens of mortar*

suitable proportion of silt improves the general grading and, hence, the working qualities of the mix as a whole. Washed sand may be rather harsh due to the absence of silt and can often be improved considerably by the addition of a suitable proportion of sand containing fines. Marine and estuarine sand is not recommended[3] unless it has been adequately washed to remove all deleterious salt.

Material	Thickness not exceeding (mm)	Maximum working stress (MN/m²)
Brick	As applicable	4.60–6.20†
Polythene	0.50	6.90
Polythene	0.75	5.52
Pitch polymer	1.15	0.69
Asbestos core bitumen	3.10	0.35
Lead core bitumen	3.60	0.07
Aluminium core bitumen	4.00	0.07

Table 2 *Recommended working stresses for damp proof course materials at temperatures of 22 °C or below**

DAMP PROOF COURSE MATERIALS

It is extremely important to select the correct damp proof course material for each individual building as to adopt the cheapest may prove false economy. This is particularly true for loadbearing brickwork where high stresses and/or relatively high temperatures can cause a degree of structural distress. Sustained loading and the resulting creep of the damp proof course materials may need careful consideration and Plowman and Smith[6] have made certain recommendations as to working stresses, Table 2.

* Where damp proof course materials are used on a south-facing wall or one which is likely to be subjected to long periods of sunshine, the above stresses may need to be reduced.

† The permissible stresses quoted for damp proof course bricks are based on those shown in Table 3 of CP 111 and should be treated as basic stresses dependent on the strength of bricks used. Damp proof course bricks have an average absorption not greater than 4.5 per cent by weight and usually crush in excess of 69 MN/m² (Plowman and Smith[6], 1970).

WALL-TIES AND THEIR FUNCTION

Cavity-wall construction offers advantages over solid construction as far as resistance to water penetration and thermal insulation are concerned but from the structural point of view, as discussed elsewhere[7], the solid wall is generally to be preferred.

When cavity walls are subjected to bending due to eccentricity of loading a tendency towards differential lateral deflexion occurs between the leaves. Wall-ties play an important function in equalizing deflexions of the two leaves, particularly when only one leaf is loaded. When ties are capable of transmitting the tensile and compressive forces without appreciable stretching or buckling, the bending stresses will be the same in the two leaves, providing they are of equal stiffness.

Davey and Thomas[7] state that the tendency towards lateral deflexion is greater for the cavity wall, and in *CP 111* allowance is made for this by taking the effective thickness of a cavity wall (for slenderness ratio purposes) as equal to two-thirds of the sum of the thickness of the two leaves. Tests at the Building Research Station have shown that for cavity walls approximately 2.5 m (9 ft) high and 1.5 m (4 ft 6 in) long, loaded eccentrically on one leaf, the unloaded leaf followed approximately the curvature of the loaded leaf regardless of whether the ties were in compression or tension. In the first series of tests Davey and Thomas record that a 1 : 3 cement : sand mortar was used, and the ties were of the butterfly type made from 9 swg steel wire. In the second series a 1 : 1 : 6 cement : lime : sand mortar was adopted, the ties being of the same form but made from 12 swg hard drawn copper wire. The ties in both series were spaced at 1 m (3 ft) centres horizontally and 0.5 m (1 ft 6 in) vertically. It was noted in the second series of tests that lateral movement of the two leaves was approximately the same when the ties were in tension but rather less when in compression, probably due to bending of the wall-ties.

Although wall-ties may ensure that both leaves of a cavity wall have the same radius of curvature, it does not follow that the ultimate strength of the wall will be greater; indeed, the reverse may be the case. In the worst case, with the load eccentrically applied to one leaf and the other leaf unloaded, the flexural stress induced in the unloaded leaf may, at high loads,

Specification	Working loads of ties engaged in dovetail slots set in structural concrete	
	Tension (N)	Shear (N)
1. Dovetail slot type ties **(a)** Galvanized or stainless steel fishtail anchors 3 mm thick in 1.26 mm slots, 150 mm long. Slots set in 1 : 2 : 4 concrete	1000	1200
(b) Galvanized or stainless steel fishtail anchors 2 mm thick in 2 mm slots, 150 mm long. Slots set in 1 : 2 : 4 concrete	700	1100
(c) Copper fishtail anchors 3 mm thick in 1.26 mm copper slots, 150 mm long. Slots set in 1 : 2 : 4 concrete	850	1000

	Working loads of ties embedded in mortar			
	Tension (N) Mortar*			Shear (N) Mortar*
	Grades I	III	IV	Grades I, III, IV
2. Cavity-wall type ties **(a)** Wire butterfly				
(1) Zinc coated or stainless steel 3.26 mm diameter	650	650	500	500
(2) 2.64 mm diameter	500	500	450	500

(b) Strip fishtail
    (1) Zinc coated 3 mm × 20 mm
    (2) Copper 3 mm × 20 mm
    (3) Bronze 3 mm × 20 mm
    (4) Stainless steel 3 mm × 20 mm
	1400	1000	700	900

Table 3 *Recommended working loads for wall-ties (* Equivalent mortar mixes/grades as indicated in Table 1)*

cause it to crack so that its contribution to the loaded leaf will disappear. Even in such conditions, however, ties are normally required, in order to stabilize the unloaded leaf.

The results of tests on wall-ties have been discussed by the author[8] and are summarized in Table 3.

MORTARS—GENERAL

The desirable properties of all mortar mixes for brickwork are said to be:

(a) workability,
(b) good water retentivity,
(c) sufficiently early stiffening,
(d) development of suitable early and final strength,
(e) good adhesion or bond, and
(f) durability.

Modern mortars usually consist basically of cement, lime, sand and mixing water, although lime is often replaced or used in conjunction with a plasticizer. Lime and cement mainly determine the rheological properties of mixes although the grading of the sand or the effect of additives may modify them. Workability, water retentivity and bond are properties which lime imparts, while the cement and sand confer strength and durability. Cement-lime mixes are generally superior to straight lime or cement mixes for most building work.

Ready-mixed mortars or, more correctly, ready-mixed lime-sand for mortar (to which cement must be gauged) are becoming popular and provide a means factory control over the constituents, which is highly desirable.

Mortar strength/wall strength

Mortar has always been one of the building industry's poor relations and unlike concrete has not until comparatively recently been subjected to quality control procedures. In 1967 a model specification[3] was issued in which minimum strengths were quoted for varying mix proportions and from that time more serious consideration appears to have been given to the science of mortar technology. Variability in mortar strength is not surprising when one considers that it is a composition of cement, sand and lime, not to mention other commonly-used additives which can have considerable effect on the ultimate performance values, also the water/cement ratio and methods of batching and mixing. The relationship between the compressive strength of mortar and that of the wall in which it is used cannot be stated precisely, but Klein[9] has indicated that for bricks in the intermediate strength range laid in cement : lime : sand mortars, the change in wall strength is related to approximately the fourth root of the change in mortar strength.

WALL STRENGTH/BRICK STRENGTH

The ratio of wall strength to brick strength cannot be given precisely as so many variables are involved. Clay bricks, for example, are produced basically by three different processes; likewise the clay, shale or brickearth used in their manufacture being indigenous has widely different characteristics. Calcium silicate and concrete bricks are entirely different products and should be treated as such when designing walls in which they are incorporated. For clay bricks the ratio of wall strength to brick strength based on a large number of tests[7,10,11,12] varies from 0.2 to 0.4.

STRENGTH OF BRICKWORK

The influence of brick and mortar strength on the loadbearing capacity of brickwork has been studied extensively at the Building Research Station and elsewhere[7,10,11,12] and tests have shown that the strength of brickwork increases as the strength of both brick and mortar increases, but in neither case in direct proportion (Fig 6).

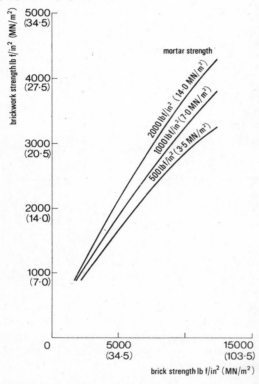

Figure 6 *Relationship between crushing strength of brickwork and strengths of bricks and mortar.*

Mechanism of failure under vertical loading

Failure of brickwork subject to axial compression is normally by vertical splitting due to horizontal tension in the brickwork (Fig 7). The reason for this type of failure is that the mortar and brick have widely differing strain characteristics. It has been suggested that the mortar between the bricks behaves like a pad of rubber and that the rubber, having a high lateral strain when compressed, forces the bricks apart and hence the tensile failure by vertical splitting. The tensile strength of the brick is therefore important.

The above analogy tends to simplify the mechanism of failure, which is influenced by other factors including Poisson's ratio, Young's modulus of elasticity, bond between bricks and mortar, coefficient of friction between mortar and brick, shear resistance and tensile strength.

Shear failure has also been observed in several wall tests (Fig 8).

Where brickwork is loaded eccentrically failure will normally take place by crushing and spalling on the compression face (Fig 9). This may be relieved to some extent by raking back the joints although this,

Figure 7 *Typical failure by vertical splitting wall panel subjected to axial loading.*

Figure 9 *Vertical splitting of wall panel accompanied by spalling in the top courses.*

Figure 8 *Vertical splitting of wall panel accompanied by shear failure.*

in turn, reduces the effective wall area.

When weak bricks and mortar are used and the slenderness ratio and eccentricity are high, failure may occur by buckling.

Mechanism of failure under lateral loading
Failure of brickwork subject to lateral loading is a function of the degree of vertical pre-load (Fig 10). As lateral load is increased incipient cracking occurs, the 'yield line' pattern depending upon the peripheral fixity.

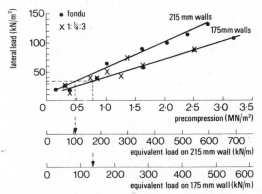

Figure 10 *Graph showing results of lateral load tests on 175 mm and 215 mm walls at various levels of precompression.*

Walls restrained top and bottom only: When walls are restrained top and bottom and subjected to lateral loading horizontal cracking occurs at the mid height. Providing the vertical load is not excessive the mode of failure is entirely geometrical. As the lateral force increases the horizontal crack opens on the tension face, thus causing a wedging action as the upper and lower sections of the wall rotate. Before complete failure of the wall can occur the test machine platens (or the concrete floors in an actual structure) must lift to increase the distance between restraints and allow the wall to fail as indicated in Fig. 11. In the test machine used for this research only the upper

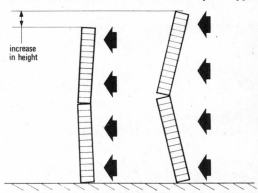

Figure 11 *Failure mechanism of brick walls under low compressive loads when subjected to lateral pressures (restrained top and bottom only).*

beam was free to move, and measurements have shown that at low pre-loads (78 kN/m), a vertical extension of about 8.5 mm occurs before failure. At low pre-loads the strengths·of the mortar and bricks appear to play little or no part in the total resistance to lateral loading.

As vertical pre-compression increases the mode of failure changes and local crushing occurs on the side of the wall subjected to the lateral forces (Fig 12). Eventually if sufficient vertical load is applied the compressive stress becomes critical and failure will

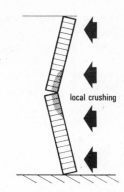

Figure 12 *Walls subjected to high compressive loads and lateral pressures (restrained top and bottom only).*

occur with little or no lateral load as described above. The precise relationship at very high levels of pre-compression has not so far been investigated, but the range indicated in Fig 12 is adequate for practical purposes.

Walls restrained on four sides: When walls are restrained on all four edges the mode of failure changes from the single hinge to the more usual 'yield line' pattern (Fig 13). Lateral loading tests using airbags and gaseous explosions in an actual test facility[13] have produced remarkably similar results and verify the modes of failure described above.

Figure 13 *Fully restrained 112.5 mm wall failing in a 'yield line' pattern. (BCeramRA gas explosions test site).*

Walls with limited pre-load subjected to wind loading: Research into the effects of wind forces on panel walls with little or no pre-load is currently under way and a five-storey experimental building at Torphin Quarry is being used for this purpose in which simulated wind forces can readily be applied using airbags; research in this field is also continuing at the BCeramRA. The results of this work are likely to be far-reaching and will almost certainly necessitate a new design concept unlike the existing one which is based on the dubious tensile bond properties of bricks and mortar. Earlier work on wind load analysis of multi-storey brickwork structures in this research series is described elsewhere.[14]

SLENDERNESS RATIO

The strength of brickwork decreases as the slenderness ratio increases although a different relationship exists for single leaf walls (stretcher bond) and bonded walls. The reduction in strength for an increase in slenderness ratio is influenced by the basic brickwork strength (Fig 14).

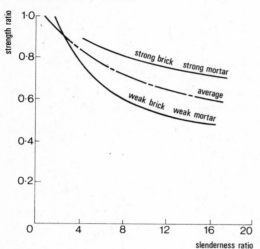

Figure 14 *Influence of slenderness on the reduction in brickwork strength for weak and strong brickwork. (After F. G. Thomas, 1953.)*

Early work at the Building Research Station included testing walls between knife edges or, more correctly, between 25 mm (1 in) square bars. For walls loaded eccentrically in this way failure was generally by buckling, and the reduction in strength was considerably greater than for the walls loaded eccentrically between reinforced concrete slabs where failure was usually by vertical splitting. Tests on 112.5 mm (4.5 in) walls axially loaded with both ends solidly bedded showed little or no reduction in strength when the wall height increased from 1 m (3 ft) to 3 m (10 ft).

A recently issued amendment to *CP 111*[15] revised the permissible slenderness ratios based on actual wall thicknesses as opposed to nominal dimensions quoted in the original code. The new ratios are much more realistic and the basic slenderness ratio is now 27 for units set in mortars gauged with cement, except for walls less than 90 mm thick, in buildings of more than 2 storeys, when the ratio becomes a maximum of 20.

ECCENTRIC LOADS

In some types of building eccentricity of floor loading is ignored, for example in buildings having stiff floors, short spans or walls stiffened by intersecting walls. This may be a reasonable assumption where mortars not stronger than $1 : \frac{1}{2} : 4\frac{1}{2}$ cement : lime : sand are used since the mortar joint will accommodate any local high stressing by readjustments within the joint.

For long spans, ie when the floor span is more than thirty times the thickness of the wall, and also for timber and other lightweight floors, an eccentricity

equal to one-sixth of the bearing width is often assumed.

When the compressive stress in the wall is greater than about 0.7 MN/m² and where a strong mortar is used (ie $1 : \frac{1}{4} : 3$ cement : lime : sand) the wall/slab junction will normally be fully fixed. The equivalent eccentricity may then be calculated after distributing the floor slab fixed end moments (due to live load only or combined with dead load, depending on the method of construction) to the wall in a similar manner to that for reinforced concrete columns in *CP 114*[16].

Where metal hangers are used to support timber joists the load is normally assumed to act 25 mm from the face of the wall.

Some guidance on assumed eccentricities for domestic construction is given in Building Research Station Digest 61 (2nd Series)[17].

BRICKWORK BONDS

A series of tests[18] carried out on $^1/_6$th scale brickwork walls indicated that no significant difference in strength occurred between the following bonds: English, Flemish, English Garden Wall and Header Bond.

Five full-size walls, one brick thick, with both leaves in stretcher bond connected by strip metal ties to provide a fair face on both sides have been loaded axially[19] and the results indicate that the wall strengths, although slightly lower than for walls of traditional brickwork bonds, are satisfactory. However, where high strength brickwork is required or where walls are loaded eccentrically, it may be necessary to carry out wall tests.

It has been established that axially loaded single-leaf walls are considerably stronger than bonded walls, other factors including brick and mortar strength, workmanship and slenderness ratio being equal. The load factors for one recent series of tests[20] on half brick thick storey-height walls ranged between 6.3 and 12.3. For one brick thick storey-height walls the load factor normally ranges between 4 and 7.

COMPOSITE ACTION OF BRICK WALLS SUPPORTED ON REINFORCED CONCRETE BEAMS

The behaviour of brickwork supported on beams can follow two patterns: (a) brickwork can bond with the supporting beam, sometimes with the aid of shear connections, and form an effective composite unit; or (b) no bond is achieved between the brickwork and supporting beam and the two materials behave in a less composite manner, ie the beam provides its normal moment of resistance but the brickwork tends to arch and, hence, subject the supporting beam to a very much smaller bending moment than is normally allowed for. Wood in his papers describes bending moments measured in experiments at the Building Research Station and elsewhere[21],[22]. Special care is necessary when designing heavily loaded brick walls supported on reinforced concrete beams, and recent tests carried out at the Building Research Station have shown that arching action clearly takes place leading eventually to the crushing of bricks close to the supports. Wood and Simms have suggested a tentative design procedure [23] based on an intuitive assessment of the way composite action of

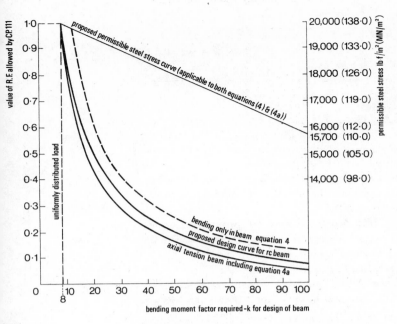

Figure 15 *Composite action with loadbearing walls.*

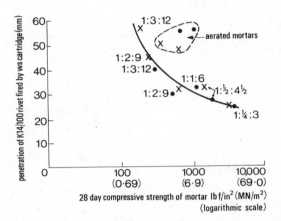

Figure 16 *Relation between mortar strength and penetration of cartridge fired rivets.*

such walls would occur and these are summarized in Fig 15.

The action described in (b) above needs careful consideration at the design stage as such arching can in some instances produce considerable lateral thrusts which must be catered for in the structure.

QUALITY CONTROL

Structural brickwork is designed as an engineering material and must, therefore, conform to the strict requirements of such forms of construction. Site and laboratory control is now accepted as standard practice for calculated loadbearing brickwork construction and comprehensive details of suitable testing have been specified elsewhere[3]. The following structural tests have been suggested.

Bricks

The average crushing strength should be determined as specified in the appropriate British Standard and it is hoped that a unified method of testing will shortly be agreed. At the moment the method of testing varies for different materials and is something of a misnomer as the number obtained is then used to determine permissible working stresses from Table 3 of *CP III*.

The recently issued *CP III* draft for comment[24] followed BS 3921[25] in recommending that bricks should be regularly tested in the factory and that a record of the results should be used to plot control charts based on the principles discussed in BS 2564[26]; indeed, the draft code[24] recommended increased permissible stresses for brickwork when quality control of bricks and mortar is practised. The crushing strength test will undoubtedly remain for some time but it should be remembered that this test may not hold the key to brickwork strength since failure under compressive loading is usually by vertical splitting. The tensile strength of the brickwork is, therefore, important as also are the other properties of the brick and mortar, such as Young's modulus and Poisson's ratio, since these influence the magnitude of the transverse forces. Some work has been carried out on the tensile properties of brick and the results have been reported elsewhere[27]. Clay bricks are sometimes wetted to reduce the initial suction by hosing or by immersion in a tank of water for a short period—the latter method being preferable due to the more uniform wetting. However, bricks should not receive this treatment unless the suction rate exceeds 2 kg/m²/min; indeed, some manufacturers do not recommend wetting with certain products. Calcium silicate bricks should never be wetted to reduce their initial suction rate and it is desirable that a highly water-retentive mortar be used as an alternative to wetting the bricks. The suction rate of concrete bricks is usually low and wetting of them is neither necessary nor desirable.

Mortar

The SCAG model specification[3] recommends minimum average compressive strengths for mortar prisms tested in the laboratory and, undoubtedly, the next edition of *CP III* will include similar requirements.

The mortar prism compressive test has many defects and may bear little relationship to the strength of the 10 mm mortar joints. However, such testing gives an indication of mortar strength and, hence, control of uniformity of strength throughout the construction. Specification of mortars has in the past been based on volume proportions with no strength requirement. However, it is preferable to specify a grade by strength in addition to volume proportions. Undoubtedly future trends will be to specify grades of mortar based on a strength requirement and, other factors being equal, mix proportions will become the concern of the mortar technologists and not necessarily that of the specifier.

On occasions when mortar cubes failed to meet the specified strength requirement engineers have inquired if an alternative method existed to assess the actual strength of mortar in the wall. A method of determining the strength of mortar in brickwork by measuring the depth of penetration of cartridge-fired steel pins has been described by Albrecht and Engelke[28] and Ryder[29] has continued this work at the Building Research Station. Fig 16 indicates the results obtained by Ryder and illustrates that the strength of cement : lime : sand mortars can be predicted with reasonable accuracy whereas the results of aerated cement : sand mortars cannot.

Brickwork prisms

Brickwork cubes have been suggested as suitable specimens for testing[3]. In the author's opinion such specimens at the moment reflect merely a method of quality comparison and are not indicative of any definite cube strength/wall strength ratio. American practice permits brickwork prisms to be used for determining working stresses, but the aspect ratio of their specimens is such that it is unlikely that the well-known 'machine platen effect' will come into play as must be the case with the squat '9-in cube'☐

REFERENCES

1 CP III : 1948, **Structural Recommendations for Loadbearing Walls**, now superseded by CP III: 1964 Metric Version: Part 2: 1970, British Standards Institution.

2 Simms, L. G., 'Frog-up and frog-down brickwork compared', *The Builder*, August 24, 1956, p. 329–331.

3 *Model Specification for Loadbearing Clay Brickwork, Special Publication No. 56*, by the Structural Ceramics Advisory Group of the Building Science Committee. BCeramRA, 1967.

4 *Draft Australian Standard Rules for Brickwork—Committee BD/30—Brick Masonry Codes, Doc. 1053*, June 1966, Standards Association of Australia.

5 Thomas, K., Coutie, M. G., and Pateman, J. 'The effect of pigment on some properties of mortar for brickwork', *Proceedings SIBMAC*, April 1970.

6 Plowman, J. M., and Smith, W. F., 'The selection of damp proof course materials for loadbearing structures', *Proceedings SIBMAC*, April 1970.

7 Davey, N., and Thomas, F. G. T., 'The structural uses of brickwork', *Structural and Building Paper No. 24*, ICE, February 1950.

8 Thomas, K., 'The strength, function and other properties of wall-ties', *Proceedings British Ceramic Society*, No. 17, February 1970.

9 Klein, A., 'Multi-storey flat buildings in calcium silicate bricks and blocks and the testing of wall panels of bricks, blocks and mortar for calculated masonry', *Proceedings Int. Symp. on Autoclaved Calcium Silicate Building Products 1965*, 239, Soc. of Chem. Ind.

10 Thomas, F. G., 'The strength of brickwork', *Structural Engineer*, Vol. 31, No. 2, February 1953, p. 35–46.

11 Simms, L. G., 'The strength of walls built in the laboratory with some types of clay bricks and blocks', *Proceedings British Ceramic Society*, July 1965, p. 81–92.

12 Prasan, S., Hendry, A. W., and Bradshaw, R. E., 'Crushing tests on storey-height panels $4\frac{1}{2}$ in thick', *Proceedings British Ceramic Society*, July 1965, p. 67–80.

13 Astbury, N. F., et al., *Gas Explosions in Loadbearing Brick Structures, Special Publication No. 68*, BCeramRA, 1970.

14 Hendry, A. W., 'Wind load analysis of multi-storey brickwork structures', *Brick Development Association, Research Note* Vol. 1, No. 3, January 1971.

15 Amendment Slip No. 1 to *CP III: Part 2: 1970.*

16 *CP 114: 1957*, **Structural Use of Reinforced Concrete in Buildings**, British Standards Institution.

17 *Digest No. 61, Second series*, 'Strength of brickwork, blockwork and concrete walls', D.o.E., Building Research Station, August 1965.

18 Sinha, B. P., and Hendry, A. W., 'The effect of brickwork bond on the loadbearing capacity of model brick walls', *Proceedings British Ceramic Society*, No. 11, July 1968.

19 Anon., 'Loading tests on brick walls built in stretcher bond', *C.P.T.B. Technical Note*, Vol. 1, No. 4, January 1964.

20 Bradshaw, R. E., and Hendry, A. W., 'Further crushingtests on storey-height walls $4\frac{1}{2}$ in thick', *Proceedings British Ceramic Society*, No. 11, July 1968.

21 Wood, R. H., *Part One*, The composite action of brick panel walls supported on reinforced concrete beams, Studies in Composite Construction, National Building Studies Research Paper, No. 13, 1952.

22 Wood, R. H., *Part Two*, Interaction of floors and beams in multi-storey buildings, Studies in Composite Construction, National Building Studies Research Paper No. 22, 1955.

23 Wood, R. H., and Simms, L. G., A tentative design method for the composite action of heavily loaded brick panel walls supported on reinforced concrete beams, Current Paper C.P.26/69, D.o.E., Building Research Station, July 1969.

24 Draft British Standard Code of Practice (Revision of CP III: 1964, Structural Recommendations for Loadbearing Walls), Document 70/1556.

25 *BS 3921: Part 2: 1969*, Specification for Bricks and Blocks of Fired Brickearth, Clay or Shale—Metric Units, British Standards Institution.

26 *BS 2564: 1955*, Control Chart Technique when Manufacturing to a Specification, with Special Reference to Articles Machined to Dimensional Tolerances, British Standards Institution.

27 Thomas, K., and O'Leary, D. C., 'Tensile strength tests on two types of brick', *Proceedings SIBMAC*, April 1970.

28 Albrecht, W., and Engelke, H., 'Prufung der Festigkeit von Maurmorteln durch Bolzenscheiversuche', *Die Ziegelindustrie*, Vol. 24, 1964, p. 914–919.

29 Private communication.

The stability of a five-storey brickwork cross-wall structure following the removal of a section of a main loadbearing wall

A W Hendry and B P Sinha

SYNOPSIS

It is now a requirement under the Building Regulations that structures of five storeys and over should remain stable following the removal of a specified length of loadbearing wall, although at a substantially reduced safety factor. Three experiments are described in this paper which had the object of providing confirmation that this could be achieved in a simple five-storey brickwork cross-wall structure. In each test a section of loadbearing wall was removed and measurements were made of applied loads, deflexions and strains. The theoretical conclusion that the structure would remain stable under these conditions was confirmed and some information was obtained concerning the strength of 114 mm (4.5 in) thick wall panels subjected to lateral loading.

INTRODUCTION

This paper describes the results of three experiments on a section of a five-storey brickwork cross-wall structure in which sections of the main cross-walls were removed at ground-floor level with a view to testing the stability of the structure in a damaged condition, as might occur following an internal explosion. The structure (Fig 1) had been constructed previously for a series of lateral loading tests and was not specially designed to resist the stresses set up by partial collapse, but prior to the present tests the stability of the structure following the removal of a section of cross-wall was assessed by the methods described elsewhere.[1] These calculations indicated that the structure would not collapse following the removal of a major loadbearing element.

In each of the present tests, a section of loadbearing wall was removed under controlled conditions and measurements of applied loads, deflexions and strains were made in the course of the tests, thus affording information as to lateral rigidity and strength of a 114 mm (4.5 in) single leaf wall.

THE TEST STRUCTURE

Plan form and elevation

The structure consisted of three pairs of 114 mm (4.5 in) cross-walls stabilized by two pairs of shear walls, as indicated in Fig. 2. The wall layout was the same on each of the five storeys of the structure.

Bricks

Wire-cut bricks with an average crushing strength of 34.6 MN/m² (5020 lbf/in²) were used for the construction of the building. The average compressive strength of six-course brick prisms was 17.2 MN/m² (2495 lbf/in²) at 28 days.

Cement and sand

A rapid hardening cement 'Ferrocrete' to BS 12: 1958 and ordinary building sand conforming to BS 1200: 1955 were used for the construction of the walls.

Mortar

A 1 : ¼ : 3 cement : lime : sand mix (by vol) was used for the construction of the building. The

Figure 1 *Test structure.*

materials were gauged before dry mixing. The average crushing strength of 102 mm (4 in) mortar cubes was 14.5 MN/m² (2100 lbf/in²) at 28 days.

CONSTRUCTION DETAILS
Walls
The walls were built according to the BCeramRA Model Specification.[2]

Reinforced concrete slabs
To save time and cost of shuttering, 50.8 mm (2 in) thick 'Omnia wide slabs' with 76.0 mm (3 in) *in situ* concrete topping were used for the floors (Fig 3). The floors were made of panels 3.16 m (10 ft 4.5 in) long and 1.2 m (3 ft 11.25 in) wide. The precast panels were lifted and kept in position by props with no bearing on the walls (Fig 4). A 1 : 2 : 4 ready-mix concrete having a minimum strength of 20.7 MN/m²

((3000 lbf/in²) at 28 days was poured on the top of the 50.8 mm (2 in) precast panels to give a 127 mm (5 in) thick slab throughout. By adopting this method of construction a good joint, similar to a cast *in situ* slab, was obtained between the finished slabs and the walls underneath. However, there is no through reinforcement (Fig 4) in the bottom of the slab over the brick wall. With this form of construction the minimum practical slab thickness is 127 mm (5 in) although it would be possible to use a thinner slab with normal *in situ* concrete. Mesh reinforcement ((16 mm ($\frac{1}{4}$ in) square twisted bars at 203 mm (8 in) c/c)) was provided in the bottom of the precast slab and in the top of the slab over supports. The details are given in Fig 3. The slab was designed to carry 1915 N/m² (40 lbf/ft²) superimposed load and self weight of 2875 N/m² (60 lbf/ft²); the design calculations are given in Appendix 1.

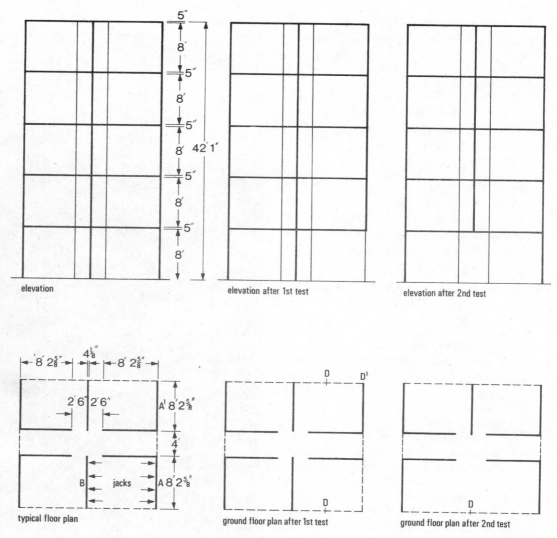

Figure 2 *Test structure before and after use.*

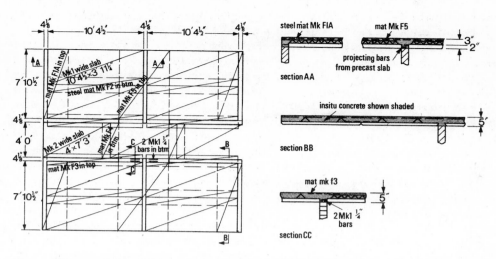

Figure 3 *Layout of precast slabs and reinforcement (in situ concrete).*

ARRANGEMENT FOR APPLYING THE TRANVERSE LOAD

The tranverse load was applied by jacking against a steel frame, which was fixed to the concrete floor of the quarry by nine projecting-type Rawlbolts going to a depth of 184 mm (7.25 in) (Fig 5). The transverse load from four 10 tonne jacks was distributed to eight point loads on a line across the width of the wall through 305 mm (12 in) span steel spreader beams 50 × 50 mm (2 × 2 in). The load from each jack was measured by a 3 tonne load-cell calibrated in the laboratory. Rubber packing was inserted between the rollers of the spreader beams and the wall for the proper distribution of the load. All four jacks were controlled by a single valve and supplied by a mobile electro-hydraulic pump unit.

TESTS ON THE STRUCTURE

The tests were carried out in the following sequence.

(1) The outer loadbearing cross-walls of the ground floor (Fig 2, either panel A or A¹) was loaded to destruction leaving no support in that region for the floor above it.

(2) The cross-wall in (1) was rebuilt.

(3) The central loadbearing wall (B, Fig 2) was similarly removed 15 days after the reconstruction of the outer wall.

The horizontal deflexions of the panel and the vertical deflexions of the floor slab were measured by a dialgauge. Demec gauges of 610 mm (24 in) and 203 mm (8 in) gauge length were used to measure the strains in the panel. Typical deflexion results are given in Figs 6, 7 and 8.

The walls in both tests 1 (Panel A or A¹) and 2 failed due to development of horizontal cracks at the centre of the panel at a brick/mortar interface. The ultimate loads were 43.1 kN (4.33 tonf—Panel A), 47.5 kN (4.77 tonf—Panel A¹) and 62.5 kN (6.27 tonf) respectively. The calculated precompressions for the walls in test 1 and 2 were 365 kN/m² (53 lbf/in²) and 510 kN/m² (74 lbf/in²). In the first test (Panel A) the maximum deflexion after removal of the wall (Fig 2) could not be obtained as the falling wall knocked out the device for recording the vertical deflexion. The maximum deflexion of the floor slab was 4.04 mm (0.159 in) after removal of Panel A¹ (Fig 9). The deflexion of the floor slab at the centre

Figure 4 *The precast Omnia slab in position.*

Figure 5 *Test arrangement.*

243

(D, Fig 2—Panel A and A¹) of the free edge was 1.45 mm (0.057 in) and 1.67 mm (0.066 in) respectively. In the second test the maximum slab deflexion (D—Fig 2) was 5.44 mm (0.214 in) after removal of the wall. No damage was noticed anywhere in the structure (Fig 10) except immediately above the wall removed where the joint between the first floor slab and wall was found to be broken and the wall appeared to be partly hanging from the second floor slab and partly supported by the first floor slab spanning between the ground-floor shear walls. From the recorded strains before and after the test this wall also appeared to have been relieved of load from floors above. In each case the structure as a whole was found safe after the tests and remained so for a week before the damaged element was replaced.

The uplift of the floor slab above the test wall was also recorded during the tests, with the results shown in Table 1.

Lateral load on panel	Test 1 (Panel A¹)	Test 2
	Slab Deflexion D¹	Slab Deflexion D
28.21 kN (2.83 tonf)	0.102 mm (0.004 in)	—
35.9 kN (3.6 tonf)	—	0.229 mm (0.009 in)
40.3 kN (4.04 tonf)	0.508 mm (0.02 in)	—
47.5 kN (4.77 tonf)	4.87 mm (0.192 in)	—
49.8 kN (5 tonf)	—	0.762 mm (0.03 in)

Table 1 *Showing the uplift of slab*

ASSESSMENT OF STABILITY

Examination of the structure shows that if one of the central cross-walls is removed at ground-floor level the weight of each section of cross-wall on storeys above plus the self weight of the slab on which it rests and any imposed loading on it will have to be resisted by bending action in that slab. As we are concerned with ultimate load behaviour it is appropriate to apply a yield-line analysis. This permits the estimation of the load factor for the structure following the removal of the section of cross-wall. A detailed calculation, shown in Appendix 2, indicates that a load factor of 1.94 on dead load plus 1915 N/m² (40 lbf/ft²) superimposed load will exist after removal of the centre cross-wall.

Where a section of the end cross-wall is removed, it is to be expected that each section of cross-wall above will be supported vertically by its connection to the corresponding shear wall and prevented from rotating by the floor slabs. The transfer of load to the shear walls will result in concentrated bearing stresses at the outer end of the shear wall on the ground floor. The actual distribution of loads is somewhat indeterminate but a safe assumption would be that the whole weight of the section of cross-wall above the one removed has to be supported by the floor slab below: in other words assuming that the bond between the cross-wall and the shear wall breaks. A calculation on this basis is shown in Appendix 2 and shows a load factor of about 2.44 after removal of the end cross-wall.

Discussion

The removal of one of the major loadbearing elements did not precipitate the collapse of the structure, hence it appears that the building is not susceptible to collapse following major local damage. Although the test was carried out without superimposed load on the floor slabs, this has been accounted for in the calculations for the worst condition. The slab appeared to be safe, even after taking superimposed

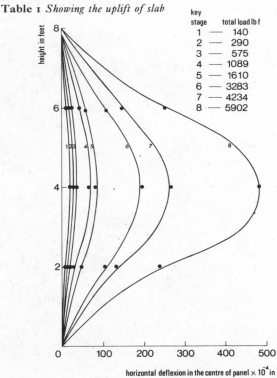

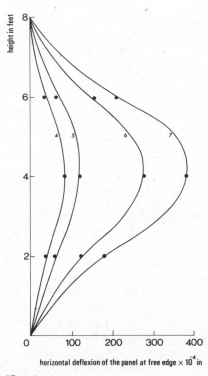

key
stage	total load lbf
1	140
2	290
3	575
4	1089
5	1610
6	3283
7	4234
8	5902

horizontal deflexion in the centre of panel × 10⁻⁴ in

horizontal deflexion of the panel at free edge × 10⁻⁴ in

Figure 6 *Panel deflexion at various stages of loading in the 1st test (Panel A).*

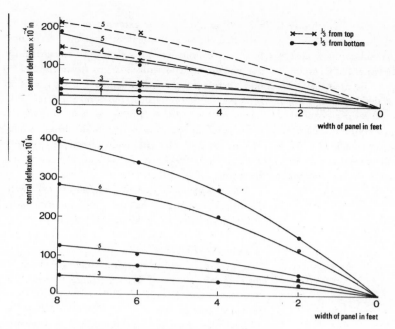

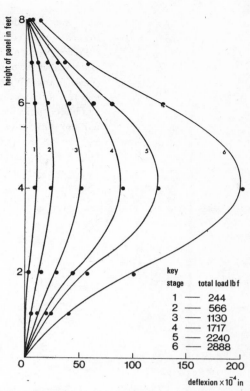

Figure 7 *Horizontal deflexion of the panel, 1st test (Panel A).*

Figure 8 *Horizontal deflexion of the panel in 2nd test at various loads.*

key

stage	total load lbf
1	244
2	566
3	1130
4	1717
5	2240
6	2888

Figure 9 *Structure after removal of the end loadbearing element.*

Figure 10 *Structure after removal of central loadbearing element.*

245

load into consideration, and following the tests showed no sign of damage.

Precompression increased the load-carrying capacity of the wall in the transverse direction and from tests 1 (Panel A and A[1]) and 2 it can be inferred that a 114 mm (4.5 in) brickwork wall of the dimensions tested and under similar conditions will resist an equivalent uniform static pressure of 15.26 kN/m² (2.21 lbf/in²) and 21 kN/m² (3.05 lbf/in²) respectively, before failure. This is far below the design static pressure of 34 kN/m² (5 lbf/in²) as required by the revision[3] to the

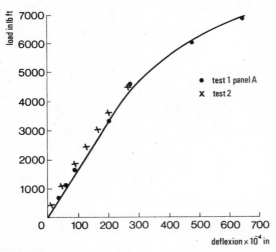

Figure 11 *Relationship between transverse load and panel deflexion.*

Building Regulations and also in one case less than the 17 kN/m² (2.5 lbf/in²) recommended by the Institution of Structural Engineers.[4] Although the panel fails to achieve the above requirements, the stability of the structure as a whole was very adequate after removal of one section of cross-wall, although it was not designed with this in mind.

The deflexion of the panel at one-third height from top is slightly more than at the corresponding distance from the bottom (Fig 7), possibly due to different support conditions. The relationship between the central deflexion of the panel and transverse load is linear up to approximately half the ultimate load (Fig 11).

CONCLUSIONS

These tests demonstrated that a cross-wall load-bearing brickwork structure could remain stable in the event of one section of a main cross-wall being

removed as a result of an explosion or other accident even though, as in the present structure, it did not have fully continuous *in situ* slabs. The test structure was not designed to withstand this treatment and it may therefore be concluded that the design of a brickwork structure of this type to meet the requirements of the *Building (Fifth Amendment) Regulations 1970* on the 'alternative path' basis would present no difficulty. In many cases, there would seem to be no necessity for additional elements to secure the safety of the structure.

The tests also indicate that a 114 mm (4.5 in) wall panel in a structure of the type tested would resist a lateral pressure of 14–21 kN/m² (2–3 lbf/in²).

ACKNOWLEDGEMENT
The work described in this paper was carried out at the request of the Director, Building Research Station. Funds in support of the experiment were provided by the Ministry of Public Building & Works, now the Department of the Environment. The experimental structure was built in connection with research work sponsored by the Brick Development Association and the British Ceramic Research Association. A grant for staff salaries was provided by the Science Research Council.

REFERENCES
1 Hendry, A. W. **Preliminary report on an assessment of the liability of a number of existing loadbearing brickwork structures to progressive collapse.** Private report to Director, Building Research Station, March 1969.

2 British Ceramic Research Association, **Model Specification for Loadbearing Clay Brickwork**, BCeramRA Spec. Publ. 56, 1967.

3 Ministry of Housing and Local Government, Circular 62/68: **Standard to avoid progressive collapse, large panel structures**, London, HMSO, 1968.

4 Institution of Structural Engineers, **Guidance on the Design of Domestic Accommodation in Loadbearing Brickwork and Blockwork to Avoid Progressive Collapse.** London, I.S.E., June, 1969.

5 Jones, L. L. and Wood, R. H. **Yield-line analysis of slabs.** London, Thames & Hudson; Chatto & Windus, 1967.

APPENDIX 1

Design of floor slabs*
Wide slab units:
    loading 40 lbf/ft² super P_{cb} = 1000 lbf/in²
    60 lbf/ft² self wt. P_{st} = 33 000 lbf/in²
             q =   100 lbf/in²
Effective span 10.8 ft
    Mid-span moment
Dead load $0.071 \times 60 \times 10.8^2 \times 12$ =   5960 lbf/in
Live load $0.096 \times 40 \times 10.8^2 \times 12$ =   5375 lbf/in
                                  11 335 lbf/in

    Support moment
Dead and live load $0.125 \times 100 \times 10.8^2 \times 12$ = 17 500 lbf/in

A_{st} for mid-span $\dfrac{11\ 335}{33\ 000 \times 4 \times 0.95}$ = 0.0905 in²/ft

Use 0.25 in square twisted bars at 8 in (0.095) centres both ways.

A_{st} for support moment $\dfrac{17\,500}{33\,000 \times 4 \times 0.95} = 0.134$ in²/ft

Use $^5/_{16}$ in square twisted bars at 8 in centres with distributed steel 0.25 in (0.146 + 0.094) square twisted bars at 8 in centres.

Shear stress $\dfrac{0.62 \times 100 \times 10.8}{12 \times 4 \times 0.89} = 15.7$ lbf/in²

Lintol section 5×4 in

   Bending moment $= 100 \times 6 \times 3^2 \times 1.5 = 8100$ lbf/in

$d_1 = \dfrac{8100}{4 \times 140} = 3.8$ in

Depth provided $= 4$ in

$A_{st} = \dfrac{8100}{33\,000 \times 4 \times 0.896} = 0.068$ in²

Use $2 \times \frac{1}{4}$ in.

---

1 in = 25.4 mm; 1 lbf/in² = 6895 N/m²; 1 lbf/in = 0.113 Nm.

APPENDIX 2

Stability calculations: yield-line analysis of slabs*

An approximate analysis of the resistance of the floor slabs to failure after removal of the walls has been carried out by yield-line theory according to normal procedures[5].

In this case (Fig 12) the slab is loaded by the weight of one storey-height panel of brickwork on the side fc.

Consider the yield-line pattern shown above. Let the slab be given a unit deflexion on line bc. To simplify calculations and since the load is distributed, the trapezoidal area B can be divided into a triangular and a rectangular area. Since bc deflects by one unit, the centre of gravity of each of the triangles abd and bde will deflect by one-third of a unit. Similarly, the centre of gravity of the rectangular area befc and the wall will each deflect by half of a unit.

Total external work $P = W \times \delta$
$$= \text{Total load} \times \text{Deflexion of C.G. of loads}$$
$$= W' \times I \times \delta_w + w\,(A_A \delta_B + A_A \delta_B)$$

where

$\delta =$ deflexion of C.G. of loads

$w =$ distributed load (100 lbf/ft²)

$A =$ area (ft²)

$W' =$ wt. of wall/ft run (328 lbf/ft run)

$l =$ length of wall (ft)

Suffix A, B, w represent sections abd, dbcf and the wall.

Hence, $P = W' \times 8 \times \frac{1}{2} + w\,[\frac{1}{2} \times X \times 8 \times \frac{1}{3} + \frac{1}{2} \times 8 \times X \times \frac{1}{3} + (10.75 - X) \times 8 \times \frac{1}{2}] = 5612 - 133.33X$

$$\dots (1)$$

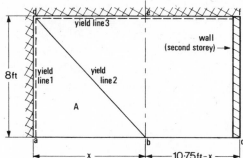

Figure 12 *Yield-line analysis of slabs: test 1.*

Internal dissipation of energy in the yield lines (1, 2 and 3)

$D = \Sigma\, ml\theta_n$

where

$m =$ normal moment/unit length on a yield line depending on the magnitude of the reinforcement.

$\theta_n =$ angle of rotation normal to yield line for the section under consideration.

$l =$ length of yield line (ft)

$D = \dfrac{0.146}{0.094}\, m \times 8 \times \dfrac{1}{X} + m \times 8 \times \dfrac{1}{X} + m \times X \times \dfrac{1}{8} + (2m \times 2.5 + m \times 6 + 2m \times 2) \times \dfrac{1}{8}$

---

1 ft = 0.305 m; 1 lbf/ft² = 47.88 N/m²; 1 lbf/ft = 1.356 Nm.

(See design calculations for reinforcements)

$$D = m\left(\frac{20.42}{X} + \frac{X}{8} + \frac{15}{8}\right) \qquad \ldots(2)$$

External work P = Dissipation of energy D, and by equating the expressions for P and D (equations (1) and (2)) we get:

$$m = \frac{5612 - 133.33X}{\frac{X}{8} + \frac{15}{8} + \frac{20.42}{X}}$$

$$\frac{dm}{dx} = 0$$

or $0 = -133.33\left(\frac{X}{8} + 1.875 + \frac{20.42}{X}\right) - (5612 - 133.33X)\left(\frac{1}{8} - \frac{20.42}{X^2}\right)$

or $X^2 + 5.72X - 120.44 = 0$

$$X = \frac{-5.72 \pm \sqrt{(5.72)^2 + 4 \times 120.44}}{2}$$

Acceptable solution $X = +8.6$

$$m = \frac{5612 - 133.33 \times 8.6}{\frac{8.6}{8} + \frac{15}{8} + \frac{20.42}{8.6}} = 838.66 \text{ ft lbf/ft} = 10\ 064 \text{ in lbf/ft}$$

This is less than the design moment at midspan and much less than the ultimate moment so that the slab will not fail when the wall under boundary fc is removed.

In this case (see Fig 13) because there was no through reinforcement over the wall in the bottom of the slab, it was assumed that a crack would develop along line cf as soon as wall support below is removed. The

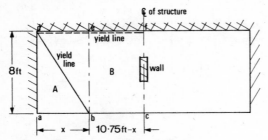

Figure 13 *Yield-line analysis of slabs: test 2.*

wall load will then be equally shared by the two sections of the slab.

External work $P = 2W' + w(8X/3 - 4X + 43)$ (Calculation similar to Test 1) $= 4956 - 133.33X$ $\ldots$ (3)

Dissipation of energy $D = m\left(\frac{X}{8} + \frac{15}{8} + \frac{8}{X}\right) \qquad \ldots(4)$

(See the design calculation, Appendix 1)

By equating P and D we get (equations (3) and (4)):

$$m = \frac{4956 - 133.33X}{\frac{X}{8} + \frac{15}{8} + \frac{8}{X}}$$

$$\frac{dm}{dX} = 0 = -133.33\left(\frac{X}{8} + \frac{15}{8} + \frac{8}{X}\right) - (4956 - 133.33X)\left(\frac{1}{8} - \frac{8}{X^2}\right)$$

or $869.4937 X^2 + 2133.28X - 39\ 648 = 0$
or $X^2 + 2.45X - 45.60 = 0$

$$X = \frac{-2.45 \pm \sqrt{(2.45)^2 + 4 \times 45.60}}{2}$$

Acceptable solution $X = +5.64$

$$m = \frac{4956 - 133.33 \times 5.64}{\frac{5.64}{8} + \frac{15}{8} + \frac{8}{5.64}} = 1051 \text{ ft lbf/ft} = 12\ 618 \text{ in lbf/ft}$$

This is greater than the design moment at midspan but still much less than the ultimate moment of the slab so that failure will not take place when the wall is removed□

INDEX